EASEMENTS RELATING TO LAND SURVEYING AND TITLE EXAMINATION

EASEMENTS RELATING TO LAND SURVEYING AND TITLE EXAMINATION

DONALD A. WILSON, LLS, PLS, RPF

WILEY

Cover Design: Anne Michele Abbott
Cover Illustration: From *Atlas of Washington County, Maine,* 1881. George N. Colby & Co.

This book is printed on acid-free paper.

Published by John Wiley & Sons, Inc., Hoboken, New Jersey
Published simultaneously in Canada

Library of Congress Cataloging-in-Publication Data:

Wilson, Donald A., 1941-
Easements relating to land surveying and title examination / Donald A. Wilson.
pages cm
Includes bibliographical references and index.
ISBN 978-1-118-34998-4 (cloth), ISBN 978-1-118-41706-5 (ebk.);
ISBN 978-1-118-42001-0 (ebk.); ISBN 978-1-118-67524-3
1. Servitudes—United States. 2. Surveying—Law and legislation—United States.
3. Land titles—United States. I. Title.
KF657.W55 2013
346.7304'35—dc23

2013007088

Printed in the United States of America
10 9 8 7 6 5 4 3 2 1

CONTENTS

PREFACE

One of the most frequently presented topics, by a variety of speakers, is that of easements. Easements, in general, is one of the most complex areas of real property law and real property ownership, and one the most frequently litigated. That, by itself should tell us two things: they can be extremely important, even necessary, to people, and they are commonly a subject of argument and dispute. To the professional, the land surveyor, the title researcher and the attorney, there is a third message, which is more on the positive side. There is an abundance of information available to us from the court systems. However, being as complex as it is, that by itself presents somewhat of a difficulty in that it would take many years to read the literally thousands of decisions that are readily available. As a quick example, consider the following:

With any [apparent] easement, there is a consideration of both easement and fee (2 possibilities), whether public or private (2 possibilities).

There are 10 ways in which an easement can be created (10 possibilities) and 13 ways in which one can be extinguished (13 possibilities),

Which translates to: $2 \times 2 \times 10 \times 13$, or 520 possible situations. Worse yet, in the case of a road, there are 8 possibilities, for a total of

$$520 \times 8 = 4160 \text{ possibilities}$$

With 4160 possibilities of getting into the correct category, the frightening part is that means that there are 4159 ways of being incorrect. Selecting the wrong category means attempting to apply the wrong rules, or the wrong law, to a given situation. This becomes exceedingly critical when considering a road, either public or private. From a legal standpoint, there are title considerations, both easement and [underlying] fee, and from a technical, or surveying, standpoint, there are considerations of location and width, along with points of origin and termination.

Taking the 4160 possibilities, and expanding on the potential of any of them occurring in any or all of our 50 states, presents the researcher with a mind-boggling number of items for consideration.

Knowing the correct category of any estate in land is a key to doing appropriate research. Only some types of easement are a matter of public record, yet people spend hours, sometimes days, searching for records that don't exist. Finding none, they often make an assumption, which can easily be incorrect. Easements arising through use are usually thought to be prescriptive, when many (I hesitate to say most) of them are permissive, again misleading the unwary into the wrong category. Evidence of location is often misidentified, for example, not all fences mark right of way lines, and not all traveled ways are in the center of their easement. In fact, occasionally one is found completely outside of its defined location.

Most easement presentations by fellow professionals involve one day of instruction, which is mostly on the general aspects of easements. To learn about the practical aspects of easements, and how the courts have decided specific problems and arguments, will demand attendance at a series of follow-up seminars. Over the years I personally have offered seminars entitled Locating Historical Rights of Way, Easements and the Land Surveyor (responsibility & liability), Locating Undescribed Rights of Way, Highways and Railroads, Roads and Streets, Easements and Reversions, and Dedication of Public and Private Rights. Even with all that, I feel that I have barely scratched the surface because of the innumerable aspects of the various topics.

Through the years a number of us have been frustrated by the fact that while there are numerous surveying and legal treatises that devote a chapter or a short discussion to easements, it has been many years since a full treatment has been published. In 1900, John Cassan Wait compiled an extensive treatment of real property rights entitled *The Law of Operations Preliminary to Construction in Engineering and Architecture: Rights in Real Property, Boundaries, Easements, and Franchises for Engineers, Architects, Contractors, Builders, Public Officers, and Attorneys at Law*. It was published by John Wiley & Sons, and contains an extensive treatment of easements, Part IV, consisting of 9 chapters containing some 123 pages of material extensively supported by court decisions. In 1989, Backman and Thomas published *A Practical Guide to Disputes Between Adjoining Landowners—Easements*, and, in 1991, I wrote *Easements and Reversions*, which was published by Landmark Enterprises. Almost daily there are many decisions rendered, and new issues addressed, making the overall field even more complex and extensive than ever before. Several professionals have suggested to me that an up-to-date treatment on the subject is badly needed. This is my attempt at bringing out a more extensive overview than has been available in the recent past. I have tried to illustrate it with real examples, and I have selected several leading court decisions dealing with practical situations, while also considering some elusive and complex situations. While Backman & Thomas is a wonderful treatment of the general aspects of the extensive field of easements, this treatment, with its examples, is directed toward land surveyors and title persons.

Newfields, New Hampshire
June 2012

A court once noted that, "Few things are as certain as death, taxes and the legal entanglement that follows a sale of landlocked real estate."

—*Cobb v. Daugherty,* 225 W.Va. 435 (2010)

ACKNOWLEDGMENTS

I would be remiss not to acknowledge my various audiences of the past several years who have supported the selection of material that I felt important to professionals. Presentations demand extensive research for supporting legal principles, and for answers to particular problems. The questions that have arisen give one cause to think about what is important to others, and updating presentations is a constant process in the evolving attempt to make them more meaningful.

Thanks to fellow professionals with whom I have worked or who have provided court decisions from their own research and from whom I have learned so much:

Kristopher Kline, PLS, who is constantly making me aware of court decisions new to me, having derived them from his relentless and untiring research. I am grateful for his willingness to share.

Wendy Lathrop, PLS, who assisted on a "roads" conference and with whom I co-presented a roads seminar.

Paul Alfano, Esq., who presented the lead-off topic at the "roads" conference, as well as giving a joint presentation with me to the New Hampshire surveyors.

John Cunningham, Esq., with whom I had the pleasure of working on the aforementioned "roads" conference, along with a stand-alone roads seminar we taught together.

Paul Gillies, Esq., who also assisted with the "roads" conference and with whom I had the pleasure of co-presenting a seminar sponsored by the Vermont Department of Transportation.

Along the way, there have been several local practicing attorneys, too numerous to mention, who have continually presented me with some tricky and challenging easement problems, and with whom I have had the honor to work in the courtroom. Such teamwork can be very rewarding, especially when the right answer is found and a successful presentation is made, win, lose, or draw. It is always a learning experience for everyone involved.

And finally, to my editor Daniel Magers and my former editor Robert Argentieri, along with members of the staff at John Wiley & Sons, who believed in this project as filling a major gap in the current literature relating to rights in real property. I appreciate their patience with me in seeing this through.

EASEMENTS RELATING TO LAND SURVEYING AND TITLE EXAMINATION

CHAPTER 1

INTRODUCTION

Easements are not real property, but they are treated like real property and are often confused by the uninitiated, who believe they are real property, when they are in fact not. Therein lies a major source of confusion among the general public, which, over time, has led to an astounding number of problems and controversies. To understand the nature and impact of easements first requires at least a basic understanding of real property law and ownership, in many cases much more than just a basic grasp of the subject.

The other common source of confusion is that an easement, being a right, or an interest, is invisible. Some characteristics of the ultimate result of an easement, such as pavement in the case of a road, for instance, may be visible, but the easement itself cannot be seen. Many would look at these characteristics and conclude that that is the easement, when in reality, it is the *result of the easement* or *evidence of the location* of the easement.

RIGHTS AND INTERESTS IN LAND; TRANSFER OF OWNERSHIP

Ownership in General

> *Land ownership.* When something is owned, it is generally thought that someone has *title* to it, or in it. Title may be in several forms, however. Title, by definition, can be regarded as the right to or ownership of property. The word is used to designate the means by which an owner of lands has the just possession of his property, the legal evidence of his ownership, or the means by which his right to the property has accrued.[1]

A fee-simple estate is the highest and greatest estate in land that one can obtain. Those who possess a fee-simple or fee simple absolute estate are, for all purposes,

[1] *Patton on Titles*, § 1.

the owners of the land. The words *fee-simple absolute* in reality are not a single term; each distinct word carries a meaning that explains the entire term. *Fee* denotes that the estate is one that can be inherited or devised by a will. *Simple* denotes that the estate is not a fee tail estate, wherein the estate must be inherited by a specific individual. *Absolute* means that there are no conditions or limitations so far as time is concerned on the estate, and this estate may continue forever (not like a fee or an estate that may be determinable upon the happening of an event).[2]

Fee tail, or an estate tail, is an early English type of estate, which in all probability was borrowed from the Romans. It is a true freehold estate limited by the grantor to the heirs of the grantee's body or to a special class of individuals, either male or female (e.g., the eldest, the youngest, or other). If the conditions are breached (no male or female is produced by the grantee), the estate reverts to the grantor or his heirs. In the United States, individual states have determined that this type of estate was never adopted as part of the English common law that a conditional fee was present and that, upon the birth of a child, it was converted to a fee-simple estate, or that statutes eliminated the estate and any reference to fee tail connotes fee-simple absolute.[3]

A *life estate* is considered a freehold estate since it can be conveyed to a third party, yet its duration is only measurable by some life. In essence the life estate lasts only for the life of some person. An "ordinary" life estate is normally worded "to Jones for life and then to Brown in fee-simple." An estate *per autre vie* has a measured life other than that of the holder of the estate and may be worded "to Jones for the life of Brown and then to Smith in fee-simple." This estate terminates with the death of Brown.

Life estates may be created by expressed provisions or words in a will or deed or by contract between heirs or parties in interest. If a question exists as to the creation of a life estate, the courts will look at the precise words used.[4]

Nature of modern estates. By law, an estate is the interest a person has in real or personal property. The word is sometimes used to mean the property or assets of a person, as in "the estate of John Doe." In general, real estates are classified by the time of enjoyment, and they are (1) estate in fee, (2) estate for life, (3) estate for years, and (4) estate at will.

An *estate in fee*, sometimes called an *estate in fee-simple*, is the most absolute interest a person can have in land. It is of indefinite duration and is freely transferable and inheritable. More than one fee can be held on a given parcel of land; for example, one person may have the fee to minerals and another the fee to the land, excepting the minerals.

[2] Brown, Curtis M., Walter G. Robillard, & Donald A. Wilson. *Boundary Control and Legal Principles.* New York: John Wiley & Sons, Inc., 1986.
[3] Ibid.
[4] Ibid.

A *defeasible fee-simple estate* is one in which a future event must be met. It is an estate in fee that is liable to be defeated by some future contingency.[5] The title is conveyed on the condition that certain things will be done within a time limit or that certain things will never be done. A fee may pass on the condition that a storm drain is installed, or that the property is never used for the sale of alcoholic beverages.

An *indefeasible estate* is one which cannot be defeated, revoked, or made void. The term is usually applied to an estate or right that cannot be defeated.[6]

A *life estate* is an estate limited to the life of the person or persons holding it, and will automatically revert upon the death of the individual. However, the holder of interest may, in some cases, transfer their interest to another, which will terminate upon the death of the original holder of the life estate. This is known as an estate *per autre vie* (for the life of another).

An *estate for years* is usually created by a lease between two parties whose relationship is that of landlord and tenant, such as a lease to use a parcel for 10 years, conditioned on payment of a given amount of money or other consideration.

An *estate at will* may be terminated at any time.[7] It is an estate less than freehold, where lands and tenements are let by one person to another, to have and to hold at the will of the lessor, and the tenant, by force of this lease obtains possession.[8]

Transfer of Title and Property Rights Whatever title or interest a person or entity has may be transferred, or acquired. The means by which this is accomplished may be categorized as follows:[9]

MEANS OF TRANSFERRING OR OBTAINING TITLE OR RIGHTS IN LAND[10]

Public grant. A public grant is defined as a grant from the public; a grant of a power, license, privilege, or property, from the state or government to one or more individuals, contained in or shown by a record, conveyance, patent, charter, or similar entity.[11]

[5] *Black's Law Dictionary.*

[6] *Black's Law Dictionary.*

[7] Brown, Robillard, & Wilson, supra.

[8] *Black's Law Dictionary.*

[9] Brown, Curtis M., Walter G. Robillard, & Donald A. Wilson. *Brown's Boundary Control & Legal Principles.* 4th Edition. Hoboken, NJ: John Wiley & Sons, Inc., 1995.

[10] Subject titles taken in part from Robillard, Wilson, & Brown, *Brown's Boundary Control & Legal Principles,* 6th edition, Hoboken, NJ: John Wiley & Sons, 2009.

[11] *Black's Law Dictionary.*

A *grant* is the generic term applicable to all transfers of real property.[12] A *charter* is an instrument emanating from the sovereign power, in the nature of a grant, either to the whole nation, or to a class or portion of the people, or to a colony or dependency, and assuring to them certain rights, liberties, or powers. Such were the charters granted to certain of the English colonies in America.[13] A *patent* is a document issued by the government to one to whom it has transferred or agreed to transfer land, in order to vest in the transferee the complete legal title, or to furnish evidence of the transfer.[14] It is frequently a conveyance by which the United States passes title to portions of the public domain.[15] It is equivalent to a deed, and often passes fee-simple title.

Often overlooked is the transfer of title by *vote of proprietors*. Proprietorships generally had the authority to transfer parcels of land, and did so by vote, there being neither further conveyance nor record of transfer other than the notation in the proprietors records. Such a transfer generally resulted in fee-simple title,[16] although necessary easements may be part of the grant, frequently by implication.

Act of a legislature. Legislative acts frequently result in fee-simple title being granted.[17] Often found as the subject of legislative acts are mill grants, turnpikes, plank roads, canals, and similar entities, along with major highways, especially when they cover long distances, or are in more than one state or colony. Post roads and similar entities are included in this latter category.

Private grant. A private grant is a grant by a public authority vesting title to public land in a private person. *United Land Ass'n v. Knight*, 85 Cal. 448, 24 P. 818 (1890).[18]

The most common private grant is a deed. Deeds may come in a variety of forms, and there are also similar instruments that may pass property rights from one person to another

Will. With a will, the decedent has the right to designate whom the next title holder will be of the items contained in the estate. The next transfer may found either from one or more of the devisees or from the executor of the will, if empowered to sell.

Intestate succession. Where there is no will, the statute current at the time of the death of owner dictates who are the legal heirs. In conducting title research or

[12] *Black's Law Dictionary*.

[13] *Black's Law Dictionary*.

[14] *American Law of Property*, § 949.

[15] *St. Louis Smelting & Refining Co. v. Kemp*, 104 U.S. 636, 26 L.Ed. 875.

[16] The proprietors of common and undivided lands may divide the same among themselves by metes and bounds, and lots and ranges. They may make partition either by a vote or deed, or they may convey their undivided interests without partition. *Corbett v. Norcross & al.*, 35 N.H. 99 (1857).

[17] A grant of land by an act of the legislature vests an actual seizin in the grantee. *The Proprietors of Enfield v. Permit*, 8 N.H. 512, 31 Am. Dec. 207 (1837).

[18] *Black's Law Dictionary*.

compiling chains of title back in time, it is important to keep track of which statute applies at any particular point in time.

Involuntary alienation. This category includes bankruptcies and foreclosures, including takings through the statutory lien process for nonpayment of taxes.

Adverse possession. This is the process whereby long-term occupation and possession can ripen into title. This also includes unwritten agreements. Each state has a statute of limitations for bringing an action for the recovery of real property when occupied by one other than the record title holder. Requirements, as well as the time(s) required under different sets of circumstances, vary considerably. The definition is the actual, open, and notorious possession and enjoyment of real property, or of any estate lying in grant, continued for a certain length of time, held adversely and in denial and opposition to the title of another claimant, or under circumstances which indicate an assertion or color of right or title on the part of the person maintaining it, as against another person who is out of possession.[19]

Title acquired by adverse possession is a new and independent title by operation of law and is not in privity in any way with any former title. Generally, it is as effective as a formal conveyance by deed or patent from the government or by deed from the original owner. In fact, it is a good, actual, absolute, complete, and perfect title in fee-simple, carrying all of the remedies attached thereto. The title acquired will pass by deed. After the running of the statute, the adverse possessor has an indefeasible title, which can only be divested by conveyance of the land to another, or by a subsequent ouster for the statutory limitation period.[20]

When the occupation is through use rather than possession, an easement generally results, although essentially the same requirements must be fulfilled. The process is known as *prescription.* With any type of occupation against another title holder, one can only acquire the extent of title that the title holder holds.

Eminent domain. Eminent domain is the power of the state to take private property for public use.[21] The process is known as condemnation, and the condemnor can only acquire the extent of the title held by the condemnee. It requires a formal procedure outlined in the law of the state.

Escheat. Escheat signifies a reversion of property to the state in consequence of a want of any individual competent to inherit. The state is deemed to occupy the place and hold the rights.[22] When a property holder dies without will or legal heirs, generally the law provides for such property to automatically become state property through the process of escheat.

Dedication. An appropriation of land to some public use, made by the owner. A dedication may be express, as where the intention to dedicate is expressly manifested by a deed or an explicit oral or written declaration of the owner, or

[19] Black's Law Dictionary.

[20] *3 Am Jur 2d, Adverse Possession, § 298.*

[21] *Black's Law Dictionary.*

[22] Black's Law Dictionary.

it may be implied, shown by some act or course of conduct on the part of the owner from which a reasonable inference of intent may be drawn. A dedication may also occur according to common law and may be either express or implied, or it may be statutory, made under and in conformity with provisions of a statute regulating the subject, making it necessarily express.[23]

Dedication is the devotion. In order for the public to acquire the dedicated rights, there must also be an acceptance by the appropriate governing authority. The acceptance is a discrete event.

Containing the element of estoppel. An estoppel is a bar or impediment raised by the law, which precludes a person from alleging or from denying a certain fact or state of facts, in consequence of his previous allegation or denial or conduct or admission, or in consequence of a final adjudication of a matter in a court of law.[24]

The doctrine of estoppel rests upon principles of equity and is designed to aid the law in the administration of justice when without its intervention injustice would result. The rule is grounded in the premise that it offends every principle of equity and morality to permit a party to enjoy the benefits of a transaction and at the same time deny its terms or qualifications.[25]

In brief, estoppel is an equitable principle utilized to prevent one who has failed to act when he should have acted from reaping a profit to the detriment of his adversary.[26]

Regarding easements, estoppels may fall into three possible categories: estoppel by deed, estoppel by record, or estoppel by conduct.

Accretion. Accretion is the increase of real estate by the addition of portions of soil, by gradual deposition through the operation of natural causes, to that already in possession of the owner.[27] It may be by water, which is the most common, or by wind, which is often subtle, long-term, and not considered by the average person. Ownership of accretion generally is in the owner(s) of the land it attaches to.

Parol gift. While generally transfers of real property fall under the English Statute of Frauds (1677) and therefore must be in writing, under certain circumstances there may be a parol transfer of property, or property rights. In addition, transfers made prior to the adoption of the statute of frauds by a particular state, are not affected, and therefore may be valid although not in writing.

Operation of law. Situations not covered by the foregoing, affected by law or legal requirements, may be included in this category. Most of them are not a matter of record and take place automatically upon the satisfaction of specified requirement. *Reversion* is a classic example, whereby an estate, or part

[23] *Black's Law Dictionary.*

[24] *Black's Law Dictionary.*

[25] *Thompson v. Soles*, 299 N.C. 484, 263 S.E.2d 599 (1980).

[26] *Sizemore v. Bennett*, 408 S.W.2d 449 (Ky., 1966).

[27] *Black's Law Dictionary.*

thereof,[28] automatically returns to its origin when a certain situation arises. It is defined as the manner in which rights, and sometimes liabilities, devolve upon a person by the mere application to the particular transaction of the established rules of law, without the act or cooperation of the party himself.

Custom. Custom is a usage or practice of the people, which, by common adoption and acquiescence, and by long and unvarying habit, has become compulsory, and has acquired the force of a law with respect to the place or subject matter to which it relates. It is a law not written, established by long usage, and the consent of our ancestors. If it be universal, it is common law; if particular to this or that place, it is then properly *custom.*

Customs result from a long series of actions constantly repeated, which have, by such repetition, and by uninterrupted acquiescence, acquired the force of a tacit and common consent.[29]

Prior appropriation. This is a doctrine developed by Sir William Blackstone in his *Commentaries on the Laws of England.* Distinct from the prior appropriation doctrine of the western United States pertaining to water rights, Blackstone labeled flowing water as "transient property," usable by the owner(s) of land against whom it touched. He stated that water was a corporeal right, a transient element to the public but subject to a qualified individual property or title during use. Title subsists only during time of use, as water cannot be possessed or appropriated in the same manner as land. The prior appropriation theory is distinct from theories of acquisition of incorporeal rights by prescriptive long user.[30]

Any one of the foregoing may play a role in the creation, termination, alteration, or transfer of one or more easements.

Corporeal and Incorporeal Hereditaments Corporeal hereditaments are substantial permanent objects that may be inherited. The term "land" will include all such.

Incorporeal rights are those rights in things that can neither be seen nor handled, that are creatures of the mind and exist only in contemplation.[31] They are *rights* and *interests* acquired by one person in and over the land of another, and include rights of way, rights to water, support, air, light, view, and a multitude of similar interests that are appurtenant to and constitute the enjoyment of land. They are known as easements or servitudes.[32]

An incorporeal hereditament has been defined to mean a right growing out of, or concerning, or annexed to, a corporeal thing, but not the substance of the thing itself.[33]

[28] *Black's Law Dictionary.*

[29] *Black's Law Dictionary.*

[30] Blackstone, *Commentaries*, 1765-1769.

[31] Blackstone op. cit.

[32] *Black's Law Dictionary.*

[33] *Huston v. Cox*, 103 Kan. 73, 172 P. 992.

CHAPTER 2

EASEMENTS IN GENERAL

DEFINITION: WHAT IS AN EASEMENT?

An easement is a liberty, privilege, or advantage without profit, which the owner of one parcel of land may have in the lands of another; or, as conversely stated, it is a service which one estate owes to another, or a right or privilege in one man's estate for the advantage or convenience of the owner of another estate.[34]

It is not ownership of the fee, or the land itself, and is called a non-possessory interest in another's land for a special purpose.[35] It is more than a mere personal interest, and since it constitutes an actual interest in the land, it is regarded as realty.[36]

An easement exists within (in, on, over or under) land of another. A person cannot have an easement in their own land, nor would they need one, having already all the rights and benefits of land ownership.[37]

[34] C.J.S. Easements § 1.

[35] Easement is a right which one person has to use the land of another for a definite purpose. *Brown v. Sneider*, 400 N.E.2d 1322, 9 Mass. App. 329 (1980).

[36] "Easement" is not estate in land, nor is it land itself, but rather is property or interest in land. *Burris v. Cross*, 583 A.2d 1364 (Del.Super., 1990).

"Easement" is interest in land created by grant or agreement, express or implied, which confers right upon holder thereof to some profit, benefit, dominion, enjoyment or lawful use out of or over estate of another; thus, holder of easement falls within scope of generic term "owner." *Copertino v. Ward*, 473 N.Y.S.2d 494, 100 A.D.2d 565 (1984).

[37] 25 Am.Jur.2d § 2. A person cannot have an easement in his own land since all the uses of an easement are fully comprehended in his general right of ownership.

An "easement" is an interest which a person has in land in possession of another, and therefore an owner cannot have an easement in his own land. *Van Sandt v. Royster*, 83 P.2d 698, 148 Kan. 495 (1938).

The reason why one may not have an easement in his own land is that an easement merges with the title, and while both are under the same ownership the easement does not constitute a separate estate. *Sievers v. Flynn*, 305 Ky 325, S.W.2d 364 (1947).

Because an easement is for a special or definite purpose, the easement holder is limited as to what can be done on the land. Usually the scope is governed by the purpose of the easement, what is reasonable for its intended purpose, and within any stated restrictions.[38]

An easement is also often defined as being a right that one has in land of another *not inconsistent* with a general property in the owner.[39] One must be cautious however, as this may not always be the case. Easements created many years ago may still be in effect and contrary to the present owner's plans for use of the land or sometimes even prevent certain uses. Some of these outstanding easements may encumber the land to the degree that they are superior rights created and conveyed away long in the past, yet are still outstanding and very much in effect.[40] Those created, or the subject of conveyance, long ago are often outside the parameters of a normal title examination and often not discovered under present title standards. Even though a matter of record, such an easement may only appear well beyond the period of search (earlier in time), yet still be very much in existence, and have as much effect as when it was created.

Since an easement is an interest in property which, though distinct from an ownership interest in the land itself, nevertheless confers upon the holder of the easement an enforceable right to use property of another for specific purposes.[41] Because the interest is itself non-possessory, the holder of an easement does not have the degree of control over the burdened property that is enjoyed by the owner of the servient estate.[42]

Identical with Servitude *Servitude* is the civil law term for *easement* in the common law, and the two terms are often used indiscriminately.[43] It has been defined as the right of the owner of one parcel of land, by reason of the ownership, to use the land of another for a special purpose of his own, not inconsistent with the general

One cannot be said to have an easement in lands the fee-simple to which is in himself. *Othen v. Rosier*, 148 Tex. 485, 226 S.W.2d 622 (1950).

While two adjoining estates are both owned by the same person, no easement can be created in one of them for the benefit of the other. *Carby v. Willis*, 89 Mass. 364, 83 Am.Dec. 688 (1863).

[38] 25 Am.Jur.2d, § 72. A principle which underlies the use of all easements is that the owner of the easement cannot materially increase the burden of the servient estate or impose thereon a new and additional burden. Though the rights of the easement are paramount, to the extent of the easement, to those of the landowner, the rights of the easement owner and of the landowner are not absolute, irrelative, and uncontrolled, but are so limited, each by the other, that there may be a due and reasonable enjoyment of both the easement and the servient tenement. The owner of an easement is said to have all rights incident and necessary to its proper enjoyment, but nothing more.

[39] 25 Am.Jur.2d § 1.

[40] Grantees take title to land subject to duly recorded easements which have been granted by their predecessors in title. *Borders v. Yarborough*, 75 S.E.2d 541, 237 N.C. 540 (1953); *Waldron v. Town of Brevard*, 62 S.E.2d 512, 233 N.C. 26 (1950).

[41] *DeReus v. Peck*, 114 Colo. 107, 162 P.2d 404 (1945).

[42] *Wright v. Horse Creek Ranches*, 697 P.2d 384 (Colo, 1985).

[43] 28 C.J.S., Easements, § 1d.

property of the owner.[44] The word *servitude* should not, however, be construed as the same as *equitable servitude*, which is a form of restrictive covenant.

An example would be a perpetual servitude of passage over their respective property for the use and benefit of the [grantees], described as a "30 foot road servitude."[45]

Different from License An easement differs from a *license* by the latter being a mere *revocable* right, privilege, or permission to enter upon or do acts upon another's land. A license is usually temporary, usually may not be either bought or sold, and is revocable at any time. However, in some cases it may be permanent and treated much the same as fee ownership.

An example of a license is the right to leave one's vehicle in a parking area when shopping, attending a sports event, or at a restaurant. Another is having verbal or written permission to walk on a neighbor's land in the performance of a service, such as carrying out a land survey or painting a building.

Different from Profit A *profit a prendre*, also called "right of common," is the right exercised by one person in the soil of another accompanied by participation in the profits of the soil. It is not an easement since one of the features of an easement is the absence of right to participate in the profits of the soil charged with it. It is similar to an easement, however, in that it is an interest in the land. It is created by grant and may be either appurtenant or in gross.[46]

Examples of rights of *profit a prendre* are the right to cut timber; the right to take gravel,[47] soil, coal, or other minerals generally from the land of another; the right to graze animals; and the rights to fish or hunt. Flowing water is not considered a product of the soil; therefore. the right to take such is an *easement* not a *profit a prendre*. A right to take ice,[48] however, has sometimes been considered a right of *profit a prendre*.[49]

Differing from Fee[50] Sometimes it is difficult to distinguish between easement and fee, depending on the wording of the document. Certain creations only result in

[44] 25 Am.Jur.2d § 1.

[45] *Elston v. Montgomery*, No. 46,262-CA, Court of Appeals of Louisiana, Second Circuit, May 18, 2011.

[46] 25 Am.Jur.2d § 4.

[47] The words "reserving the gravel" in a deed created in the grantor a *profit a prendre*. *Beckwith v. Rossi*, 157 Me. 532, 174 A.2d 732 (1961).

[48] The grant of a right to harvest ice amounts to more than merely a revocable license; it constitutes a right in the nature of an easement appurtenant to the land or a *profit a prendre*. *Gadow v. Hunholz*, 160 Wis. 293, 151 N.W. 810.

[49] 25 Am.Jur.2d § 4.

[50] The following cases are by no means intended to be comprehensive but only to illustrate the differences among various courts regarding the descriptive language used in various conveyances. Researchers should employ all of the tools at their disposal and thoroughly study the variety of decisions relative to the particular question at hand. *136 ALR 379, Deed as conveying fee or easement*, is a logical place to begin.

an easement, depending on the law at the time of creation.[51] Others, which can go either way, may require legal research to develop an opinion, or, in the extreme, a court ruling.

For example, ordinarily, absent modern statutes, the creation of a highway results in an easement right in favor of the public. However, the New Hampshire court decided in 1871, in the case of *Coburn v. Coxeter*,[52] that a conveyance of a strip of land, in explicit terms, with a restriction that it shall be used only for a road, was a grant of the fee, and not of a mere easement. Their reasoning was that "it is inconceivable that land should have been granted in explicit terms, when only an easement was intended."

Three key elements in the granting document are[53]:

- Does it recite "land" or "right"?
- The amount of the consideration
- Later treatment and wording
- The wording in the document(s)

  When documents use the word "land," there is an *indication* that it is fee being spoken of, whereas if the word "rights," "use," or similar vernacular is used, that is an *indication* that an easement is being described. When the word "damages" appears, that is a *strong indication* that only an easement is involved since damages are paid where there is a taking, and most takings, absent a statutory provision to the contrary, result in an easement. Even if there is a statute allowing the fee to be taken, the taking authority may still acquire an easement if that is all they need or want.
- The consideration.

  It would seem that the purchase of the fee in a parcel would be more than purchasing merely an easement. Comparing land values at the point of time in question may provide insight into what was being purchased. However, a word of caution: sometimes the property acquired was so small that the difference between the value of the fee and the value of an easement may not be a reliable guiding factor.
- Subsequent treatment or description of the property.

  Subsequent documents, especially those of the servient estate, may lend insight into whether fee or easement was conveyed. For instance, if the grantor later stated "subject to rights conveyed to . . . " would indicate an easement, whereas "excepting the land sold to . . ." would indicate the prior conveyance of fee.

[51] For instance, common law dedication results in an easement, as does condemnation, absent a special statute providing otherwise.

[52] 51 N.H. 158 (1871).

[53] See the subsequent decision of *Ray v. King County* for a more extensive list.

Selected Examples — Private Way Deeds given for the creation of a right of private passage across a designated parcel are similar to those given to railroad companies or for railroad purposes. If the granting clause purports to convey a strip, piece, or parcel of land rather than a right or privilege with respect to land, the conveyance, in the absence of other information regarding the intent of the parties, will be construed as passing title in fee. If the granting clause conveys a right or privilege with respect to the land rather than the land itself, the conveyance, absent other information regarding intention, will be construed as passing only an easement. Generally, there will be other information from which to ascertain the intent of the parties.

Wording	Decision	Case
"hereby grant to the [grantee], his heirs and assigns for the sole purpose of an alleyway, to be used in common with the owners of other property adjoining said alleyway, all that tract of land," etc.	Easement	*Pellisier v. Corker* (1894) 103 Cal. 516, 37 P. 465
"grant, bargain and convey: All of that certain lot or parcel of land," and, after a description of the land in question, providing: "This grant is for the purpose of granting to party of the second part a right of way from his premises to the country road."	Easement	*Parks v. Gates* (1921) 186 Cal. 151, 199 P. 40
"also grant to" the grantee "a strip of land described as follows, to wit . . . for a road to and from said premises first above described"	Fee	*Soukup v. Topka* (1893) 54 Minn. 66, 55 N.W. 824
"And also all that tract or parcel of land......being a strip of land three rods wide excepted and reserved" from the land previously conveyed, "which land was reserved from said conveyance . . . and is hereby conveyed in order that cattle and horses may have the use of water at said spring."	Fee	*Walls v. Rodman* (1937) 252 App.Div.826, 299 NYS 314
"grant, release and confirm unto" the grantee "all that certain strip of land hereinafter described, of the width of 24 feet for a private road,"	Fee	*Kilmer v. Wilson* (1867) 49 Barb (NY) 86

Selected Examples — Public Way Most cases involving the interpretation of deeds to public entities for public street or highway purposes almost exclusively involve instruments which purport to convey "land" rather than a "right." Language in these cases nevertheless indicates that such a deed meant to convey only a "right" or "privilege," the broad principle that deeds conveying a right rather than land would be applicable and, in the absence of further evidence, the instrument would be construed as passing merely an easement.

Wording	Decision	Case
"the same to be used as a street or highway and for no other purposes"	Easement	*Bainton v. Clark Equipment Co.* (1920) 210 Mich. 602, 178 N.W.
"for highway purposes"	Easement	*Mueller v. Schier* (1926) 189 Wis. 70, 205 N.W. 912
"grant . . . all that certain tract and parcel of land . . . described as follows . . . together with . . . the reversion and reversions, remainder and remainders, rents, issues and profits thereof," and containing an habendum clause reading: "to have and to hold all and singular, the said premises, for the purposes of a public road of said city"	Fee	*Cooper v. Selig* (1920) 48 Cal App. 228, 191 P. 983
"do hereby grant, bargain, sell and convey . . . the said parcel of land herein above last described, being the width of thirty feet. To have and to hold the same for their use and behoof for a . . . public roadway"	Fee	*Mt. Olive Stave Co. v. Hanford* (1914) 112 Ark 522, 166 S.W. 532

Selected Examples — Railroads A railroad right of way is an interest in the realty. Such a right of way usually is an easement merely. Under a statute providing that real estate received by a railroad shall be held and used for the purpose of such grant only, the rule has been stated that where a railroad acquires a right of way, whether by condemnation, voluntary grant, or conveyance in fee upon a valuable consideration, it takes by a mere easement over the land, not a fee. Even though a right of way acquired by a railroad company is ordinarily designated an "easement," it is well settled that it is an interest in land of a special and exclusive nature and of a high character, and it has also been held that a railroad right of way may be a fee, and so far as the right of possession for railroad purposes is concerned, it has most of the qualities of an estate in fee, including perpetuity and exclusive use and possession. Whether it is a fee or an easement generally depends upon the character of the acquisition.[54]

The fact that the deed contains covenants of warranty, or that the right acquired is designated as "a fee," is not necessarily controlling. But an interest in fee, and not merely an easement, has been held to be conveyed by a deed to a railroad company, its successors and assigns, forever, to have and to hold for all purposes mentioned in the act of incorporation. The general rule is that conveyances to railroads which purport to grant and convey a strip, piece, parcel, or tract of "land," and do not contain additional language relating to the use of purpose to which the land is to be put, or in other ways cut down or limit, directly or indirectly, the estate conveyed, are usually construed as passing an estate in fee.[55]

It has been held that where a deed contained additional language referring in some way to a "right of way," it operated to convey a mere easement rather than a title in fee. The mention in such a conveyance of rights, members, and appurtenances belonging and appertaining to the strip of land conveyed, or the use of the words, "forever in fee-simple," has been held not to enlarge such an easement into a fee. A railroad to which is conveyed a "right of way" for "all the uses and purposes of said railroad company so long as the same shall be used for the construction, use and occupation of said road company" acquires an easement only.[56]

NOTE

There may be a difference between land for right of way purposes and land for other purposes. Courts have frequently regarded land for right of way as easement, but land for material and improvements as fee (see Figure 2.1).

---

[54] 65 Am Jur 2d Railroads, § 73.
[55] 65 Am Jur 2d Railroads, § 76.
[56] 65 Am Jur 2d Railroads, § 77.

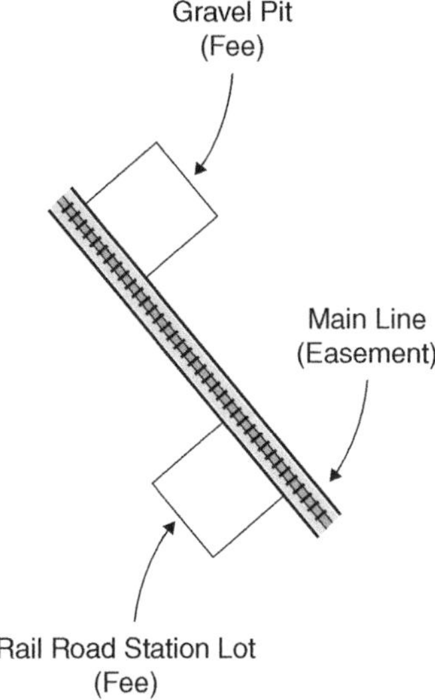

Figure 2.1 Section of railroad track illustrating both easement and fee ownership.

Wording	Decision	Case
All that certain lot, piece or parcel of land . . . bounded and particularly described as follows……..	Fee	*Moakley v. Blog* (1928) 90 Cal. App. 96
A strip of land 100 feet wide . . . being 50 feet each side center line	Fee	*Radetsky v. Jorgensen* (1921) 70 Colo 423
Right of way then "also two acres of land for purposes of a railroad depot	Easement and Fee	*Askew v. Vicksburg, S. & P.R. Co.* (1931) 171 La. 947
To be used for railroad purposes only: A parcel of land 100 ft. in width, lying 50 ft. on each side of the center line of the [railroad], as located and established upon and across the lands of said [grantor]	Fee	*Quinn v. Pere Marquette R. Co* (1931) 256 Mich. 143
a strip, tract, or parcel of land for a railroad right of way . . . 100 ft. in width, being 50 ft. on each side of the centerline of the . . . railroad as surveyed	Easement	*Sherman v. Petroleum Exploration* (1939) 280 Ky. 105

(continued)

Wording	Decision	Case
deed headed "Right of way"	Easement	*Knoxville v. Phillips* (1931) 162 Tenn. 328
"hereby bargains, sells and conveys unto the said [company] and their successors to be forever used enjoyed by the said company for railroad purposes only,"	Fee	*Nashville, C. & St. L.R. Co. v. Bell* (1931) 162 Tenn. 661
granting, bargaining, selling etc., for a small consideration, to a railroad, its successor and assigns, a "piece or parcel of land, being a strip of land 100 feet wide, for the purpose of constructing and maintaining a railroad thereon,"	Easement	*Chicago, R.I. & P.R. Co. v. Olson*, (1953) 222 Ark. 828

Illustrative Case The case of *Ray v. King County*[57] takes this discussion to a higher level and demonstrates the difficulty the court system sometimes faces. This difficulty and frustration arise whenever a similar conveyance is encountered. The issue in the case was whether a grant to a railroad resulted in the grant of the fee or a mere easement.

The court discussion is as follows:

"Where a deed conveys a right of way to a railroad, the conveyance may be fee-simple or may be an easement only. The interpretation of such a deed is a mixed question of fact and law. It is a factual question to determine the intent of the parties. Courts must then apply the rule of law to determine the legal consequences of that intent. Whether a conveyance is one of fee title or an easement is a conclusion of law as to the effect of a deed.

The description in the deed in question:

Hilchkanum to S.L.S. and E.R.Y. Co.; Right of Way Deed

In consideration of the benefits and advantages to accrue to us from the location construction and operation of the Seattle Lake Shore and Eastern Railway in the County of King, in Washington Territory, we do hereby donate grant and convey unto said Seattle Lake Shore and Eastern Railway Company a right of way one hundred (100) feet in width through our lands in said County described as follows to-wit

Lots one (1) two (2) and three (3) in section six (6) township 24 North of range six (6) East.

[57] 120 Wn.App. 564, 86 P.3d 183 (Wash.App. Div. 1 2004).

Such right of way strip to be fifty (50) feet in width on each side of the center line of the railway track as located across our said lands by the Engineer of said railway company which location is described as follows to-wit.

Commencing at a point 410 feet West from North East corner of Section six (6) township 24 N R 6 East and running thence on a one (1) degree curve to the left for 753 3/10 feet thence South 16 degrees and 34 minutes West 774 2/10 feet thence with a 3 degree curve to the right for 700 feet thence with an 8 degree curve to the right for 260 4/10 feet thence South 58 degrees and 24 minutes West 259 6/10 feet thence with an 8° curve to the left for 564 4/10 feet thence South 13° 15' W 341 4/10 feet thence with a 6° curve to the right for 383 3/10 feet thence S 36° 15 W 150 feet to South boundary of lot 3 of said Sec 6 which point is 1320 feet North and 2170 feet west from SE corner of said Sec 6

And the said Seattle Lake Shore and Eastern Railway Company shall have the right to go upon the land adjacent to said line for a distance of two hundred (200) feet on each side thereof and cut down all trees dangerous to the operation of said road.

In *Brown v. State,* 130 Wn.2d 430, 924 P.2d 908 (Wash. 1996), our supreme court most recently articulated the principles governing resolution of the mixed questions of fact and law before us. There, the court resolved a dispute between property owners abutting the railroad right of way, who claimed reversionary interests in it, and the state, which purchased the right of way from a successor in interest to the original grantees of the strip under some 37 deeds. The deeds, which were dated between 1906 and 1910, were on preprinted forms with blank lines containing handwritten descriptions of the specific properties conveyed. The court ultimately held that the deeds conveyed fee-simple title because they were "in statutory warranty form, expressly convey fee-simple title, and contain no express or clear limitation or qualification otherwise." The court began its analysis by noting that the decisions dealing with conveyancing of rights of way to railroads in various jurisdictions "are in considerable disarray" and "turn on a case-by-case examination of each deed." In Washington, the general rule is that when construing a deed, "the intent of the parties is of paramount importance and the court's duty to ascertain and enforce."

The court then identified the following factors for determining intent: (1) whether the deed conveyed a strip of land, and did not contain additional language relating to the use or purpose to which the land was to be put, or in other ways limiting the estate conveyed; (2) whether the deed conveyed a strip of land and limited its use to a specific purpose; (3) whether the deed conveyed a right of way over a tract of land, rather than a strip thereof; (4) whether the deed granted only the privilege of constructing, operating, or maintaining a railroad over the land; (5) whether the deed contained a clause providing that if the railroad ceased to operate, the land conveyed would revert to the grantor; (6) whether the consideration expressed was substantial or nominal; and (7) whether the conveyance did or did not contain a habendum clause, and many other considerations suggested by the language of the particular deed. In addition to the language of the deed, we will also look at the circumstances surrounding the deed's execution and the subsequent conduct of the parties.

The court also noted the special significance that has been accorded the term "right of way" in Washington deeds.

The court, in Ray, analyzed the conveyance in light of the seven criteria previously enumerated along with the eighth item, the subsequent conduct of the parties The conclusion was that the 1887 deed from the Hilchkanums conveyed fee title to the Railway Company.

Selected Examples — Miscellaneous *Cemeteries.* As pointed out in 10 Am Jur 503, Cemeteries, § 22, an interest in a burial lot is of a somewhat peculiar nature, and although the courts usually, but not invariably, hold that the owner of such a lot does not have a fee title thereto, they are not in agreement as to the precise nature of the interest held, some calling it an "easement," some a "license," and others a "privilege."[58]

Wording	Decision	Case
"One who purchases a lot in a public cemetery for burial purposes, though the right there in be exclusive, does not acquire any title to the soil, but only an easement or license for the use intended."	Easement or license	*Nicolson v. Daffin* (1914), 142 Ga. 729, 83 S.E. 658
"As it is a public cemetery, the sale of lots therein does not pass to the grantees the title in fee to such lots, but thereby assures to the grantee a license or easement therein for burial purposes, so long as the cemetery shall be used for cemetery purposes. . . . The fee of the lots in such case remains in the trustees of the cemetery."	Easement or license	*McWhirter v. Newell* (1903) 200 Ill. 583, 66 N.E. 345

Illustrative Decision The case of *American Trading Real Estate Properties, Inc. v. Town of Trumball*,[59] serves as an illustration. In this case, the court noted: "The trial

[58] Most of the writings speak in terms of public cemeteries, private cemeteries being a different situation. Title to private cemeteries may be fee, easement, license, or in trust, so an investigation is necessary to determine not only the status of the title, but also in whom it rests.

[59] 215 Conn. 68; 574 A.2d 796 (Conn., 1990).

court found the following facts. The property at issue in this case is an overgrown and impassable roadway running between Route 111 and Old Mine Park in the town of Trumbull. The roadway was created by grant in 1867, when the defendant's predecessor in title purchased "a narrow strip of land for the purpose of a road or passage way . . . being twenty feet in width and about nine hundred & fifty feet in length. . . ." Immediately following this provision, the 1867 deed reserved to the grantors "the privilege of passing over, crossing and recrossing said tract whenever our convenience or necessity may require."

The precise description is noteworthy:

> *a certain narrow strip of land for the purpose of a road or passage way, commencing at the Monroe Turnpike about Two Hundred and fifty five feet East from a certain Elm stump which constitutes said Hubbell's Western boundary point and running through land of the said Mary J. Turney in a line with a certain apple tree with dead top, till it intersects an old road or cart path a little to the East of a large chestnut tree, from thence following said path till it reaches land now belonging to said Grantee, at a certain pair of bars, said East post being said Eastern boundary point, said tract being twenty feet in width and about nine hundred & fifty feet in length—reserving to ourselves and our heirs and successors the privilege of passing over, crossing and recrossing said tract whenever our convenience or necessity may require.*

"Our discussion of this issue must begin with the case of *Chatham v. Brainerd*, 11 Conn. 60 (1835), which, despite its venerable age, is the controlling precedent. In *Chatham v. Brainerd*, this court was called upon to interpret the language of a resolution by which a town granted one acre of land 'for a burying ground.' Although the court recognized the possibility that a grantor could limit a conveyance to an easement while retaining fee-simple title to the property, the court concluded that the provision 'for a burying ground' did not evince an intent to create an easement. Instead, the court held, 'it was the land, which was granted, though the object was designated.' The court further emphasized that 'the word land, comprehends any ground, soil or earth whatsoever.' [E. Coke, *A Commentary upon Littleton*, Book 1, § 1, p. 4 (1703)].' Like the language at issue in *Chatham v. Brainerd*, supra, the deed in the present case made specific reference to 'a certain narrow strip of land' as the subject of the conveyance, and then went on to state the object of the conveyance without expressing any intention to limit the grant to an easement."

Canals A canal may be a particular type of estate, or it may comprise a combination of several different estates. Sometimes particular wording is interpreted the same as that for highways, since, in a sense, a canal may be thought of as a type of highway.

Restrictive covenants. Restrictive covenants concerning the use of the land or the location or character of improvements on the land have been said to create easements. Such covenants, being limitations on the manner in which one may use his land, do not create true easements, but rather rights in the nature of servitudes or easements, some of which are characterized as "negative easements" or "reciprocal negative easements."[60]

Natural rights. Certain rights, such as lateral and subjacent support of the land in its natural condition, or the right to receive the benefit of the flow of a natural watercourse and the discharge of surface water from a higher to a lower plane, are said to be inherent in the land and are termed "natural rights." They have been said to be similar to easements, but they are not true easements, since they have not been created by an act of man, but strictly by nature.[61]

EASEMENT TERMINOLOGY

In order to fully grasp the nature of easements, and to study or research them, it is necessary to be familiar with their unique terminology.

Dominant and Servient Ownership The land benefited by an easement is known as the *dominant tenement* or *dominant estate*, while the land burdened or encumbered by the easement is known as the *servient tenement* or *servient estate*. Consequently, the respective owners are known as the *dominant owner* or the *servient owner* (See Figure 2.2).[62]

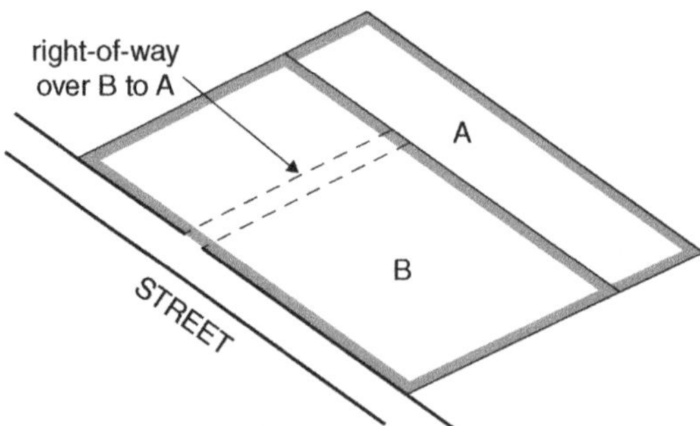

Figure 2.2 The owner of parcel A, benefitted by the right of way, is known as the *dominant tenement*, parcel B is burdened by the right of way and is known as the *servient tenement*.

[60] 25 Am.Jur.2d § 5.

[61] 25 Am.Jur.2d § 6.

[62] An easement consists of two separate estates: the dominant estate, which has benefit of easement and to which it is attached, and the servient estate on which easement is imposed or rests. *Edgell v. Divver, 402 A.2d 395 (Del.Ch., 1979).*

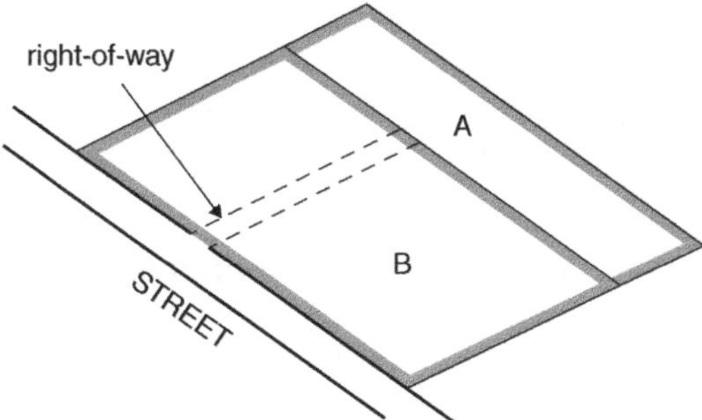

Figure 2.3 The aforementioned right of way in the former diagram over Lot B is *appurtenant* to Lot A.

Appurtenant Easements vs. Easements in Gross If an easement was created for the purpose of benefitting *other land* owned by the holder of the easement, it is *appurtenant*—that is, appurtenant to the land it benefits (see Figure 2.3).[63]

The important characteristic of an appurtenant easement is that it *runs with the land* and is a part of a conveyance of land, whether or not they are specifically mentioned in the conveyance.[64] An old saying is, "once an easement, always an easement," which is essentially true, since easements do not automatically disappear if they are not described in a conveyance or if they are not used.[65] Easements, in general, continue until properly terminated. Extinguishment, or termination, of an easement is a distinct process and is the subject matter of Chapter 5, where it is discussed in detail.

An easement *in gross* does not benefit any other land; it exists independently of other land and is a mere personal interest in or right to use the land of another.[66] In

[63] *Davis v. Briggs*, 117 Me. 536, 105 A. 128 (1918).

[64] A deed that conveyed the dominant estate passed an easement that ran with the land, though the deed did not mention such easement. *Khori v. Dappinian*, 125 A. 268, 46 R.I. 163 (1924).

When an easement, though not originally belonging to land, has become appurtenant to it by grant or prescription, a conveyance of the land will carry with it such easement, whether mentioned in the deed or not, though it may not be necessary to the enjoyment of the land by the grantee. *Cole v. Bradbury*, 29 A. 1097, 86 Me. 38 (1894); *Dority v. Dunning*, 6 A. 6, 78 Me. 381 (1866).

[65] As a general rule, an easement acquired by grant or reservation cannot be lost by mere nonuse for any length of time, no matter how great. 23 Am.Jur.2d § 105.

[66] An easement in gross cannot be acquired except by grant or by prescription. *Branan v. Wimsatt*, 298 F. 833, 54 App. D.C. 374 certiorari denied 44 S.Ct. 639, 265 U.S. 591, 68 L.Ed. 1195 (1924); *Miller v. Lutheran Conference & Camp Ann'n*, 200 A. 646, 331 Pa. 241, 130 A.L.R. 1245.

The general rule in the United States today with respect to easements in gross is that if they are commercial in character, such as railroad or public utility easements, they are transferable, noncommercial easements in gross are transferable only if such intention is indicated in its creation, such as by the use of the words "heirs and assigns." 2 Am.L.Prop. § 5; RESTATEMENT § 489, 491.

other words, there is a servient tenement but no dominant tenement.[67] Examples of easements in gross are utility rights-of-way, major highways, and the right to flow land with water by damming a stream or river. Easements in gross do not impart specific benefit to other land owned by the holder of the easement and exist independently.

Determining Whether an Easement Is Appurtenant or in Gross Occasionally it is difficult to identify whether a particular easement is appurtenant (running with the land it benefits) or in gross (existing independently). Whether a particular easement is classed as appurtenant or in gross depends mainly on its nature and the intention of the parties creating it.[68] If the easement by its nature is an appropriate and useful adjunct to the land conveyed, and there is nothing to show that the parties intended it to be a more personal right, it should be an easement appurtenant and not an easement in gross.[69] Easements in gross are not favored in the law, and an easement will never be presumed as personal when it may fairly be construed as appurtenant to some other estate. If doubt exists as to its true nature, an easement will be presumed to be appurtenant, rather than in gross.[70]

Whether an easement is appurtenant or in gross is to be determined by a fair interpretation of the grant or reservation creating the easement, aided, if necessary, by the situation of the property and surrounding circumstances.[71] Chapter 6 is devoted to the interpretation of language found in appropriate documents.

Affirmative as Opposed to Negative Easements may also be classed as either *affirmative* or *negative*. An affirmative easement is one that *allows* the holder of the easement to do certain acts on land of another (see Figure 2.4).[72]

[67] For example, a public utility company's easement to place and maintain power lines across another's land.

Easement in gross usually has no dominant estate; it is purely servient. *DeSon v. Parker*, 361 N.E.2d 457, 3 O.O.3d 430.

[68] There is no incompatibility in the grant of a right of way to several persons, which grant confers a right appurtenant as to some of the grantees and a right in gross as to the others; such a grant may be partly appurtenant and partly in gross. *Louisville & N.R. Co. v. Koelle*, 104 Ill. 455; *Davis v. Briggs*, 117 Me. 536, 105 A. 128 (1918).

An "easement in gross," because of its personal nature, is not assignable or inheritable, but an "easement appurtenant" runs with the land. *Davis v. Briggs*, 105 A. 128, 117 Me. 536 (1918).

In determining whether there was an easement in gross or an appurtenant easement, intent of parties to be gathered from the nature of the subject matter and language used in deed, was required to control. A construction that an easement is on appurtenant rather than in gross is favored. *Sabins v. McAllister*, 76 A.2d 106, 116 Vt. 302 (1950).

[69] *Greenwalt v. McCardell*, 178 Md. 132, 12 A.2d 522 (1940); *Smith v. Dennedy*, 224 Mich. 378, 194 N.W. 998 (1923); *American Rieter Co. v. Dinallo*, 147 A.2d 290.

[70] 25 Am.Jur.2d § 13.

In case of doubt, presumption favors easement being appurtenant. *Sullivan Granite Co. v. Vuono*, 137 A. 687, 48 R.I. 292 (1927).

[71] 25 Am.Jur.2d § 13.

[72] Maine Real Estate Law, Chapter 6.

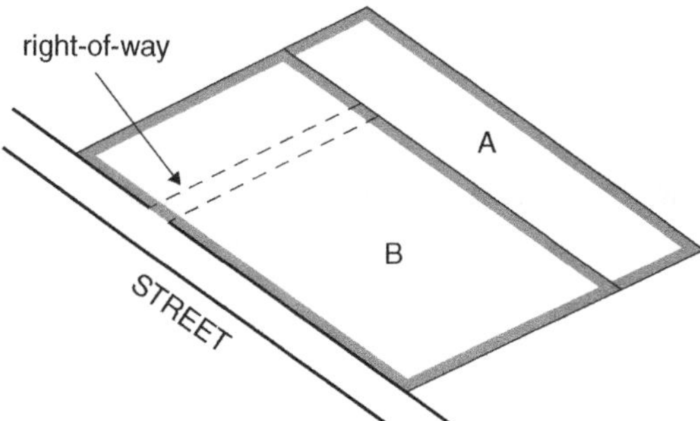

Figure 2.4 A right of way across Lot B as an access to Lot A is an affirmative easement for Lot A.

A negative easement is one that allows the holder of the easement to prevent an owner from doing certain acts on his own land (see Figure 2.5).[73]

A common modern example of a negative easement is a *conservation easement*, whereby development rights of a parcel are permanently controlled by other than the landowner. (see the section on conservation easements in Chapter 3).

Apparent vs. Nonapparent Easements may be known as apparent or nonapparent. An *apparent easement*, as the name implies, is one that is obvious, and one that is understood to be open and visible, such as a pathway or a road. The word "apparent," however, does not necessarily mean *actual* visibility, but rather "susceptibility of ascertainment on reasonable inspection by persons ordinarily conversant with the

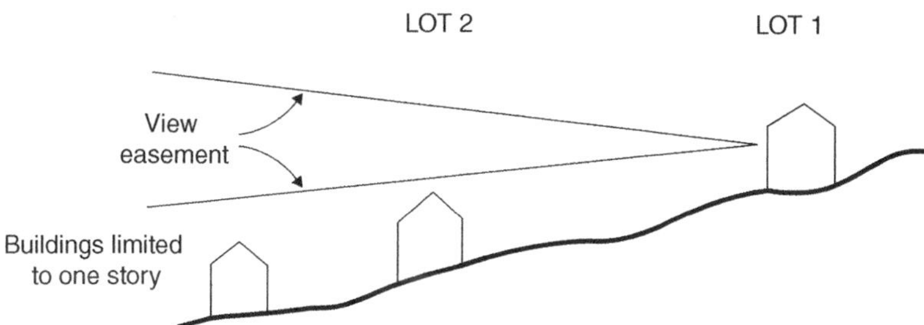

Figure 2.5 A view easement for the owner of Lot 1 is a negative easement over Lot 2, since it prevents 2 from blocking 1's view.

[73] Maine Real Estate Law, Chapter 6.

subject." An underground drain may be an apparent easement, even though not visible from the surface.[74]

Continuous vs. Noncontinuous Easements may also be *continuous* or *noncontinuous*, also known as *discontinuous*. A continuous easement is one in which the enjoyment is, or may be, continuous without the necessity of any actual interference by man,[75] or as one depending on the natural formation of, or some artificial structure upon, the servient tenement, obvious and permanent, such as the bed of a running stream, an overhanging roof, a pipe for conveying water, a waterspout that discharges water whenever it rains, a drain, or a sewer.[76]

A noncontinuous, or discontinuous, easement is one that can only be enjoyed through the interference of man, that is, one that has no means specially constructed or appropriated to its enjoyment and that is enjoyed at intervals, leaving between those intervals no visible sign of its existence, such as a right of way or a right to draw water.[77]

Perpetual vs. Temporary Easements are presumed to be perpetual unless terminated by their own terms or extinguished in one of the formal manners outlined in Chapter 5. Easements are often created to endure forever. They are sometimes called *perpetual* or *permanent servitudes* because they are without limitation as to duration. These are sometimes referred to as *easements in fee*. This term is not recommended for use, since it can be misleading. An easement in fee should not be confused with ownership in fee-simple or fee-simple absolute.

Example: "As a condition of each such payment by the Authority, the petitioner to whom it is made shall grant to the Authority a permanent easement in fee, free

[74] 25 Am.Jur.2d § 9.

An easement is "apparent" if its existence is indicated by signs, which must necessarily be seen or which may be seen or known on careful inspection by persons ordinarily conversant with the subject. *Slear v. Jankiewicz*, 54 A.2d 137, 189 Md. 18 (1947), certiorari denied 68 S.Ct. 453, 333 U.S. 827, 92 L.Ed. 1112 (1948).

Where licensed surveyor, whom landowner had retained to survey tract for, *inter alia*, visible evidences of interests not appearing of record, did not discover such use of right of way by trucks going to and from loading platform on adjoining property as to warrant notice, such use of way for that purpose lacked those qualities of apparency and permanency essential to creation of easement by implied grant. *A.J. and J.O. Pilar, Inc. v. Lister Corp.*, 123 A.2d 536, 22 N.J. 75 (1956).

[75] The test of continuousness is that there is an alteration or arrangement, permanent in nature, that makes one tenement dependent in some measure on another. *German Sav. & L. Soc. v. Gordon*, 54 Or. 147, 102 P. 736 (1909).

[76] 25 Am.Jur.2d § 10.

[77] Ibid.

A way is said to be a noncontinuous easement, because in the use of it there is involved the personal action of the owner in walking or driving on it. *Hoffman v. Shoemaker*, 69 W.Va. 233, 71 S.E. 198 (1911).

A cement driveway between houses on adjoining lots, though a visible easement, is not a continuous one. *Milewski v. Wolski*, 314 Mich. 445, 22 N.W.2d 831, 164 A.L.R. 998 (1946).

of all encumbrances over and affecting the grantor's land (which easement shall be appurtenant to the locus) to maintain the project as against the properties of the several petitioners."[78]

Easements for construction purposes are generally temporary, since they are needed only during the construction project for which they are created. Occasionally they are called "temporary construction easements."

An easement for a particular purpose, while not expressly limited as to time, may prove to be temporary if that purpose disappears. Easements so limited may not be perpetual.[79]

Public or Private Easements Easements may be classified as private or public, depending on whether they are in favor of an individual or the public at large. A private easement is defined to be "a privilege, service or convenience which one neighbor has of another, by prescription, grant or necessary implication, without profit."[80]

Private easements exist independent of public easements[81]; the distinguishing feature is that in case of private easements there must be two distinct tenements, one dominant and the other servient, whereas, public easements are easements in gross, and in this class there is no dominant tenement.[82]

INTERMITTENT EASEMENTS

An intermittent easement is one that is usable or used only at times, and not continuously.[83] An example is a winter road (see Figure 2.6).

New Hampshire statute:

> § 231:24. Winter Roads
>
> The selectmen, upon petition, may, in any case where, in their judgment, the public good requires it, layout a public road exclusively for winter use, such public road to be open only from November 15 until April 1, and they shall assess the damages to the owners of land over which such road may pass in the form of yearly rentals. Hearings

[78] *Kelloway v. Board of Appeal of Melrose*, 361 Mass. 249, 280 N.E.2d 160 (1972).

[79] It appears to have been the intent of the parties to continue the right of passage only so long as either of the buildings then in existence continue to stand, and not to create a perpetual easement attaching to the land of the dominant estate. *Cotting v. City of Boston*, 201 Mass. 97, 87 N.E. 205 (1909).

[80] *Cleveland, C., C. & St. L. Ry. Co. v. Munsell*, 94 Ill.App. 10.

[81] A common example is dedicated streets in a subdivision. If accepted, they become public; whether the streets are accepted or not, lot owners have private easements in the streets independent of any public rights. This is true whether or not streets have been constructed or improved.

[82] 28 C.J.S. § 3e.

[83] *Eaton v. Railroad Co.*, 51 N.H. 504, 12 Am. Rep. 147.

shall be had upon 7 days' notice to landowners. In all other respects such laying out shall be subject to the provisions for laying out a class V highway.

The existence of gaps in time between a claimant's use of another's land will not necessarily destroy the continuity of use. Easements that are seasonal or periodic may be acquired by prescription. For example, one may obtain a prescriptive easement by driving cattle to and from a summer range, by using a beach or a driveway only during the summer, by traveling a roadway in the haying season, or by making seasonal use

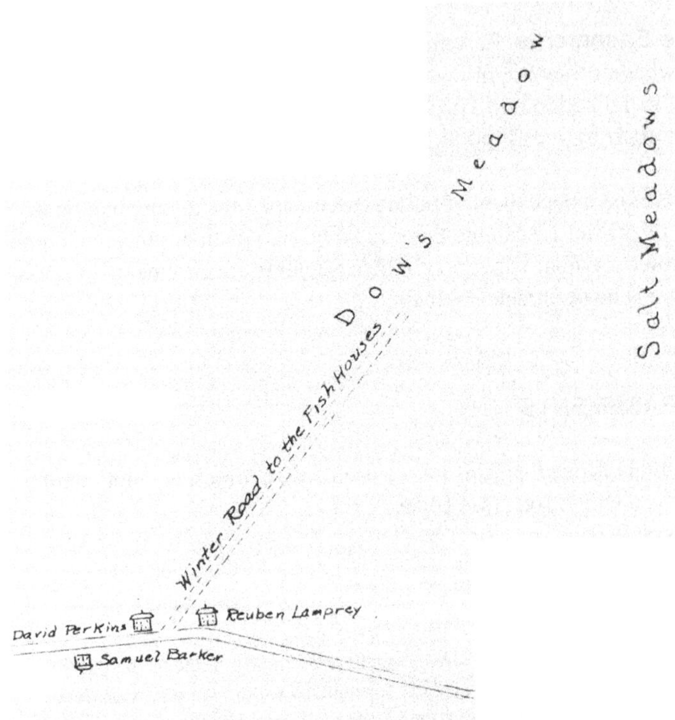

Figure 2.6 Example of a "winter road" shown on an 1841 map.

of a path. Likewise, intermittent but recurring use of a rural roadway for hauling wood and other purposes constitutes the required continuity. Further, in an unusual case, . . . a county club acquired a prescriptive easement to use adjacent land as a rough when several poorly hit golf balls landed on the servient estate each day. In an equally unusual circumstance, the continuity requirement was satisfied by the existence and infrequent use of a fire escape between two buildings. It has also been held that infrequent use of a roadway to visit lots purchased for eventual retirement living was continuous. Moreover, in assessing the continuity requirements courts will consider use by visitors, service providers, and family members attributable to a claimant.[84]

Quasi-Easements Rights in the nature of easements, but not possessing all the essential qualities, are sometimes mentioned as "quasi easements" or "seeming servitudes." They are existing conditions in the land retained, the continuance of which would be so clearly beneficial to the land conveyed that they would be presumed to be intended and must be such as are apparent in the sense of being indicated by the objects that are necessarily seen or that would be ordinarily visible by persons familiar with the premises. They may arise where the dominant and servient estates are severed after having been unified, and the use by the owner of one part of his land for the benefit of another part, is spoken of as a quasi-easement, and the land benefited is referred to as the *quasi-dominant tenement*, which the part utilized for the benefit of land benefited is called the *quasi-servient tenement*.[85]

An easement, in the proper sense of the word, can only exist in respect of two adjoining pieces of land occupied by different persons, and can only impose a negative duty on the owner of the servient tenement. Hence an obligation on the owner of land to repair the fence between his and his neighbor's land is not a true easement but is sometimes called a "*quasi*-easement."[86]

Reciprocal Negative Easements If the owner of two or more lots, so situated as to bear the relation, sells one with restrictions of benefit to the land retained, the servitude become mutual, and, during the period of restraint, the owner of the lot or lots retained can do nothing forbidden to the owner of the lot sold, this being known as the doctrine of "reciprocal negative easement."[87] (See Figure 2.7.)

Secondary Easements A secondary easement is one that is appurtenant to the primary or actual easement; every easement includes such "secondary easements,"

[84] *O'Dell v. Stegall*, 703 S.E.2d 561 (W.Va., 2010) citing Bruce & Ely, *The Law of Easements and Licenses in Land*, § 5:15 (Footnotes omitted).

[85] 28 C.J.S., § 1d; *Van Sandt v. Ryoster*, 83 P.2d 698, 148 Kan. 495 (1938).

[86] Gale, Easements, 516, cited in Black's Law Dictionary.

[87] *Sanborn v. McLean*, 233 Mich. 227, 206 N.W. 496, 497 (1925), cited in *Black's Law Dictionary*.

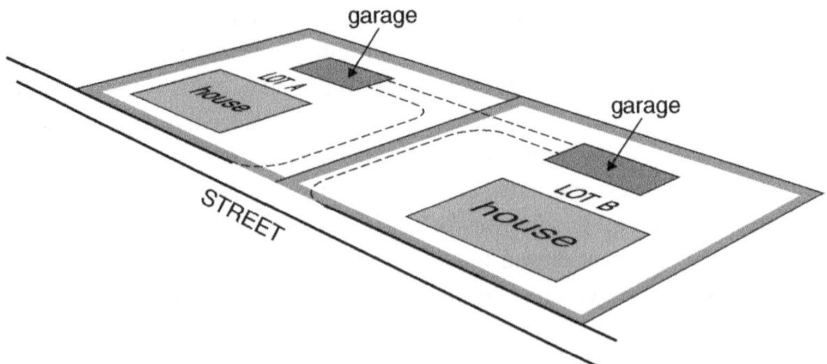

Figure 2.7 An example of a reciprocal negative easement.

that is, the right to do such things as are necessary for the full enjoyment of the easement itself.[88]

The extent of an implied or secondary easement is dependent on the purpose and extent of the grant of the primary easement. The right is limited and must be exercised in such reasonable manner as not to injuriously increase the burden on the servient tenement.[89]

Easement by Substitution In one case, the defendants [further] alleged that the trial justice erred in failing to address their claim for an easement by substitution over the driveway. The court stated "we have held that when the owner of a servient estate closes with a wall or other structure the original way and points out another way which is accepted by the owner of the dominant estate, the new way may become a way by substitution."[90] Again, our review of the record reveals unaddressed issues that were raised in the pleadings and testified to at trial. Salvatore Scavello testified that Mr. James "took us out there and he said this is where I'm going to give you the right of way, pointing to approximately the middle of his driveway, and said it's going to go north along the road 50 feet and to a certain wall. . . . He [said] we're going to put it so you've got part of my driveway because it would be rather difficult to move the wall."

"This testimony indicates that perhaps plaintiffs' predecessor granted defendants an easement by substitution. However, the trial justice failed to determine whether sufficient factual support existed to conclude that an easement by substitution was granted. Thus, because the trial justice failed to credit or reject this testimony, and failed to make any findings of fact or conclusions of law on this issue, this Court has no choice but to remand this case to the Superior Court for a new trial, to determine whether defendants acquired an easement by substitution over plaintiffs' property".

[88] *Toothe v. Bryce*, 50 N.J. Eq. 589, 25 A. 182; *North Fork Water Co. v. Edwards*, 121 Cal. 662, 54 P. 69.

[89] 28 C.J.S., § 76 b.

[90] *Ondis v. City of Woonsocket*, 934 A.2d 799, 803 (R.I. 2007) (quoting *Hurst v. Brayton*, 43 R.I. 378, 381, 113 A. 4, 5 [1921]).

CHAPTER 3

TYPES OF EASEMENTS

There are nearly an infinite number of types of easements, as an easement, or right in another's land, may be created for almost any purpose. Throughout the years, the laws relating to easements have developed from a wide variety of uses by one person in land of another person. Almost anything conceivable may be the subject of an easement, provided it falls under the category of a true easement and not some other right or doctrine.

There are several kinds of easement that are commonplace and often appear within the records, public and nonpublic.

RIGHT OF WAY

Perhaps the most common of all easements, and one that comes readily to mind when the word "easement" is used, is a *right of way*. Rights of way may be in any one of a number of different forms, however, such as a right of way for ingress and egress, a power or other utility line, a pipeline, or a highway. (See Figure 3.1.)

A right of way is the privilege that one person or class of people has of passing over the land of another in some particular line. It is an easement, but the term is used to describe either the easement itself or the strip of land that is occupied for the easement.[91]

A right of way may be public or private, and public ways as applied to ways by land are usually termed "highways" or "public roads," and every citizen has the right

[91] 25 Am.Jur.2d, § 7.

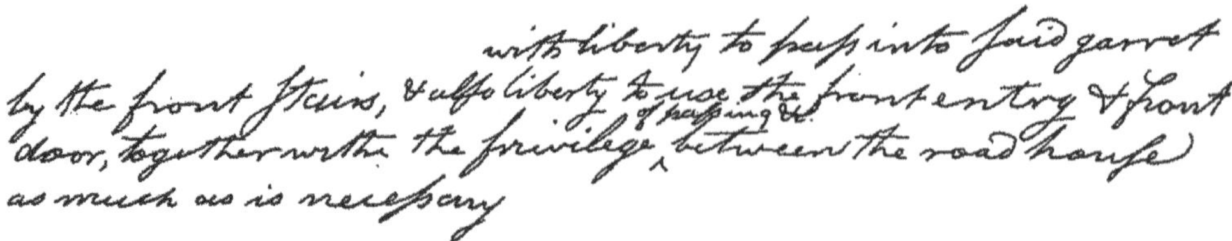

with liberty to pass into said garret by the front stairs, & also liberty to use the front entry & front door, together with the privilege [of passing &c.] between the road house as much as is necessary

Figure 3.1 Copy from an 1803 document giving several rights in the premises, including the right to use the front stairs.

to their use. A private way, to the contrary, relates to that class of easements in which a particular person or particular description or class of people has an interest or right, as distinguished from the general public. Generally, a private way is a way for travel over the surface of land, but there may also be a right of way over a passageway or stairway.[92]

For example, rights of way for ingress and egress, or to pass over a person's land, may not be obvious. They may not be readily visible on the surface of the earth, and even though on the public record, may be well outside the period of time examined in a normal title search.[93] Those not mentioned in title documents, but created in other ways, are particularly troublesome, since detection is often shown by evidence of use or parol testimony. These would likely never appear in a title examination but might be shown on a survey of the premises. A researcher should be on the alert for such evidence, as it indicates an encumbrance on the land.

The highway right of way, or easement, is the most frequent type subject to the process of *reversion*. Chapter 8 is devoted to this important subject.

[92] Ibid.

[93] In searching titles for the existence of a way, it is necessary to search both the dominant and servient tenements. Usually the servient tenement makes no mention of being burdened by the easement. An owner of the servient tenement conveyed the easement to an owner of the dominant tenement at some point in time. It is necessary to identify that conveyance and that point in time, as it may never be mentioned in any of the subsequent conveyances of the dominant tenement. A common example is when an easement is created in a will, or more often, in a partition of the real estate among heirs. Frequently, subsequent deeds to the various parcels do not mention the easement(s), but the right(s) exist(s) and "runs with the land" unless and until its proper termination.

RIGHT OF WAY LINE

A common misunderstanding is in the use of the term *right of way*. As stated previously, the term may be used to describe both the easement itself and the strip of land that is occupied for the easement.[94]

The *right of way line* is the line defining the edge of the easement, denoting its extent. Generally, there are two, one on either side. Often the term *right of way* is used when the land itself is not an easement but a strip owned in fee-simple by another. A common example of this is a railroad "right of way." Many, but not all, railroad beds are owned in fee by the railroad company. Some are easements, but they may also be based on a lease or other type of estate. Many of the larger and longer railroad beds are a "checkerboard" combination of rights. (See Figure 3.2.)

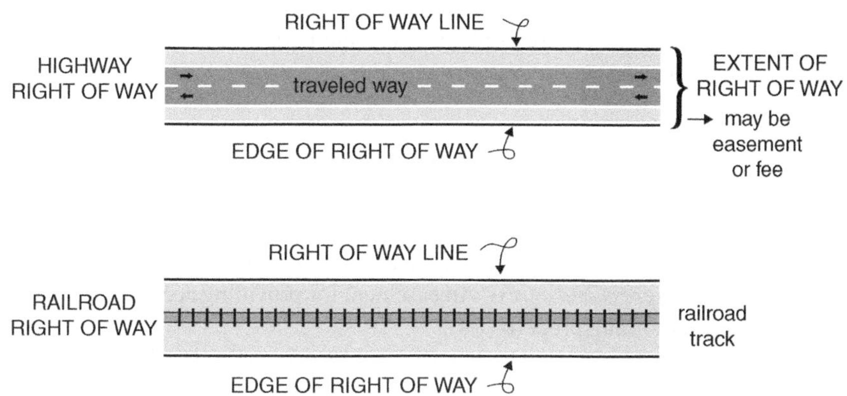

Figure 3.2 Examples of rights of way—highway and railroad.

[94] The term *right of way* has a twofold signification. It sometimes is used to describe a right belonging to a party—a right of passage over any tract, and it is also used to describe that strip of land that railroad companies take upon which to construct their roadbed. *Maysville & B.S.R. Co. v. Ball*, 56 S.W. 188, 108 Ky. 241 (1900).

Term *right of way* in deed is frequently used to describe not only easement but also the strip of land occupied by such use. *Moakley v. Los Angeles Pac. Ry. Co.*, 34 P.2d 218, 139 Cal. App. 421.

A "right of way" may consist of either of the fee or merely of a right of passage and use or servitude. *Brightwell v. International – Great Northern R. Co.* (Tex.) 41 S.W.2d 319 (1931).

In railroad parlance, "right of way" means either the strip of land on which track is laid or legal right to use such land, and in the latter sense may mean easement. *Quinn v. Pere Marquette Ry. Co.*, 239 N.W. 376, 256 Mich. 143 (1931).

The following description is from a deed conveying the fee to a 20-foot strip to be used as a right of way to a lake.

Example: "Also a certain piece or parcel of land shown on said plan as a 20-foot passway located between said Wentworth Cove Road and Lake Winnipesaukee and situated southerly of Lot # 1 as shown on said plan and between said Lot # 1 and an unnumbered lot owned by one O'Callaghan."

Highways Highways, particularly public ways, are probably the most common type of easement, although some may be held in fee-simple ownership. Since they are so commonplace, so extensive, and so troublesome, they are treated throughout the text with numerous examples and sample situations.

Easements Associated with Highways There are several types of easements commonly associated with [public] highways, some permanent, others used during the construction process:

- *Construction and maintenance of slopes*: An easement for constructing and maintaining the slopes necessary for supporting the roadway.
- *Drainage easement*: An area necessary for constructing and maintaining a drainage system or for directing natural drainage across a parcel of land.
- *Easement for ingress and egress*: An easement for providing access over, across, and through adjoining property(ies).
- *Utility easement*: An easement on a parcel of land for providing for the installation and maintenance of public utilities
- *Retaining wall easement*: An easement for the installation and maintenance of a retaining wall adjacent to the roadway.
- *Sight distance easement*: An easement across a parcel of land that allows the highway authority to clear it in order to maintain horizontal or vertical sight distances along the roadway.
- *Guardrail easement*: An easement over private property for the installation and maintenance of a guardrail adjacent to a roadway, sidewalk, or trail.
- *Sidewalk easement*: Frequently sidewalks are constructed within the designated right of way, but not always. For a variety of reasons, it is sometime necessary to divert a sidewalk outside of the right of way.
- *Bike path, or walking path, easement*: Alongside some major highways, particularly in remote areas, areas are set aside for the use of bicyclists or walking pedestrians.
- *Dead-end easement*: This involves setting aside a path for the use of pedestrians on a dead-end street to access the next public way. Such an easement could be found within the covenants of a homeowners' association, notes on a subdivision plat, or within the deeds of affected properties.

- *Temporary construction easement*: An easement leased from the property owner for the duration of the construction period.
- *Temporary driveway easement*: An easement not necessary for the construction of the roadway. This type of easement is to connect private driveways to the newly constructed roadway.

Private Roads Private ways are also very numerous and may be particularly problematical, since they are not always well described, or are not described at all when created by unwritten means, such as through prescription or by verbal agreement. The following text will contain numerous examples and situations. Some of them may also be fee-simple ownership. Alleys are often a special case.

As a general rule, courts have construed the term "private road," as used in the statutes, as meaning a public road in the sense that it is open to all who see fit to use it, although the principal benefit inures to the individual or individuals at whose request it was laid out. If a road has never been dedicated and accepted, laid out by public authority, or established by prescription, such a road is private. The term *private road*, it has been said, is used merely for the purpose of classification and to distinguish a class of public roads benefiting private individuals who, instead of the public at large, should bear the expense of their establishment and maintenance.[95]

Driveways A driveway is a type of private way. It may be easement or fee, may be entirely within the parcel it services or may also be across one or more other parcels of land.

Also known as an *easement of access*, not all lots abut a road, so an easement through one or more other lots must be provided for access. Sometimes adjacent lots have "mutual" driveways that both lot owners share to access their respective garages in the backyard. The houses are so close together that there can only be a single driveway to accommodate both backyards.

The same can also be said for walkways to the backyard: houses are so close together that there is only a single walkway between them, so the walkway is shared. Even when the walkway is wide enough, an easement may exist to allow for access to the roof and other parts of the house close to a lot boundary. Such an easement should be part of each property deed.

Alleys The word *alley* when used in a deed between individuals, frequently means a "private way."[96] Since by their very nature they are often between two or more structures, they are often very important to landowners for service and for parking.

[95] 72 C.J.S. Private Roads, § 2.

[96] Ibid.

Railroads Rail lines are another very common example of widespread easements, although railroads may be easement, fee, a combination of both, or some other estate in land, depending on the railroad and its mode of creation. One of the biggest problems encountered is the mistaken belief that railroads are all fee-simple ownership, or, in the alternative, are easements simply because the phrase "right of way" is used in conjunction with railroad descriptions.

A grant of land to a railroad for right of way purposes is substantially different from any like grant for other purposes. The character of the contemplated use makes it different. It is intended that the use by a railroad company will be perpetual and continuous. A railroad company performs a public service and is burdened with a public duty. In the performance of that duty, it is held to the exercise of the highest degree of care, and the complete, convenient, and safe use of its right of way requires that its possession be exclusive—a possession not shared with another—that it have complete dominion over its right of way, and that it enjoy all those rights that usually attend the fee.

Railroad rights of way are distinguished from other ways principally, if not entirely, by the difference in use. There is not the same necessity for exclusive possession of a right of way by canal companies as by railroads. The reasons for according to railroads the right to the exclusive possession are not applicable to canal companies. The use of right of way for a ditch or canal does not require the exclusive possession of, or complete dominion over, the entire tract which is subject to the secondary," as well as the principal, easements.

As stated earlier, because of the character of the contemplated use, the grant of a railroad right of way requires that the railroad company's possession be exclusive. "[A]n exclusive easement is an unusual interest in land; it has been said to amount to almost a conveyance of the fee."[97] "Because an exclusive grant in effect strips the servient estate owner of the right to use his land for certain purposes, thus limiting his fee, exclusive easements are not generally favored by the courts."[98]

Unlike a railroad right of way, an easement for a street, road, or highway is not granted for the exclusive use of a private entity. It is likewise not even granted for the exclusive use of the public entity owning the easement. "[A]ll roads, streets and highways are held in trust by the state and its political subdivisions for use by the public.[99]" An easement for a highway right of way is not sufficiently similar to an easement for a railroad right of way to hold that the *Bowman* opinion applies to highway easements.[100]

Flowage A flowage easement is the right held by a person, or a group of persons, to flood water on land of another, or others.[101] Flowage easements are common and

[97] *Latham v. Garner,* 105 Idaho 854, 856, 673 P.2d 1048, 1050 (1983).

[98] Ibid.

[99] *State ex rel. Rich v. Idaho Power Co.,* 81 Idaho 487, 506, 346 P.2d 596, 606 (1959).

[100] *Lake CDA Investments, LLC v. Idaho Dept. of Lands*, 149 Idaho 274, 233 P.3d 721 (Idaho 2010).

[101] Right of flowage is an easement. *Gager v. Carlson*, 150 A.2d 302, 146 Conn. 288 (1959).
The right of flowage is only an "easement." *Great Hill Lake v. Caswell*, 11 A.2d 396, 126 Conn. 364 (1940).

can be troublesome, as many of them are very old, having been created many years ago. Especially troublesome are those based on mill rights or mill privileges, some of which were created in the eighteenth century, even in the seventeenth century, and may still be existing rights, possibly owned by others, even though they have not been utilized for a century or even longer. Flowage may be for the purpose of storing water for mill purposes, hydroelectric power, flood control, or irrigation.

Another troublesome problem with flowage easements is that they were frequently granted for huge tracts, extending far inland from a lake or river. Some of these tracts have subsequently been subdivided without their technically, or legally, having been released from the easements.

Examples:

> "Reserving to the Salisbury Manufacturing Company the right to flow so much of said premises as may be by a dam seven and one-half (7-1/2) feet high at Trickling Falls and also reserving highway that passes over said land."
>
> "Also a certain tract of land situated in Kingston, described as follows: Lying at the Northerly side leading from the New Boston Road so-called to Trickling Falls, and bounded: South by said road; West by S.S. Crafts land; North by Amos Currier's land, and East by Samuel L. Blaisdell's land. Containing twenty-two acres, more or less. Said land is sold subject to the right to flow the low land by the Amesbury and Salisbury Mfg. Co."

The following description is a flowage easement over a 4-acre parcel:

> The full and free right to flow at all times by means of a dam to kept up and maintained at Trickling Falls in East Kingston near the house of the late Jacob Gale at the full height of seven and one half feet measuring from the flowing of the waterway as it now is, one undivided half part of the following described piece of land owned jointly with Thomas Gould and bounded Easterly on Powow River, southerly on land of Thomas Gould westerly on land of William Webster and northerly on land of Thomas Gould, containing four acres be the same more or less.

A recent example of outstanding flowage rights was the case in Dover, New Hampshire where an individual leased rights from the state to operate a long-abandoned hydroelectric power dam. In the process, the state did a normal title search for 35 years, which was the standard for marketability of title. But the purchaser, desirous to know the extent of his rights with his purchase, searched the records back for a period of 135 years. As a result, he found that he had the right to erect flashboards on the dam and raise the level of the water an additional 4 feet. Contemplating development for hydropower, this was important to him, since it would enable him to have considerably more power output.[102] However, to raise the water level was not without

[102] The output would have increased from 900,000 KW-hours to about 1.2 million KW-hours

consequences. Raising an additional 4 feet would result in leaving a municipal road entirely under water, as well as flooding acres of farmland and putting several new and expensive homes in danger.[103]

Normal title examinations usually do not discover these ancient, often outstanding, easements unless the easement language is carried forward in recent conveyances. Since title examination standards are merely *standards* suggesting the *minimum* examination that must be undertaken, further search and investigation should be considered to identify such outstanding rights so that they do not become a conflicting use of the land at some time in the future. The more that hydropower becomes developed, the more important these outstanding rights will become. (See further discussion on standards for title examination purposes in Chapter 13).

Private Use of Water A common right in the form of an easement in another's land, particularly an adjoiner, is that connected with the use of, or access to, water. Mostly this is for domestic purposes, but there are a number of examples of using water on another's land for watering animals or using water from another's land for irrigation. Typical domestic uses are well rights, with access and maintenance; rights to use a spring on nearby premises; and the right to lay and maintain pipes to draw water from a nearby source.

Examples:

> Together with the right to draw water from a well located at the rear of the premises herein conveyed.
>
> Reserving and excepting, however, the right to lay, maintain and repair a water pipe from said spring across the premises herein granted, for the purpose of supplying water to other property of the said Herbert A. Manning on the other side of Summit Avenue. It is understood and agreed that the said George L. and Ruth E. Cutting do not undertake to assure the said Manning the right to use a part of the water supply from said spring, or to run a pipe from said spring to the premises herein granted; this reservation extending only to the sharing of the water from said spring and the right to run a pipe across the premises herein granted.
>
> This conveyance is subject to, and includes all rights, if any, in the water supply for the described premises situated on property.
>
> With the further right to lay, maintain and repair any and all pipes necessary over the grantor's premises to Lake Winnipesaukee at such location as may be agreed upon for the purpose of furnishing the grantee's premises with lake water.

Drainage Easements for drainage are common in recent subdivisions and along highways. Runoff has to be dealt with, and often drainage from streets is routed

[103] *New York Times*, 1985.

across subdivision lots and onto adjacent land. Without the right, the runoff would be considered a trespass and could result in damages to land of another.

Example:

> This conveyance is also made subject to whatever rights of drainage and flowage, if any, the Winnipisseogee Lake Cotton & Woolen Manufacturing Co., or any others may have.

Slope A slope easement is an easement in land that will contain or be used for the construction of a slope generally to be used to accommodate the elevation difference between one property, usually a roadway, and the adjoining property. The adjoining property, or servient estate, may be lower or higher than the roadway. Typically, a slope easement is taken in eminent domain proceedings where the sovereign is taking a portion of your property for a road widening.

Aerial Easements through space for bridges and walkways between buildings are examples of aerial easements. They are three-dimensional in nature, having a planimetric description and elevations that control the vertical limits as a complementary description.

Avigation Also three-dimensional are avigation easements, or the rights for airplanes to utilize the air space over land parcels in their flight path approach to or takeoff from a runway. This is known as the glide path. Ordinarily, such an easement would be variable in the third dimension, being at 0 elevation nearest the end of the runway and becoming greater in height and elevation as one proceeds away from the runway. This is also known as an *aviation easement* (see Figure 3.3).

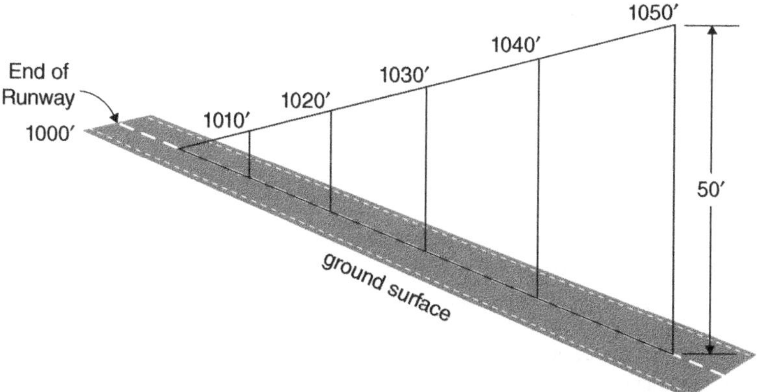

Figure 3.3 Three-dimensional easement for use of air space at end of runway.

Natural Support The natural right of an owner of land to lateral and subjacent support of that land from contiguous landowners is frequently referred to an easement of natural support. This is known as an easement *ex jure natural* (natural rights), rather than a true easement. Both lateral support and subjacent support of one's land are inherent rights under the common law.

Light and Air[104] Air space, or air rights, includes so much of the space above the ground as can be occupied or made use of in connection with enjoyment of the land. The right is not fixed or constant, and it varies with varying needs and is coextensive with them. The owner of land owns as much of the space above him as he uses, but only so long as he uses it. Everything that lies beyond belongs to the world.[105]

Airplanes have a right to use air space but since an owner of a tract of land has unlimited air rights, he may convey his interest above the surface or above a specified elevation. Condominium ownership consists of a specialized type of airspace ownership, having upper and lower horizontal limits.

The doctrine of ancient lights, recognized by English common law, protects a landowner's access to sunlight across neighboring property. When a dwelling has received sunlight across adjoining land for a specific period of time, the right to continue to enjoy such sunlight becomes part of the fee-simple ownership of the property. Under the doctrine of ancient lights, the owner acquires a negative easement under the principle of prescription. Since 1832, the prescriptive period has been 20 years.[106] Some courts however, have decided to the contrary.[107] This is a complex area of the law that needs to be studied in each individual case. With concerns for solar and wind energy, this area is becoming exceedingly important. (see following sections).

Utility Easements Easements for utility purposes may include the following:

- *Storm drain or storm water easement*: An easement to transport rainwater to a river, wetland, detention pond, or other body of water

[104] Easement of light and air is acquired only by express grant, by covenant, or by implication. *Novello v. Caprigno*, 176 N.E. 809, 276 Mass. 34 (1931).

[105] Hand & Smith, Chapter 5. *United States v. Causby*, 328 U.S. 256 (1946).

[106] Hand & Smith, Chapter 5.

Prescription Act 1832, § 3.

See also *Mathewson St. M.E. Church v. Shepard*, 46 A. 402, 22 R.I. 112 (1900): An easement inlight cannot be acquired by prescription.

Landowner has no right to light and air coming to him across neighbor's land. *Musumeci v. Leonardo*, 75 A.2d 175, 77 R.I. 255 (1950).

It is well settled that by the common law a deed of land passes no right of light and air over other lands with express words. *Brooks v. Reynolds*, 106 Mass. 31 (1870).

[107] No prescriptive right to use of light and air through widows can be acquired by use and enjoyment. *Richardson v. Pond*, 81 Mass. 387 (1860); *Keats v. Hugo*, 115 Mass. 212 .

Easement of light and air can exist only by express grant, covenant, or absolute necessity, and cannot be created by prescription. *Tidd v. Fifty Associates*, 131 N.E. 77, 238 Mass. 421 (1921).

Easements of unobstructed ocean view, breezes, light or air do not exist. *Schroeder v. O'Neill*, 184 S.E. 679, 179 S.C. 310 (1936).

- *Sanitary sewer easement*: An easement to transport used water to a sewage treatment plant
- *Electrical power line easement*: An easement for electrical service
- *Telephone line easement*: An easement for telecommunications
- *Fuel gas pipe easement*: An easement for supplying fuel

Energy Easements Easements for the development of energy projects are becoming quite common. They exist in a variety of forms, access for construction as well as maintenance, transportation of power, and protection of energy developing apparatus being among the most common.

Solar Access Because of its importance for energy, solar access is a right that has received attention in recent years. Recent court decisions have held that, if sunlight is being utilized for energy, adjoining landowners may not interfere with that use.[108] Easements protecting neighboring air space for access to sunlight will most certainly become more common.

Example: "Reserving an appurtenant easement for light, air and scenic view of the beach and ocean for the benefit of [the 116 20th Avenue property] (the 'dominant tenement')."

Some state statutes provide for easements for solar access. For example, Iowa defines such as "an easement recorded under section 564A.7, the purpose of which is to provide continued access to incident sunlight necessary to operate a solar collector." Also included in the definitions provided by this section[109] are *solar collector*, which means a device or structural feature of a building that collects solar energy and that is part of a system for the collection, storage, and distribution of solar energy. For purposes of this chapter, a greenhouse is a solar collector, and *solar energy* means energy emitted from the sun and collected in the form of heat or light by a solar collector.

The procedure for creation is as follows:

Solar Access Easements[110]

1. Persons, including public bodies, may voluntarily agree to create a solar access easement. A solar access easement whether obtained voluntarily or pursuant to the order of a solar access regulatory board is subject to the same recording and conveyance requirements as other easements.

[108] *Prah v. Maretti*, 108 Wis.2d 223, 321 N.W.2d 182 (1982) is the leading case. See also *Tenn v. 889 Associates*, 127 N.H. 321, 500 A.2d 366 (1985).
[109] Title XIV, Chapter 564A, § 564A.2.
[110] § 564A.7.

2. A solar access easement shall be created in writing and shall include the following:
   a. The legal description of the dominant and servient estates.
   b. A legal description of the space which must remain unobstructed expressed in terms of the degrees of the vertical and horizontal angles through which the solar access easement extends over the burdened property and the points from which these angles are measured.
3. In addition to the items required in subsection 2 the solar access easement may include, but the contents are not limited to, the following:
   a. Any limitations on the growth of existing and future vegetation or the height of buildings or other potential obstructions of the solar collector.
   b. Terms or conditions under which the solar access easement may be abandoned or terminated.
   c. Provisions for compensating the owner of the property benefiting from the solar access easement in the event of interference with the enjoyment of the solar access easement, or for compensating the owner of the property subject to the solar access easement for maintaining that easement.

Wind Energy Windmills have become very popular in some parts of the world in recent years as a means of generating power. Some states have enacted legislation for the planning process. For example, South Dakota has the following statute[111]:

Wind Energy Easements A property owner may grant a wind easement properly created and recorded; however, the maximum term of such easement is 50 years. Any such easement is void if no development of the potential to produce energy from wind power associated with the easement has occurred within five years after the easement began.

Since windmills can sometimes be rather noisy, approvals may require "noise easements."[112]

Communications Easement This easement can be used for wireless communications towers, cable lines, and other communications services. It is a private easement and the rights granted by the property owner are ordinarily for the specific use of communications.

View Easement A scenic view may enhance the value of a tract of land. In order to preserve and protect a view, an easement may be in order. This type of easement, commonly known as a *view easement,* may prevent someone from blocking the view of the easement owner, or permit the owner to cut blocking vegetation on land of another.

[111] Chapter 43-30S-9.

[112] Noise easements are often found with regard to wind farms, airports, and, occasionally, loud machinery.

Example:

The grantors include a servitude in favor of the grantees over other land owned by the grantors and lying Southeasterly of the granted premises so as to restrict to one and one-half stories the size of structures which may be erected Southerly of the line of sight from the Northeasterly corner of the granted premises to the Southeasterly corner of the remaining premises of the grantors at the Junction of the Lamprey River and land of Gallant.

Conservation Easements A conservation easement is a voluntary formal agreement between a landowner and a qualified land trust or government entity (referred to as the "holder" of the easement) that permanently limits uses of the land in order to protect ecological, historic, or scenic resources. A conservation easement allows a landowner to retain ownership while restricting some of the rights in order to protect the property's conservation values. Such easements are custom designed and negotiated to meet the personal and financial needs of the landowner. A conservation easement may cover portions of a property or the entire parcel. The easement should identify the rights the landowner wishes to retain, limit, or relinquish.

Conservation easements are frequently held by land trusts, which are nonprofit corporations qualified as public charities by the Internal Revenue Code. They are an important conservation tool, as well as a mechanism for individual financial planning.

Recreational Easements Some U.S. states offer tax incentives to larger landowners, if they grant permission to the public to use their undeveloped land for recreational use (not including motorized vehicles). If the landowner posts the land (i.e., posts No Trespassing signs) or prevents the public from using such an easement, the tax abatement is revoked and a penalty may be assessed. Recreational easements also include such easements as equestrian, fishing, hunting, hiking, trapping, biking, and other similar uses.

Historic Preservation Easements Similar to a conservation easement, this type typically grants rights to a *historic preservation* organization for enforcing restrictions on the alteration of a historic building's exterior or interior.

Drip[113] A species of easement or servitude obligating one person to permit the water falling from another person's house to fall upon his own land.[114] The right to cast water upon land of an adjoining owner may be acquired by deed as in the following example,[115] or by prescription.

[113] See 1 Am.Jur.2d, Adjoining Landowners, §§ 3o, et seq.

[114] 3 Kent, Comm. 436.

[115] *First Baptist Society v. Wetherell*, 34 R.I. 155, 82 A. 1061 (R.I. 1912).

Plaintiff and defendant owned adjoining estates in Newport, Rhode Island. The defendant's title derived from Sandford Bell, who, in 1846, owned the land immediately north of that on which stands the church edifice of the plaintiff. The plaintiff society so located and constructed its building, presumably through inadvertence, such that its foundation encroached upon the land of Bell a distance of 2 feet; its eaves extending over a further distance of about 2 feet 8 inches. On May 12, 1846, Bell, who happened to be a member of the plaintiff society, conveyed thereto, by quitclaim deed, all his right, title, and interest in and to the strip of land that had been appropriated and built upon by the church. Bell also included in this deed the grant of an easement or "right of drip" to the plaintiff society upon his remaining land.

The Bell deed contained the following descriptive language and habendum:

> A certain strip of land about two feet in width, running easterly and westerly along and under the northerly side of the meeting house of said society, which said meeting house lately built by said society, stands as aforesaid about two feet on the land of the grantor. It being the intention of the grantor to release to said society all of his said land adjoining the northerly side of the land of said society in said town of Newport, on which the northerly side of said meeting house stands, with the privilege for the roof of said meeting house to drip on my land forever.
>
> To have and to hold the same with the said privilege of the drippings to the said First Baptist Society of Newport and their assigns forever.

Beach Access Some jurisdictions permit residents to access public water or beach by crossing adjacent private property. Similarly, there may be a private easement to cross a private lake to reach a remote private property, or an easement to cross private property during high tide to reach remote beach property by foot.

Stairways/Access For at least 70 years the users or occupants of the second-story apartments in the two buildings have accessed those spaces through a common stairway/hallway. The stairway/hallway is approximately 4 feet wide. It was apparently built as part of the Stoner building, as it is physically part of that building. However, the Stoner building extends two feet beyond Lot 4, that is two feet onto Lot 3, on which the Alger building is located. Thus, only 2 feet of the 4-foot stairway/hallway lies on that portion of Lot 4 to which plaintiff Suzanne Stoner holds record title.[116]

Miscellaneous Easements and Servitudes As previously recognized, there is an almost infinite number of examples of, or uses for, easements. Grazing, utilities, protection from water, and, particularly, slopes for highways are very common.

[116] *Stoner v. Alger*, 670 N.W.2d 430 (Iowa App. 2003)

Some are unusual but must be recognized and honored, as they are rights that either attach to land (appurtenant) or are for benefit of others independent of land (in gross), or burden land as a superior right conveyed away and owned by another, or others.

Examples:

> Also a privilege to dry clothes in said yard and to use the necessary in said yard in common with others and to pass and repass in the usual path from said brick building to said necessary.
>
> ". . . reserve the right to pick blueberries on the granted premises so long as they do not interfere with the operations of the grantee.
>
> This deed is subject to a fifty (50) foot access easement and a beaver dam easement as shown on said plan.[117]

Some other uncommon examples are the following:

Easement to pile logs for the use of a mill
Easement for placing boxes, etc., in using a store
Custom of turning teams on land in plowing
Right of dockage and use of wharf
Easement for carrying away iron ore, etc.
Easement for taking seaweed on a beach
Reservation of grass and herbage, a servitude
Easement of a town to dig stone on another's land
Easement of a town to use parish buildings
Right to lay gas pipe, an easement in a gas company
Pew rights and burial rights

Implied Easements for Special Use Some special uses, when they are not benefitted by granted easements, may have implied easements of access and maintenance at the least. Included in this category may be such things as canals, cemeteries, dams, mills, ferry landings, and the like.

Researching Easements As mentioned earlier in the chapter, in researching easements, *both* estates must be examined,[118] as documentation regarding creation,

[117] In this particular case, the easement was for the purpose of getting to a beaver dam located on the land of another, and being able to control the height of the dam to avoid excess flooding on the land of the easement holder.

[118] Sometimes more than just two estates (ownerships) must be researched, such as when an easement burdens more than just one property and there are several servient estates even though there is only one dominant estate.

restrictions, whether easement or some other estate, as well as details about width, and the like, may appear in either chain of title. If not specific in either chain of title, each title must be examined carefully for details as to creation, alteration, relocation, extinguishment, reversion rights, and more.

This is where many people fail, as they (1) only examine one of the estates instead of both, or (2) undertake a normal title search, which, in many cases, will not take them far enough back into the records to procure all the necessary documentation regarding the easement. Chapter 10 details proper research procedure.

CHAPTER 4

CREATION OF EASEMENTS

There are 10 ways by which an easement may be created:[119]

1. Express grant
2. Reservation or exception
3. Agreement or covenant
4. Implication, including necessity
5. Estoppel
6. Prescription
7. Dedication
8. Eminent Domain
9. Custom
10. Vote of a governing body

Easements may be classified as express, implied, or prescriptive.[120] An express easement is created if the intent of the parties has been specifically evidenced by language in declaration or in documents.[121]

Courts are also willing to recognize an easement without evidence of express intent if certain requirements are satisfied by the circumstances and conditions of affected properties.[122] Implied easements may be created due to compelling circumstances, and prescriptive easements may be created through long-term use without permission or authorization.

[119] Different persons may have a right of way over the same place by different titles, one by grant, another by prescription, and a third by custom. *Kent v. White*, 27 Mass. 138 (1830).

[120] An easement of a right of way can be created only by grant, express or implied, or by prescription, or by exception. *Childs v. Boston & M.R.R.*, 99 N.E. 957, 213 Mass. 91, 48 L.R.A., N.S. 378 (1912).

[121] *A Practical Guide of Disputes Adjoining Landowners – Easements*, § 1.01 (2f).

[122] Ibid.

EXPRESS GRANT

Often called a "deeded easement," an easement may be created by *express grant.* It may be created by a deed or some related instrument, but it also may be created in a probate document, wills being the most common. However, both express and implied easements can arise through probate partitions or divisions of the estate among heirs.

By far the most commonly used method of creating an easement is with a deed, either directly or indirectly (by reference). Sometimes the easement is the sole subject matter of the deed itself, which is called an *easement deed* or a *deed of easement.* The deed must describe the interest correctly and must comply with the formalities necessary for the transfer of any interest in land. The conveyance should include words of inheritance and a statement of duration of the easement. (see Chapter 6 for suggestions on compiling easement descriptions).

Examples (see Figures 4.1 to 4.3):

> The Town of Kingston, a New Hampshire Municipal Corporation, located in the County of Rockingham and State of New Hampshire, by its Selectmen, Michael Priore, John Reinfuss, and Ralph Southwick, all of Kingston in the County of Rockingham, and State of New Hampshire, for consideration paid, Grant to Charles H. Haughey and Marguerite E. Houghey, as joint tenants with rights of survivorship, and not as tenants in common, their heirs, successors, and assigns, both of Center Street in the Town of Center Tuftonborough, County of Carroll, and State of New Hampshire, with Quitclaim Covenants,
>
> An easement in, to, upon and over a certain parcel of land, situated Northerly of the New Boston Road, which Road is also known as Rowell Road, and Westerly of that portion of the Pow Wow Pond known as Rowell's Cove and being a portion of the former right-of-way of the Boston and Maine railroad in the Town of Kingston, State of New Hampshire.
>
> The easement is to travel in a generally Northeasterly direction across the land of the grantors from a parcel located Southeasterly of the land of the grantors, said parcel being currently owned by the grantees to the land of the grantees located *nunc et future* on the Easterly side of that portion of Pow Wow Pond known as Rowell's Cove.
>
> Said easement is also given to permit the grantees, their heirs and assigns, to construct, operate and maintain one or more electrical transmission and distributive lines, including wires, poles, anchors, guys, telephone and telegraph lines and poles and appurtenant equipment in, over, upon, and across the subject premises and is also given to permit the grantees their heirs, and assigns, to construct and maintain a road for the purpose of ingress and egress and it is not to be construed as an easement given to the exclusion of the grantor, its assigns, or to others later granted a similar right.

As previously noted, easements may be created in probate documents, including wills,[123] dowers, and partitions, or divisions among heirs or others (see Figure 4.1). This is not nearly as common as it was in the past. In searching records for access to property, especially tracts that appear to be landlocked, frequently a right of way will

[123] An easement may be created by reservation in a will. *Bolomey v. Houchins*, 227 S.W.2d 752 (Mo. App., 1950).

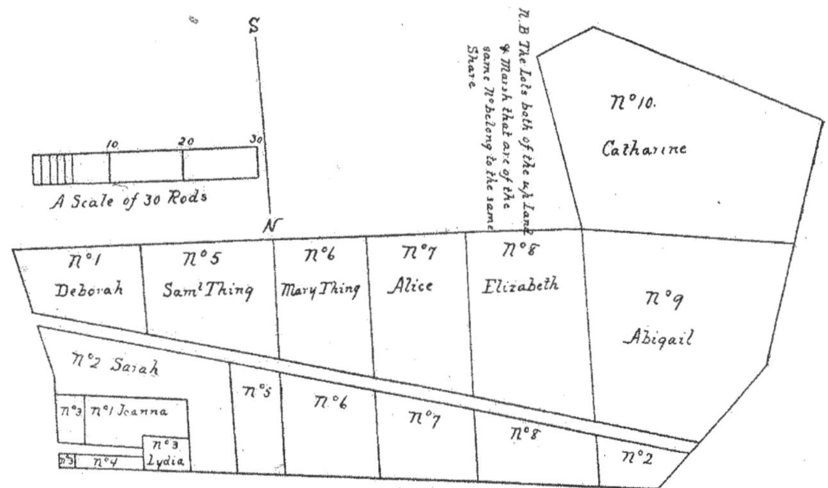

Figure 4.1 Example of a probate division.

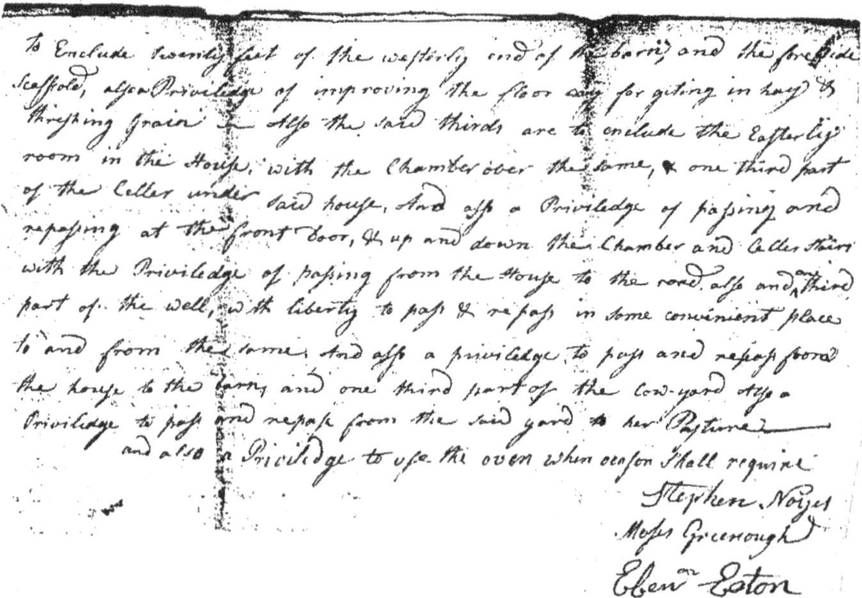

to Enclude twenty feet of the westerly end of the barn, and the fore side Scaffold, also a Priviledge of improving the floor any for giting in hay & thrashing grain — Also the said thirds are to enclude the Easterly room in the House, with the Chamber over the same, & one third part of the Celler under said house, And also a Priviledge of passing and repassing at the front door, & up and down the Chamber and Celler Stairs with the Priviledge of passing from the House to the road, also one third part of the well, with liberty to pass & repass in some convenient place to and from the same, And also a priviledge to pass and repass from the house to the barn, and one third part of the Cow-yard also a Priviledge to pass and repass from the said yard to her Pasture — and also a Priviledge to use the oven when occasion shall require

Stephen Noyes
Moses Greenough
Ebenr Eaton

Figure 4.2 Copy from a 1786 widow's allowance, or dower, giving easements in certain parts of the estate.

be found in a probate record if the particular tract in question originated in a partition of the land in the estate. Many of these have been omitted from subsequent transfers of the property, but they still legally exist.

> And also a privilege of passing and repassing at the front door, & up and down the chamber and cellar . . . with the privilege of passing from the house to the road. Also and one third part of the well, with liberty to pass and repass in some convenient place to and from the same. And also a privilege to pass and repass from the house to the barn, and one third part of the cow yard. Also a privilege to pass and repass from the said yard to her pasture. (See Figure 4.2.)
>
> . . . and reserving a right of way by my buildings where I now live to the other part of my "Great Woods so called through all the paths opened to the same, for the purpose of carrying off whatever it may be necessary to remove from said wood lot. (See Figure 4.3.)

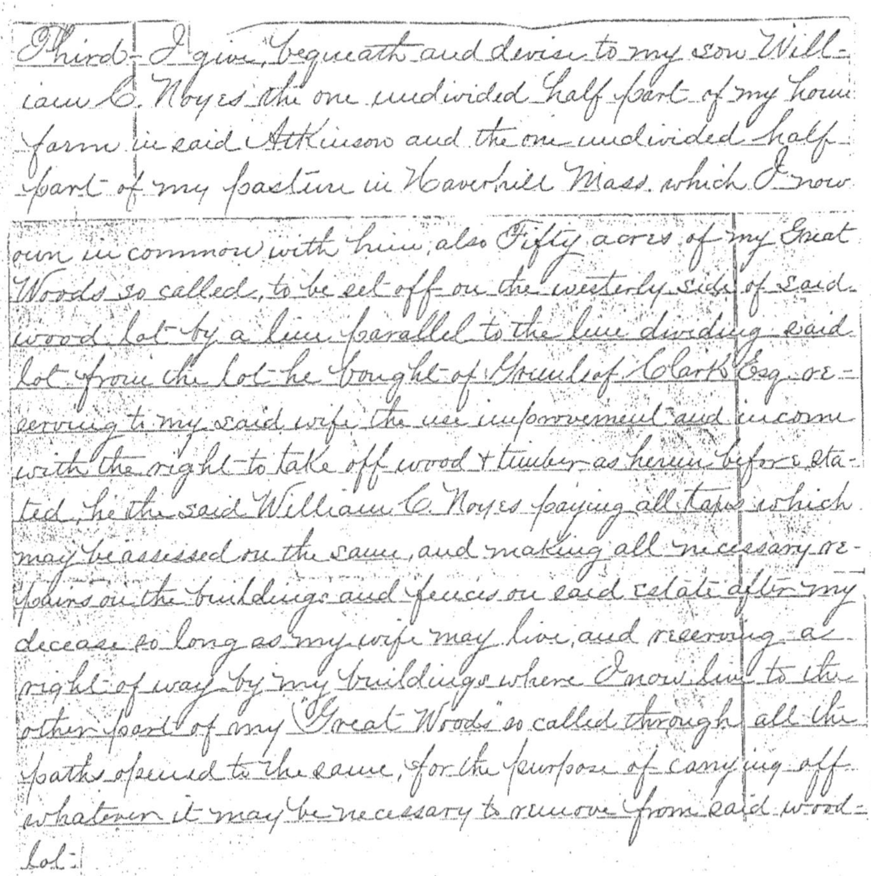

Third – I give, bequeath and devise to my son William C. Noyes the one undivided half part of my home farm in said Atkinson and the one undivided half part of my pasture in Haverhill Mass. which I now own in common with him; also Fifty acres of my Great Woods so called, to be set off on the westerly side of said wood lot by a line parallel to the line dividing said lot from the lot he bought of Greenleaf Clark Esq. reserving to my said wife the use improvement and income with the right to take off wood & timber as herein before stated, he the said William C. Noyes paying all taxes which may be assessed on the same, and making all necessary repairs on the buildings and fences on said estate after my decease so long as my wife may live, and reserving a right of way by my buildings where I now live to the other part of my "Great Woods" so called through all the paths opened to the same, for the purpose of carrying off whatever it may be necessary to remove from said wood lot.

Figure 4.3 Copy from a 1784 partition of land in an estate. A right of way is appurtenant to the parcel set off.

RESERVATION OR EXCEPTION

Also very common are easements created by *reservation* or *exception* (see Figure 4.4).[124]

Example of a reservation of a right of way over a deed parcel:

> The premises are conveyed subject to the perpetual right and easement hereby reserved in the grantor, for the benefit of all remaining land of the grantor, to pass and repass at all times over the granted premises by foot and by vehicle between said remaining land of the grantor and Exeter Road; the location of said right and easement may be made by the grantee, provided, however, that it shall be over a strip of land not less than ten (10) feet in width and shall be so located as to provide convenient access to and from said remaining land of the grantor and Exeter Road.

Example of several rights of way being excepted from previously described tracts of land in a deed:

> Said premises are conveyed subject to such rights as may be retained by Rockingham County Light and Power Company or its successors in interest to a presently abandoned power line running through the parcel first above described.
>
> Said premises are further conveyed subject to any rights of way which may continue over the abandoned woods road mentioned in the description of the first parcel.
>
> Said premises are conveyed subject to any other rights of way of record.

Reserving and *excepting*, although strictly distinguishable, are often used interchangeably or indiscriminately, as well as together, such as "reserving and excepting . . ." The use of either term is not conclusive as to the nature of the provision.[125]

An exception in a deed, as distinguished from a reservation, is some part of the boundary described in the deed which the grantor *retains* title to and does not convey by the deed.[126]

[124] There is a technical distinction between an *exception* and a *reservation*, although the two terms are often used indiscriminately. An *exception* is the process by which a grantor withdraws from the conveyance land that would otherwise have been included; the grantor merely retains or keeps the part excepted. However, a *reservation* vests in the grantor a *new* right or interest that did not exist in him before; it operates by way of an implied grant.

Mass. 1906. A right of way cannot be excepted from the operation of a deed unless either the right of way or the way itself is in existence when the deed is made. *Bailey v. Agawam Nat. Bank*, 76 N.E. 449, 190 Mass. 20, 3 L.R.A., NS 98, 112 Am. St. Rep. 296 (1906).

[125] *Florida East Coast R. Co. v. Worley*, 49 Fla. 297, 38 So. 618 (1905); *Keeler v. Wood*, 30 Vt. 246; *Roberts v. Robertson*, 53 Vt. 690, 38 Am.Rep. 710; *Stockwell v. Couillard*, 129 Mass. 231; *Martin v. Cook*, 102 Mich. 267, 60 N.W. 679 (1894); 13 Cyc. 672; 34 Cyc. 1641.

[126] *Hicks v. Phillips*, 146 Ky. 305, 142 S.W. 394, 47 L.R.A., N.S., 878 (1912).

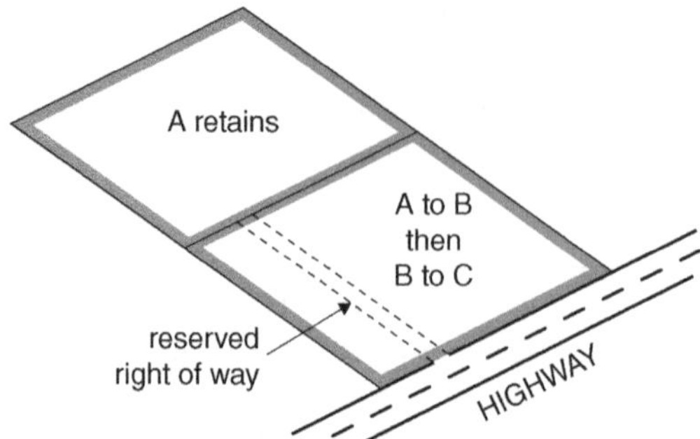

Figure 4.4 Owner A conveys to owner B *reserving* a right of way to himself over it. B later sells to C *excepting* the right of way over it to land owned by A, or A's successors in title.

A reservation is a clause in a deed, whereby the grantor reserves some new thing to himself issuing out of the thing granted and not in esse before, but an exception is always a part of the thing granted, or out of the general words or description of the grant.[127] In general, a reservation is like an exception—something to be deducted from the thing granted, narrowing, and limiting what would otherwise pass by the general words of the grant.[128] However, a reservation is not always created because the "reserves" or "reserving" is employed. "Reserving" sometimes has the force of "saving" or "excepting."[129]

Subject to. Occasionally the words "subject to" are used to indicate a reservation, or an exception.

Example:

[parcel description], followed by: "in Township 38, Range 20 (subject to existing roads and pole line easements) and situate in Hickory County, Missouri."

The words "subject to", used in their ordinary sense, mean subordinate to, "subsequent to," or limited by.[130]

[127] 4 Kent. 468; Brown, etc., v. Anderson, 88 Ky. 577, 11 S.W. 607 (1889); 11 Ky.Law Rep 107.

[128] *Dyer v. Sanford*, 9 Metc. (Mass.) 395, 43 Am.Dec. 399.

[129] 2 Coke Litt. 413; *Whitaker v. Brown*, 46 Pa. 197; *Dee v. King*, 77 Vt. 230, 59 A. 839, 68 L.R.A. 860.

[130] *Flower v. Town of Billerica*, 87 N.E.2d 189, 324 Mass. 519 (1949).

AGREEMENT OR COVENANT

An easement may be created by *agreement* or *covenant*.[131] Such agreement or covenant operates as a grant and is construed the same as an express grant. Whether an agreement or covenant runs with the land depends on whether the act that it embraces concerns or relates to the land.[132]

Some jurisdictions cite the general rule that an easement, being an interest in land, may be created only by grant, express or implied, or by prescription, and cannot be created by parol. "Despite the general rule, however, it has been recognized by the courts that an easement may exist by virtue of an estoppel, or an agreement." 26 Am. Jur.2d 430, Easements and Licenses, § 17.[133]

IMPLICATION

The general rule is when an owner of a tract of land conveys part of it to another, he is said to grant with it, by implication, all easements that are apparent and obvious, and that are reasonably necessary for the fair enjoyment of the land granted (see Figure 4.5).[134]

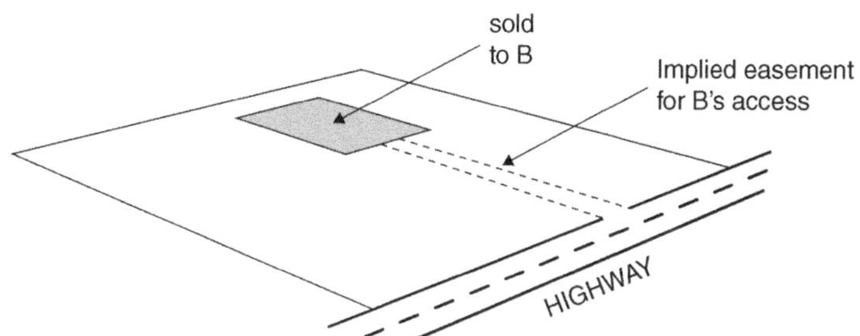

Figure 4.5 Owner A conveys to owner B without granting an easement for access. Likely, B would have an implied right of way across the remaining land of A.

[131] An easement may be created by agreement or covenant as well as by grant. *Town of Paden City v. Felton*, 66 S.E.2d 280, 136 W.Va. 127 (1951).

[132] If a covenant benefits the land to which it relates and enhances its value, the easement created by it becomes appurtenant to the land an passes with it. *Hennen v. Deveny*, 71 W.Va. 629, 77 S.E. 142 (1913).

[133] *Foldeak v. Incerto*, 6 Conn.Cir.Ct. 416, 274 A.2d 724 (Conn.Cir.A.D. 1970).

[134] Implied easements, whether by grant or by reservation, do not arise out of necessity alone, and their origin must be found in presumed intention of parties to be gathered from language of instruments when read in the light of circumstances attending their execution, physical condition of premises, and knowledge which parties had or with which they are chargeable. *Joyce v. Devaney*, 78 N.E.2d 641, 322 Mass. 544 (1948).

There are three types of implied easement, according to the three patterns in which they arise (see Figure 4.6):[135]

1. Easements implied from prior use
2. Easements implied by necessity
3. Easements implied from a subdivision plat

The first category is the implied easement based on *prior use* of the property:

In the figure, a parcel of land has a garage at the further end with a driveway leading to it. The parcel is divided into two tracts with the rear portion being sold and no mention made of the driveway, or any right of way to the garage. The driveway is implied in the conveyance and may be inferred by the purchaser.

Generally, the following requirements are necessary to justify an easement of this type.[136]

There are five elements to establish an easement by prior use:

1. Common ownership of both properties at one time
2. Followed by a severance
3. Use occurs before the severance and afterward notice
   1. Not simply visibility, but apparent or discoverable by reasonable inspection (e.g., the hidden existence of a sewer line that a plumber could identify may be notice enough)
4. Necessary and beneficial
   1. Reasonably necessary
   2. Not the "strict necessity" required by an easement by necessity

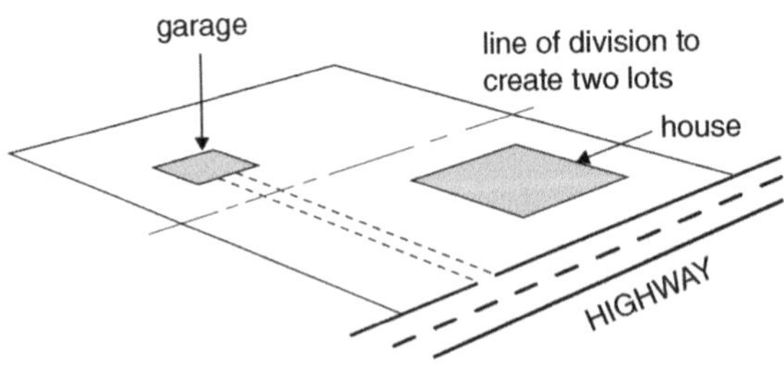

Figure 4.6 Example of an implied easement based on prior use of a driveway to a garage.

[135] *A Practical Guide to Disputes Between Adjoining Landowners – Easements*, § 2.02(3).

[136] *A Practical Guide to Disputes Between Adjoining Landowners – Easements*, § 2.02(3a).

First, the property must initially be owned as a single parcel prior to its division into a dominant and a servient portion. This element of unity of title is also a requirement for an easement by necessity, which is a type of implied easement.

Second, during the time that the unity of title exists, the sole owner must use part of the parcel in a manner beneficial to the other portion of the property. Courts often refer to this use prior to division into two or more parcels as a preexisting quasi-easement.

Third, for the conveyance to give rise to the implied easement the existing use must be apparent so that an inspection of the property would result in notice of the use. The use does not have to be visible; however, as courts have concluded that underground sewers, pipes, and drains are apparent so long as the plumbing fixtures are readily apparent.

Fourth, the original owner must convey a portion of the property to another person, either retaining the remainder himself or conveying it to a third party.

Fifth, the easement across the burdened portion must be reasonably necessary for the use of the benefited, dominant parcel to justify the conclusion that the parties had the implicit but unexpressed intent to create and easement. Generally, a use is considered reasonably necessary if it contributes reasonably to the convenient use of the benefited property. If necessity is the only factor, the courts will generally require a higher degree of necessity, often stated as strict or absolute necessity.

Other factors may affect a court's determination of whether an easement based on prior use should be implied. For example, a stricter measurement of proof is generally necessary when the easement is favor of the grantor rather than the grantee. This is because of the rule of construing a deed against a grantor, and in favor of a grantee.[137]

The Supreme Court of Ohio has established the following four elements of an easement implied from prior use[138]:

1. A severance of the unity of ownership in an estate;
2. that, before the separation takes place, the use which gives rise to the easement shall have been so long continued and obvious or manifest as to show that it was meant to be permanent;
3. that the easement shall be reasonably necessary to the beneficial enjoyment of the land granted or retained;
4. that the servitude shall be continuous as distinguished from a temporary or occasional use only.[139]

[137] Use of particular way by grantor before conveyance and by grantee thereafter may operate an assignment or designation of way by necessity. *Davis v. Sikes*, 151 N.E. 291, 254 Mass. 540 (1926).

[138] See also *Cobb v. Daugherty*, 225 W.Va. 435, 693 S.E.2d 800 (W.Va. 2010).

[139] *Ciski v. Wentworth* (1930), 122 Ohio St. 487, 172 N.E. 276.

The unity of title requirement accords with the principles of implied easements. Implied easements are easements read into a deed. "An implied easement is based upon the theory that whenever one conveys property he includes in the conveyance whatever is necessary for its beneficial use and enjoyment and retains whatever is necessary for the use and enjoyment of the land retained." *Tratt[a]r,* supra, at 291, 95

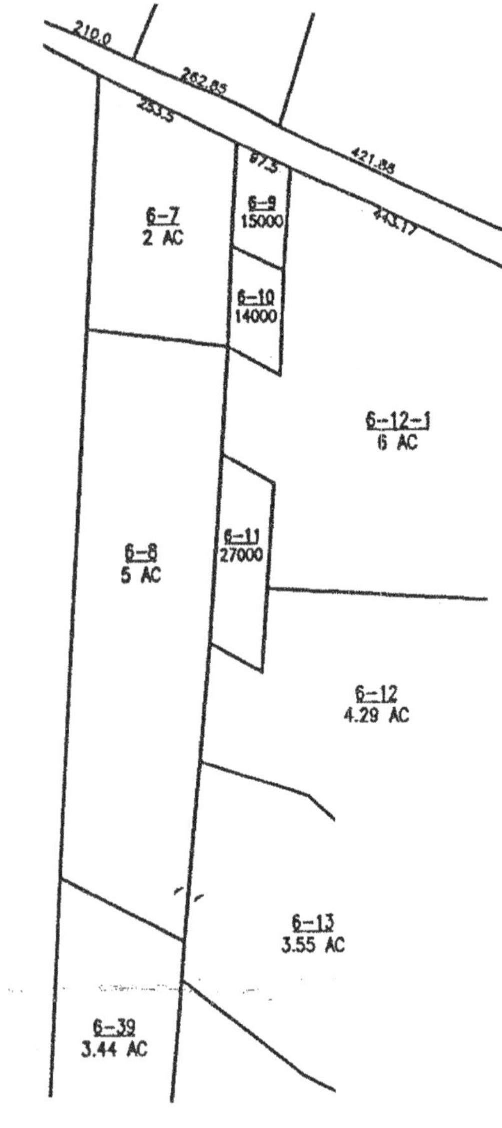

Figure 4.7 Section of assessor's plat, showing potential implied right of way to a remainder parcel having no road frontage. From Town of Seabrook, NH Assessor's Map.

N.E.2d 685. In other words, implied easements are those easements that a reasonable grantor and grantee would have expected in the conveyance, and a court will read the implied easement into a deed where the elements of that implied easement exist. However, if there is no unity of title, there is no grantor who may give an easement to the grantee. It does not matter whether a reasonable grantor would have conveyed an easement or a reasonable grantee would have expected to receive an easement. A grantor simply cannot convey what is not possessed.[140]

Implied easements, being unexpressed, are not easements of record, although later deeds may cite them as an exception. Often, the caveat, "subject to easements [affecting] the property," will be found. When searching records, one should be on guard against this possibility. Assembling surveys and assessors' maps to get an idea of the source of various titles may be of assistance in recognizing possible implied easements (see Figure 4.7).

The second category is an *easement by necessity*. The crucial factor in this type is the claimed easement is *necessary* for the use of the property owned by the person claiming the easement.[141] The basis of the concept is the promotion of usability of the land. If a parcel were actually landlocked it would lose its usefulness and, therefore, be reduced considerably in value.[142]

Usually a right of way comes to mind whenever an easement by necessity is mentioned. However, another example is where minerals are involved and they cannot be reached without an easement over (under or through) the surface owner's land.[143]

As with an implied easement based on prior use, it is necessary to establish unity of title. Courts have been willing to recognize a unity of title from the distant past even though the necessity did not exist at the time, and even though the easement has never been used.[144]

Frequently an easement of necessity cannot be claimed when conditions causing a parcel to be landlocked are caused by the claimant, but this may depend on the jurisdiction or the particular circumstances. For example, an owner with access to a public highway who blocks that way by constructing a building across it is not entitled to an easement by necessity across neighboring property, even if owned by his grantor. Likewise, an owner who creates his own landlocked parcel through subdivision cannot make a claim of easement by necessity based on the previously nonexistent

[140] *Dunn, et al. v. Ransom, et al.*, Defendants-Appellants. 2011-Ohio-4253; No. 10CA806 Court of Appeals of Ohio, Fourth District, Pike; August 18, 2011.

[141] Presence of a second or alternate way onto property is to conclusive proof that an implied easement is unnecessary. *McGee v. McGee*, 233 S.E.2d 675, 32 N.C.App. 726 (1977).

[142] *A Practical Guide to Disputes Between Adjoining Landowners – Easements*, § 2.02(3b).

Easement of absolute necessity is predicated upon strong public policy that no land may be made inaccessible and useless. *Old Falls, Inc. v. Johnson*, 212 A.2d 674, 88 N.J. Super. 441 (1965).

[143] *A Practical Guide to Disputes Between Adjoining Landowners – Easements*, § 2.02(3b).

[144] *A Practical Guide to Disputes Between Adjoining Landowners – Easements*, § 2.02(3b,i).

necessity to cross land of another.[145] Confronted with this possibility, it is important to examine the court's treatment of the problem in the governing jurisdiction.

An easement by necessity will terminate when the necessity ceases.[146] However, mere nonuse of an easement will not result in termination.[147]

Some jurisdictions require proof of *actual necessity*.[148] For instance, Maine courts have long held that no right of way by necessity exists across remaining land of the grantor where the land conveyed borders on the ocean, even though access by water may not be as convenient as access by land. The test is *necessity* and not mere convenience.[149]

The third category is an implied easement *from a subdivision plat*. When an owner of a tract subdivides in accordance to a plat, a purchaser of any lot in the subdivision also acquires an easement over the streets as laid out on the plat.[150]

Even though the existence of an easement is agreed upon, its scope may vary according to different courts. Three theories exist:[151]

1. The necessary (or narrow) rule, where the extent of the easement is *limited to the adjoining street* and connecting streets necessary to give access to public streets.
2. The beneficial (or intermediate) rule, where an easement is granted in those streets that are *reasonably beneficial* to the purchaser.
3. The broad rule, where the purchaser acquires an easement over *all the platted streets* shown on the subdivision plat.[152]

The implied easements exist over the appropriate subdivision streets, regardless of whether such streets have been improved. Obviously, this guarantees access to any and all purchasers of lots in the subdivision.

Ways of Necessity By definition, a way of necessity is an easement founded on an implied grant or implied reservation. It arises where there is a conveyance of a part

[145] *A Practical Guide to Disputes Between Adjoining Landowners – Easements*, § 2.02(3b,iii).

[146] *A Practical Guide to Disputes Between Adjoining Landowners – Easements*, § 2.02(3b,iv).

[147] 25 Am.Jur.2d, § 105.

A way of necessity ceases to exist when necessity for it ceases. *Condry v. Laurie*, 41 A.2d 66, 184 Md. 317 (1945).

[148] A "right of way by necessity" does not arise if there be already another mode of access to the land, though less convenient or more expensive to develop. *Jennings v. Lineberry*, 21 S.E.2d 769, 180 Va. 44 (1942).

[149] Right of way by necessity never exists as mere matter of convenience. *Stein v. Bell Telephone Co. of Pennsylvania*, 151 A.2d 690, 301 Pa. 107 (1930).

Convenience alone cannot give right of way of necessity. *Littlefield v. Hubbard*, 128 A. 285, 124 Me. 229, 38 A.L.R. 1306 (1925).

[150] *A Practical Guide to Disputes Between Adjoining Landowners – Easements*, § 2.02(3c).

[151] Ibid.

[152] A sale of a platted lot with reference to the plat will, as between the grantor and the grantee, gives the latter a right to use all of the streets delineated on the plat, even though the plat is unrecorded. *Robidoux v. Pelletier*, 391 A.2d 1150 (1978).

of a tract of land of such nature and extent that either the part conveyed or the part retained is shut off from access to a road to the outer world by the land from which it is severed or by this land and the land of a stranger. In such a situation, there is an implied grant of a way across the grantor's remaining land to the part conveyed, or conversely, an implied reservation of a way to the grantor's remaining land across the portion of the land conveyed. The order in which two parcels of land are conveyed makes no difference in determining whether there is a right of way by necessity appurtenant to either.[153]

A way of necessity results from the application of the presumption that whenever a party conveys property he also conveys whatever is necessary for the beneficial use of that property and retains whatever is necessary for the beneficial use of land he still possesses. Such a way is of common-law origin and is presumed to have been intended by the parties. A way of necessity is also said to be supported by the rule of public policy that lands should not be rendered unfit for occupancy or successful cultivation.[154]

Although the rule is most frequently applied to ways, it is not limited to that category of easement. Drains[155] and gas lines,[156] as well as other forms of pipe line, are common examples. Ways of necessity cannot exist where there was never any unity of ownership of the alleged dominant and servient estates, for no one can have a way of necessity over the land of a stranger. Necessity alone without reference to any relationship between the respective owners of the land is not sufficient to create the right.[157]

Rationale of the Court System "Land without means of access is practically valueless. No reasonable use can be made of it, and it has no market value. The presumption of intent on the part of the parties to the conveyance to provide a means of access is so strong . . . that the contrary thereof can hardly be supposed."[158] To ensure that land can be reasonably used, "the law implies an easement over the servient estate where the grantor owns and conveys a portion of the original lands without expressly providing a means of ingress and egress."[159] ("The way of necessity arises when the strong public policy against shutting off a tract of land and thus rendering it unusable gives rise to a fictional intent defeating any such restraint.").

An easement implied by necessity "is not established by the mere fact that one's land is surrounded by the lands of others cutting him off from public ways."[160] The

[153] *City of Whitwell v. White*, 529 S.W.2d 228 (Tenn.App., 1974).

[154] 25 Am Jur.2d, pp. 447, 448.

[155] *York v. Golder*, 151 A. 558, 129 Me. 300 (1930).

[156] *Tong v. Feldman*, 136 A. 822, 152 Md. 398, 51 A.L.R. 1291 (1927).

[157] 28 C.J.S. Easements, § 35.

[158] *Crotty v. New River & Pocahontas Consol. Coal Co.*, 72 W.Va. 68, 78 S.E. 233, (1913).

[159] *Berkeley Development Corp. v. Hutzler*, 159 W.Va. 844, 229 S.E.2d 732, (1976). *See also*, *Wolf v. Owens*, 340 Mont. 74, 172 P.3d 124, (2007).

[160] *Proudfoot v. Saffle*, 62 W.Va. 51, 57 S.E. 256, (1907).

easement arises only when a unified tract is severed—if there is no evidence that the purported dominant estate and servient estate were ever a part of the same tract of land, there cannot be a way of necessity.[161] The easement also arises only if, at the time of severance, there was no reasonable access to the landlocked property except by way of the claimed easement. This Court has stated that a party claiming an easement of necessity "can go back beyond the deed of the immediate grantee to the common source of title, however remote it may be, and claim a way by necessity, as appurtenant to the land[.]"[162] Put another way: "A way of necessity springs out of the deed at the date of the grant, and becomes appurtenant to the granted estate."[163]

"An easement may be acquired by express grant, implied grant (implication), or prescription, which presupposes a grant to have existed."[164] An easement by necessity arises by implied grant when a part of a commonly-owned tract of land is severed in such a way that either portion of the property has been rendered inaccessible except by passing over the other portion or by trespassing on the lands of another.[165] The fact that the dominant and servient estates originated from the severance of a "commonly-owned tract of land" is undisputed in the present case. The parties agree that the Hugginses are entitled to an easement by necessity across the Wrights' five acre tract.

An easement by necessity requires no written conveyance because it is a vested right for successive holders of the dominant tenement and remains binding on successive holders of the servient tenement.[166, 167]

A "way of necessity" is an "easement" arising from an implied grant or implied reservation, and it is the result of the application of the principle that whenever a person conveys property, he conveys whatever is necessary for the beneficial use of that property, and retains whatever is necessary for the beneficial use of the land he still possesses. (citations omitted).[168]

Sale or Lease of Land Surrounded by That of Grantor or Lessor. When there is a conveyance of a tract of land and there is no means of access thereto or egress therefrom except over the remaining land of the grantor, a way of necessity over such land is ordinarily granted by implication of law. Besides that, a right of way is implicitly granted by a lease of premises that cannot be reached except across the lessor's land.[169]

[161] *Derifield v. Maynard,* 126 W.Va. at 755, 30 S.E.2d at 13.

[162] *Crotty v. New River & Pocahontas Consol. Coal Co.,* 72 W.Va. at 70, 78 S.E. at 234.

[163] *Id.,* 72 W.Va. at 71, 78 S.E. at 234.

[164] *Dethlefs v. Beau Maison Dev. Corp.,* 511 So.2d 112, 116 (Miss., 1987).

[165] *Taylor v. Hays,* 551 So.2d 906, 908 (Miss. 1989).

[166] *See Broadhead v. Terpening,* 611 So.2d 949, 954 (Miss. 1992) (implication that the owner of the larger tract would not want to create a landlocked parcel by conveying an interior portion, so it is conveyed whether described or not when the dominant estate is deeded; easements by necessity run with the land and are deeded with each conveyance regardless of description).

[167] *See also Pitts v. Foster,* 743 So.2d 1066, 1068 (Miss.Ct.App.1999).

[168] *Huggins v. Wright,* 774 So.2d 408 (Miss., 2000).

[169] 28 C.J.S. Easements, § 36.

Sale of Land Surrounded by That of Grantor and Third Persons. If a parcel is granted partly surrounded by land of the grantor and partly by land of strangers, a right of way over the remaining land of the grantor exists by necessity. If the land retained is surrounded by the land conveyed and land of strangers, the grantor is entitled to a way of necessity over the land conveyed.[170]

Sale of an Item to Be Removed from Land. Where a landowner grants some product of the soil, such as timber, stone, iron ore, etc., the grant necessarily carries with it such a reasonable way over the land to allow a right to enter and take away that which was granted. The location of the right of way, however, depends on the necessity.[171]

Ways Created by Sale by Reference to Map or Plat. Where a landowner creates a subdivision of land into streets and lots delineated on a map or plan and sells lots bounded by the streets, the legal effect of the grants is to convey to any grantee an easement of way over the streets. This easement is independent of the dedication to public use of the platted streets.[172] Therefore, a purchaser in a subdivision where the streets have been accepted by public authority has two easements, one as a member of the traveling public and one through implication. This guarantees a lot owner access to the lot even if the dedication is not accepted or in the case where a street is later vacated.

A grantee who acquires property by reference to a plan acquires a private right of way in proposed streets delineated in the plan.[173] "The right of way in the proposed streets acquired by a grantee under such circumstances is an 'easement by implication based upon estoppel.'"[174] "Such an easement is not based upon a dedication to the public. It is a private right in no way dependent upon a prospective public use."[175] Id.[176]

Whether the streets have been improved, or are suitable for travel, is immaterial.

Paper Streets Bona fide streets shown on a map or plat but never constructed are known as *paper streets*. They share the same characteristics and status as other streets in that, if accepted, they become public ways, and if not accepted, they are implied ways in favor of lots purchased in accordance with the map or plat (see Figure 4.8). The concept is not unique, nor is it limited to, streets. It may also apply to pathways or other entities set aside for use by either the lot owners or the public in general. This means that ancient plats may have dedicated areas shown on them that have certain title characteristics. These areas may be developed at a later date, and if a change in use of certain aspects of the original subdivision is contemplated, the title issues affecting any of the dedicated areas must be taken into account.

[170] 28 C.J.S. Easements, § 37.

[171] 28 C.J.S. Easements, § 38.

[172] 28 C.J.S. Easements, § 39.

[173] *Callahan v. Ganneston Park Dev. Corp.*, 245 A.2d 274, 278 (Me.1968).

[174] *Id.* (quoting *Arnold v. Boulay,* 147 Me. 116, 120, 83 A.2d 574, 576 (1951)).

[175] *Id.* (quoting *Arnold,* 147 Me. at 120-21, 83 A.2d at 576).[166]

[176] *Murch v. Nash*, 2004 ME 139, 861 A.2d 645 (Me. 2004).

Consider the following:

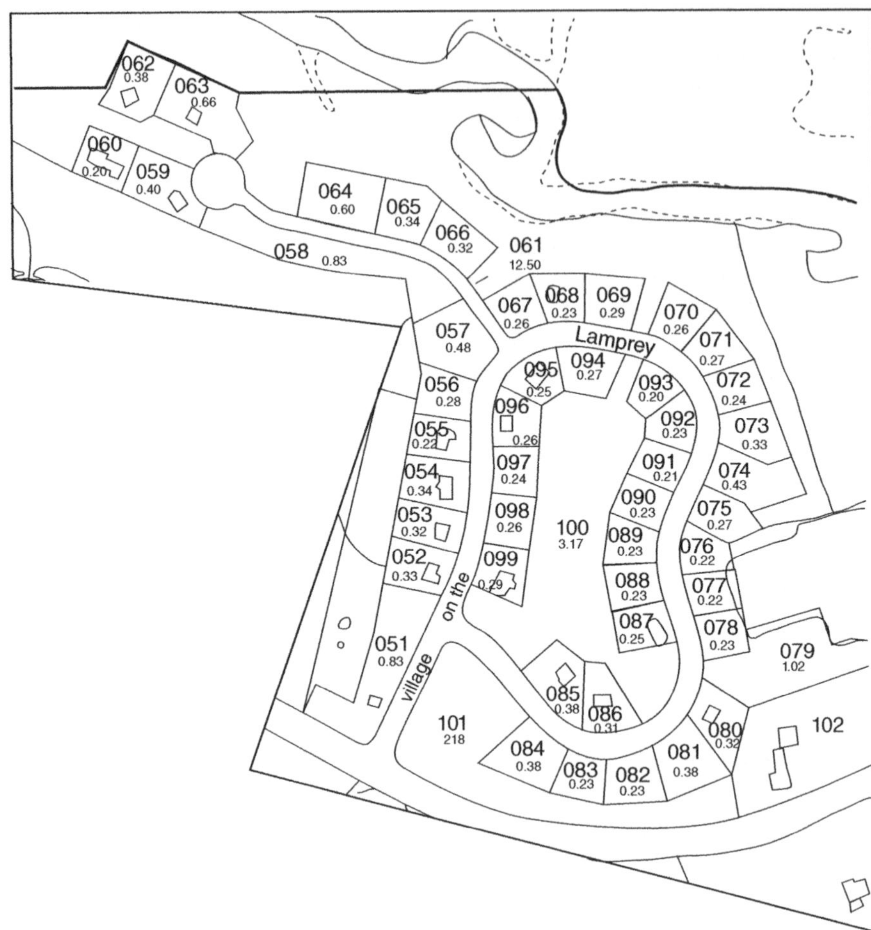

Figure 4.8 Illustration of a subdivision plat. Courtesy of Town of Epping, NH.

Another common problem is illustrated in Figure 4.9.

Ways Created by Sale of Land Bounded by Road, Street, or Alley. When a landowner conveys land by a deed that describes it as bounded by a road, street, or alley, the fee of which is in the grantor, the grantee acquires a right of way over the road, street, or alley. This rule has been held to apply regardless of whether the road is in existence, or whether the easement is necessary.[177]

Ways Created by Sale of Land without Reference in Conveyance to Roads, Streets, or Alleys. Where land is conveyed on a representation that streets exist

[177] 28 C.J.S. Easements, § 40.

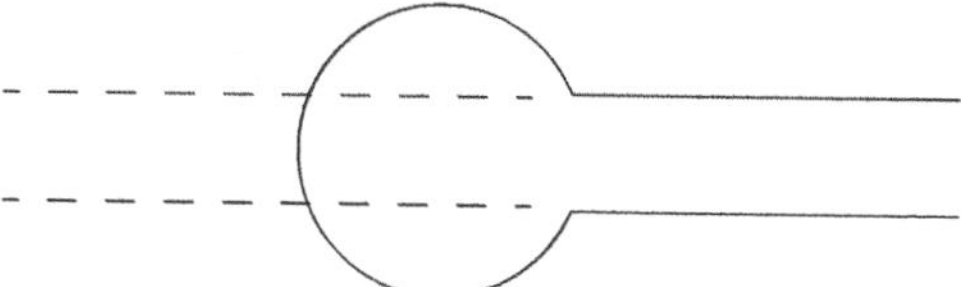

Figure 4.9 Illustration of change in cul-de-sac. As development was extended, a dead-end road was converted to a through street.

adjoining the parcel conveyed, the grantee acquires an easement in such streets even though the conveyance is by metes and bounds with any reference to the streets. The easement does not pass through the grant, but by estoppel.[178] See Chapter 14, Case # 7.

Drains and Ditches. If an owner of land makes an artificial open ditch or drain across it for the purpose of drainage and afterward sells either the upper or lower portion of the land without reference to the ditch, there is an implied grant or reservation of an easement for the use of the drain, with some exceptions.[179]

Light, Air, View, and Privacy. Easements of light and air may be created by implication, but since such easements are not favored as they unduly burden property and prevent adequate development, they will not be implied unless an intention to create them clearly appears.[180]

Rights to Use Halls and Stairways. General rules regarding implied easements apply to halls and stairways. Where a building is constructed such that occupants enter and depart by the same hall and stairway, an easement for the use of the hall or stairway is impliedly granted on the sale or lease of part of the building.[181]

Rights to Use of Park or Open Square. Where land is sold with reference to a map or plat showing a park, beach, or open square, the purchaser acquires an easement that such area shall be used in the manner designated, and an easement in the streets that provide access to such area.[182]

Strict Necessity Some jurisdictions state, or at least imply, that *strict necessity* is a requirement. The North Carolina court, however, has taken a slightly more conservative approach, in keeping with the underlying concept of land benefitted by what allows it to be useful, or enjoyed. "It is not necessary that the party claiming the easement show absolute necessity. An easement by necessity may arise even where other inconvenient access to the parcel in question exists."[183] (see discussion on quasi-easements in Chapter 2).

[178] 28 C.J.S. Easements, § 41.
[179] 28 C.J.S. Easements, § 33.
[180] 28 C.J.S. Easements, § 42.
[181] 28 C.J.S. Easements, § 43.
[182] 28 C.J.S. Easements, § 44.
[183] *Boggess v. Spencer*, 620 S.E.2d 10 (N.C.App., 2005).

Implication vs. Necessity[184] In closing, it is important to emphasize that an implied easement is distinct from an implied easement by necessity; an implied easement by necessity allows for the establishment of a right of way where one previously did not exist. An easement by necessity is a type of implied easement based upon the premise that wherever one conveys property he also conveys whatever is necessary for its beneficial use and enjoyment, including access to one's property. The party claiming the right of way bears the burden of proving the three elements:

1. Unity of title,
2. Unity severed by conveyance of one of the tracts, and
3. Necessity, both at the time of severance and the time of exercise of the easement.

ESTOPPEL

An easement may be created by under the *doctrine of estoppel*.[185] The word *estop* means to stop, prevent, or prohibit. Legally, an estoppel is a bar raised by the law which precludes a person, because of his conduct, from asserting rights that he might otherwise have—rights as against another person who, in good faith, relied upon such conduct and was led thereby to change his position for the worse.

Estoppels are of three kinds:

1. Estoppel by deed
2. Estoppel by record
3. Estoppel *in pais*, (estoppel by conduct)

Estoppel *in pais* is an *equitable* estoppel and arises when one represents by word of mouth, conduct, or silent acquiescence that a certain state of facts exists, thus inducing another to act in reliance upon the supposed existence of such facts, so that if the party making the representation were not estopped to deny its truth, the party relying thereon would be subjected to loss or injury.

[184] *Cellico Partnership v. Shelby County*, 172 S.W.3d 574 (Tenn. Ct. App., 2005).

[185] Easement may be created by estoppel. *Monroe Bowling Lanes v. Woodsfield Livestock Sales*, 224 N.E.2d 762, 17 Ohio App.2d 146, 460 O.O.2d 208 (1969).

Right of way created by estoppel is appurtenant to land conveyed and is not an easement in gross. *Uliasz v. Gillette*, 256 N.E.2d 209, 357 Mass. 96 (1970).

A way of estoppel is not a way of necessity, and the right exists even if there are other ways, either public or private, leading to the land. *Casella v. Sneierson*, 89 N.E.2d 8, 325 Mass. 85 (1950). Where property is conveyed in a deed and one or more of the calls is an abuttal on a private way, there is a grant or at least a presumption of a grant of an easement in such way when the way is owned by the grantor; and it is of no consequence that the fee to the roadway or passageway remains in the hands of the original grantor or his assigns, or that the grantor did not intend to grant an easement, or that the easement is not one of necessity, because the grantor, and all claiming under him, are estopped by deed from denying such an easement exists. *700 Lake Avenue Realty Co. v. Dolleman*, 121 N.H. 619 (1981).

It is the species of estoppel that equity puts upon a person who has made a false representation or a concealment of material facts, with knowledge of the facts, to a party ignorant of the truth of the matter, with the intention that the other party should act upon it, and with the result that such party is actually induced to act upon it, to his damage.[186]

Case #7 in Chapter 14 is an example of a right of way by estoppel.

PRESCRIPTION

An easement may be acquired by *prescription*, that is, by long continued *use* of another's land for purposes in the nature of an easement. Such an easement stands in all respects on the same footing as an easement acquired by grant.[187]

Historically, the doctrine of prescription was based on the presumption of a lost grant, the idea being that, since the use had continued for such a long period of time, it must have been based on a grant that had become lost or destroyed.

In *Crawford v. Matthews*,[188] the court noted, "[p]rescription is, in essence, a form of adverse possession. They differ in that prescription grants the adverse user an easement or incorporeal rights in the property, while adverse possession grants the adverse user legal title."

The court further noted, "Prescriptive easements are not favored in law, because they deprive the legal property owner of rights without compensation." "Thus, '[o] ne who claims an easement by prescription has the burden of proving by clear and convincing evidence all the elements essential to the establishment thereof." [189]

Additionally, easements created in other ways may be enlarged by prescription (see Figure 4.10).[190]

In most states, the requirements for establishing an easement by prescription are the same as for acquiring title by adverse possession,[191] including the following:

1. Use must be adverse
2. Open and notorious
3. Continuous
4. Exclusive
5. Under claim of right
6. For the statutory period

[186] *Black's Law Dictionary*.

[187] 25 Am.Jur.2d, § 39.

An easement in light cannot be acquired by prescription, but only covenant or grant. *Mathewson St. M.E. Church v. Shepard*, 46 A. 402, 22 R.I. 112 (1900).

[188] (Sept. 21, 1998), Scioto App. No. 97CA2555, unreported.

[189] Quoting *McInnish v. Sibit* 114 Ohio App. 490, 183 N.E.2d 237 (1953).

[190] Easements created by reservation or grant may be enlarged by prescription. *Atkins v. Bordman*, 37 Mass. 291 (1838).

[191] Adverse possession and easement by prescription depend on same elements but differ fundamentally in that one is based on claim of possession and the other claim of use. *Rasmussen v. Sgritta*, 305 N.Y.S.2d 816, 33 A.D.2d 843 (1968).

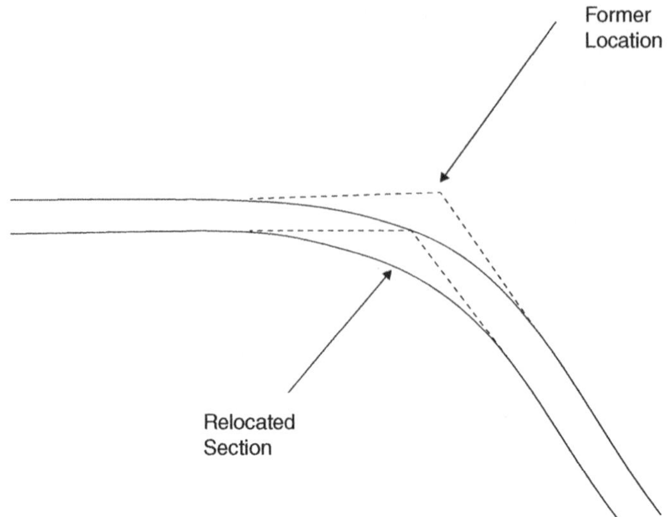

Figure 4.10 An existing easement relocated and widened through continuous use.

The basic difference between adverse possession and prescription is that with adverse possession the *possession* ripens into title (ownership), while prescription, being *use*, develops into an easement (right of use).

The first requirement is that the use must be *adverse* as distinguished from *permissive*. If the latter were the case, then the easement would be based on some form of agreement, and the user would not gain title.[192] To be adverse, the use need not be hostile in the strict sense.

The adverse use must have been with the owner's knowledge and acquiescence or it must have been so open and notorious that knowledge and acquiescence would be presumed. This requirement is for the protection of the landowner and assures them a reasonable notice of the need to preserve their rights against the adverse user.

To be *open*, the use must be made without secrecy or concealment. A concealed use will not suffice unless the owner has actual knowledge of it. To be *notorious*, it is not necessary that the use be actually known to the owner; it is only necessary that the use be such that a reasonable inspection of the premises would have disclosed its existence.

For a use to be *continuous*, it is not necessary that it be constant or made every day; it may, depending on the nature of the use, be periodic or occasional. What is essential is that there can be no break in the adverse attitude of the user.

For a prescriptive right to arise the use must not only be continuous—it must also be uninterrupted. This means that there must be no break in the use brought

[192] Use by express or implied permission or license, no matter how long continued, cannot ripen into an easement by prescription. 25 Am.Jur.2d, § 54.

about by an act of the owner; if the owner causes a discontinuance of the use, even for a brief period, there is an interruption and the prescriptive period must begin all over again if there is eventually a right to be created by prescription. In addition, there cannot be a voluntary abandonment by the party claiming the easement.[193]

Some states include an *exclusive* requirement. This means that no one else can share the use or substitute their use for the adverse user's. Use that is not exclusive but in participation with the true owner, or others, is not sufficient. However, the requirement as applied to an easement is not as strict as for adverse possession of land. It does not mean that all others be excluded from the property; it simply means that other uses cannot exclude or interfere with the easement rights. It also means that the claim cannot be shared by the public generally. If it is, the use may ripen into a public easement rather than a private one.[194] This does not mean, however, that all public easements are highways.

The word *exclusive* in reference to a prescriptive easement does not mean that there must be use only by one person, but, rather, means that the use cannot be dependent upon a similar right in others.[195]

Strictly speaking, courts do not recognize a prescriptive easement in favor of the public because there is no identifiable party for the burdened property owner to challenge. Usually, if an owner has knowingly permitted the public at large to use the property in an adverse and continuous manner, courts find an implied offer on the part of the owner to dedicate the property for public use. Custom is an alternative doctrine used in some states to achieve the same result.[196]

The public as well as an individual may, however, acquire an easement by prescription.[197] In fact, many of the very old highways were established in this manner.

As with adverse possession, the prescriptive easement doctrine does not apply against the sovereign as the servient property owner.[198] Although, in many instances, the opposite may be possible, depending on the circumstances.

In some jurisdictions, it is essential that the adverse use be under a *claim of right*.[199] It is not necessary, in order to establish a claim of right, however, that an actual claim be made; an intent to claim adversely may be inferred from the acts and conduct of the dominant users.[200]

[193] 25 Am.Jur.2d, § 56.

[194] *A Practical Guide to Disputes Between Adjoining Landowners – Easements*, § 2.03(3b).

[195] *Werner v. Schardt*, 222 Neb. 186, 382 N.W.2d 357 (1986).

[196] Ibid., § 2.03(4b).

[197] *Piper v. Voorhees*, 155 A. 556; 130 Me. 305 (1931).

[198] Unless allowed by statute, an easement, ordinarily, cannot be acquired by prescription against the government, federal or state, or a subdivision thereof. 28 C.J.S., § 9.

[199] The term "claim of right" means no more than a user "as of right," that is, without recognition of the rights of the owner of the servient estate. *Andrezejczyk v. Advo System, Inc.*, 146 Conn. 428, 151 A.2d 881 (1959).

[200] 25 Am.Jur.2d, § 52.

Claim of right is not to be confused with, or equated with, *color of title*.[201] In most states, statutes of limitation requiring color of title in order to obtain title by adverse possession, have been held not to apply to the acquisition of easements by prescription.[202]

Finally, the use must comply with the foregoing requirements for the *period required by statute*. The time requirement varies considerably depending on the jurisdiction and the attending circumstances of the use and nature of the right. It may be extended in cases of disability, such as when the owner is a minor, is mentally incompetent, is out of the country, is in prison, or is a reversioner.[203]

An easement acquired by prescription is frequently called a prescriptive easement, separating it from easements acquired by other means.

Public Highway by Prescription The Kansas case of *Brownback v. Doe*[204] dealt with this issue in some detail, and stated, among other things, the following.

To establish a road or highway by prescription, the land in question must have been used by the public with the actual or implied knowledge of the landowner, adversely under claim or color of right, not merely by the owner's permission, and continuously and uninterruptedly, for the period required to bar an action for the recovery of possession of land or otherwise prescribed by statute. When these conditions are present a highway exists by prescription; otherwise, it does not.

In designating a road as a public road by prescription, public maintenance of the road is most significant.

In order to designate a public road by prescription, there must be some action, formal or informal, by the public authorities indicating their intention to treat the road as a public road.

In a claim of title by adverse possession, every presumption is in favor of the holder of the legal title and against the claimant. The law will not allow the property of a person to be taken by another upon slight presumptions or probabilities. The facts relied upon to establish adverse possession cannot be presumed, and presumptions will not be indulged in to establish a claim of title.[205]

In *Kratina v. Board of Commissioners*,[206] the Kansas Supreme Court discussed the elements required to establish a public prescriptive easement:

[201] Color of title is a paper writing that purports to convey land but fails to do so. *Carrow v. Davis*, 248 N.C. 740, 105 S.E.2d 60 (1958).

[202] 25 Am.Jur.2d, § 53.

[203] 25 Am.Jur.2d, § 40.
Pentland v. Keep, 41 Wis. 490.

[204] 241 P.3d 1023 (Kan.App. 2010).

[205] *Chesbro v. Board of Douglas County Comm'rs*, 39 Kan.App.2d 954, 186 P.3d 829, *rev. denied* 286 Kan. 1176 (2008).

[206] 219 Kan. 499, 502, 548 P.2d 1232 (1976).

> To establish a highway by prescription the land in question must have been used by the public with the actual or implied knowledge of the landowner, adversely under claim or color of right, and not merely by the owner's permission, and continuously and uninterruptedly, for the period required to bar an action for the recovery of possession of land or otherwise prescribed by statute. When these conditions are present a highway exists by prescription; otherwise not.

The *Kratina* court stated: "Mere use by the traveling public is not enough to establish . . . that the use is adverse. . . . There must in addition be some action, formal or informal, by the public authorities indicating their intention to treat the road as a public one."

States sometimes have a statutory requirement applicable to a prescriptive easement. For example, New Hampshire requires that a road must have been used as such for public travel, for 20 years prior to January 1, 1968.[207] In litigation, the standard of proof is "by a balance of probabilities."

Dedication Easements may be created by *dedication*, such as in the case of streets in a subdivision. Dedication may be broadly defined as "the devotion of land to a public use by an unequivocal act of the owner of the fee manifesting an intention that it shall be accepted and used presently or in the future for such public use." While dedication gives the public the right of passage, it does not burden the municipality with the duty of maintenance unless the municipality accepts the dedication. Acceptance is a separate event.

Dedication has been said to be an exceptional and unusual method of passing title to an interest in land. The same quality of evidence that is needed to demonstrate the existence of a dedication must be presented to show the extent thereof. The public is not bound to accept a dedication, but when it does, the acceptance is to be limited territorially to the property dedicated and accepted.[208]

Unlike a conveyance in fee, the owner of dedicated property retains ownership of the fee under the dedicated land. However, an offer of dedication is not inconsistent with an intent to convey a fee out to the center line of a road. In other words, a property owner may dedicate a portion of his land for use by the travelling public and, by conveying lots abutting the dedicated road, convey the fee underneath the road.[209] (see Chapter 10 on reversion of highways for a discussion on the presumption of law and its caveats).

Dedications may be express, as when made by oral declaration, or by deed or note, or they may be implied as when there is an acquiescence in a public use. Implied dedications can be troublesome, since there usually is no written record, at

[207] NHRSA, Title 20, § 229:1.

[208] *Schmidt v. Town of Battle Creek*, 188 Iowa 869, 175 N.W. 517.

[209] *Creation and Termination of Highways in New Hampshire.*

least of its creation. Implied acceptance, in order for the dedication to become public, is also elusive and often requires research into whether a governing authority acted on the issue with any votes, discussion, maintenance, or appropriation or expenditure of public funds.

Two kinds of dedication of private land to public use exist: those made under the controlling principles of common law and those that result from the compliance with specific statutes.[210] Even in the same jurisdiction a dedication of land to public use may be made either according to the common law or pursuant to a statute. Moreover, a defective attempt to dedicate under a statute may, on acceptance and user by the public, be effective as a common-law dedication.

The distinction between the two is that a statutory dedication operates by way of *grant*, while common-law dedication is an *equitable estoppel*. Another important distinction is that the right conferred by common-law dedication is a mere easement, while most statutory dedications result in the fee being in the public authority to which the dedication was made.[211] This distinction can be exceedingly important at a later date when alterations or vacations are considered.

A common-law dedication does not pass legal title to the land, but it is sufficient to defeat an action at law for the recovery of the possession of the property as against those who are using it in accordance with the object and purpose of the dedication. The mere use, however, of a road over private land by the public will, by itself, not make it a public road.

In the case of *Dicken v. Liverpool Salt & Coal Co.*,[212] one of the issues was whether a particular road was a public way. The court found that the road was a private way and stated that

> in this state a private way cannot be converted into a public highway by the use thereof by the public, no matter how long that use may be continued; but it must be dedicated by the owner and accepted for the public [by the county court], which acceptance can only be shown by its record. It has long been the settled law in this state that the mere user of a road by the public for however long a time will not make it a public road. On the contrary, the mere permission by the owner of the land to the public to pass over the road is, without more, to be regarded as a license revocable at pleasure. A road dedicated to the public must in some way, directly or by inference, be accepted by the county court upon its record, or by the municipal corporation, before it can become a public road."[213]

[210] Particular care should be exercised to examine the law in existence at the time of the event(s) as statutes governing dedication have undergone radical changes over time in some jurisdictions.

[211] 23 Am Jur 2d, Dedication, § 3.

[212] 41 W.Va. 511, 23 S.E. 582 (W.Va., 1895).

[213] *Boyd v. Woolwine (W.Va.)* 21 S.E. 1020; *Com. v. Kelly, 8 Grat.* 632; *Hall v. McLeod, 2 Metc. (Ky.)* 98; *Davis v. Ramsey, 5 Jones (N.C.)* 236; *Speir v. Town of Utrecht*, 121 N.Y. 420, 24 N.E. 692; 19 Am. & Eng. Enc. Law, 107.

As can be seen from quoted authority, this concept is not unique to West Virginia. The court further stated

> (1) The owner must in some way indicate an intention to appropriate the private way to the public use. See Elliott, Roads & S. p. 92. Here there was nothing of that sort. It was the ordinary private way, along the side of which ran the private tram road indispensable to a salt-furnace property. (2) Even if intended to be dedicated, it has never been accepted. You cannot bind the public or the county without its assent, and in this state such assent must appear directly or indirectly by the record. It can speak in no other way. See Elliott, Roads & S. p. 87.

By the early common law dedication was confined to highways, as the need for the dedication of public parks, squares, and the like, was not an issue. In this country, however, the doctrine has a wider application and its limit has been judicially extended so as to include parks, public squares, school-lots, burying-grounds, lots for church purposes, and other charitable uses generally. The dedication of public squares, parks, school-lots, and lots for charitable purposes is usually made by the public authorities for the benefit of the whole community. The same rules of law apply to the dedication of squares, parks, school-lots, and the like, as apply to the dedication of highways. See the N.H. school house example in Chapter 8 on reversion.

When a developer sells residential lots in a subdivision by reference to a recorded subdivision plat which divides the tract of land into "streets, parks and playgrounds,' a purchaser of one of the residential lots "acquires the right to have the streets, parks and playgrounds kept open for his reasonable use, and this right is not subject to revocation except by agreement."[214]

Private Dedication The West Virginia court addressed this issue as recently as 1984 in *Bauer Enterprises, Inc. v. City of Elkins*.[215] In this case, a landowner owned land with streets or alleys on each side. The landowner used a portion of one of the streets as a private way, which the city never maintained. When the landowners sought a building permit to replace a demolished building, the city rejected the permit because the building would be upon (encroaching) a public street. The court found that there had been a private dedication of the street by the recordation of the plat, but that the street had not been made public by the city's acceptance of the dedication. The court determined that the evidence was insufficient to extinguish the private easement and that the city's right to accept the dedication was not terminated by its delay in acting or the landowner's adverse possession. Because the city retained a right to accept the dedication, the court concluded that the landowner was properly denied a building permit and relief by the trial court. This decision was affirmed.

[214] *Harry v. Crescent Resources, Inc.*, 136 N.C.App. 71, 523 S.E.2d 118 (1999).

[215] 173 W.Va. 438; 317 S.E.2d 798 (1984).

In the Supreme Court decision, it was noted that there are two types of dedication, public and private. There mere recording of a plat showing streets and alleys constitutes only a private dedication. The law provides that in this case all lots sold and conveyed with reference to the plat carry with them only the right to the use of the streets appearing on the plat, which is necessary to the complete enjoyment and value of the lots.

The general public acquires no proprietary rights to the platted ways in the case of a private dedication. On the other hand, the interest of the public becomes proprietary when the offered dedication is accepted by a public authority. Acceptance may be by an express act, as indicated by notation on the recorded plat, or implied, as when the public authority assumes responsibility for the maintenance and repair of the platted streets and alleys. When the plat is accepted by the public authority, the act of dedication becomes a public dedication and the streets and alleys therein become public ways.

Private easements differ from public ways in two regards. First, private easements can be used by their owners only for the purpose of reasonable ingress and egress to their property, whereas public ways can be used by the general public without reference to their destination. Second, public authorities have no obligation to maintain private easements, whereas they do have an obligation to maintain public ways. In addition, there can be no adverse possession of a public way, but if the elements of adverse possession are present a private easement may be extinguished.

Ancient Rights Extreme caution should be exercised with regard to ancient rights created by early plats and related documents. Case #17, "Proprietors Way," is an example of an ancient right created by an early plat found to be as valid in modern times as when it was created.

A further example is the Pennsylvania case of *Bruker v. Burgess and Town Council of Borough of Carlisle.*[216] While the court found some of the evidence confusing, and incomplete in its wording, it did state that ancient rights as established in favor of the public continue; however, in the absence of more complete definition, uses may change over time to meet the needs of the community.

Two things in this case are noteworthy. The first is the court's statement that "there is no doubt that the mere fact of the use of [the Square] by the public for now more than 200 years is sufficient to raise a conclusive presumption of an original grant for the purpose of a public square; such is an ancient and well established principle of the law. Nor can it be denied that, where such a dedication has been established and the public has accepted it, there cannot be an diversion of such use from a public to a private purpose, and it is also true that, where a dedication is for a limited or restricted use, any diversion therefrom to some purpose other than the one designated is likewise forbidden." (citing several cases).

The second item of note is the dissenting opinion (of two of the judges) which states, in part, "Evidence of dedications and titles based upon ancient documents or events arising out of antiquity cannot possible by as strong or clear as would be required in matters arising or titles created in modern times; and we must not

[216] 376 Pa. 330; 102 A.2d 418 (Pa., 1954).

lightly strike down rights or public uses which have existed for more than a century. Counsel for the Borough frankly admits that if the Borough can change or destroy this market place it can also change or destroy the church which was built on this same public square and the like the market place has been used for more than 200 years. This church and this market place were dedicated by Thomas Penn in 1751; and this dedication and use as a market place were thereafter frequently ratified. The Borough merely contends that there exists today no clear evidence that this square was the 'spot' which was dedicated for these purposes.

> We are living in exciting and rapidly changing times. We rode from the horse and buggy age to the automobile age and then flew too rapidly to the airplane age and the atomic age. The tremendous changes which have occurred and are still daily occurring have necessarily produced uncertainty, unrest and confusion-not only in the minds of men, but in many phases of man's life. As a consequence, "change" is on every man's lips, and unrest and uncertainty in many a man's heart. In the craving for change, in the restless quest for a Utopia of riches and ease, haven't we too often forgotten the things of the spirit, as well as the history of our Country and the immemorial customs of our people? Haven't we rushed frantically and heedlessly after false goods-material prosperity and political panaceas? Should not our wonderful farm people be allowed to preserve in a farming community a few of their ancient privileges, customs and practices, even though more income would be produced by a different (public or private) use?

Title to this market place or square is not in the Borough of Carlisle; it is in the Commonwealth of Pennsylvania, with a reversionary interest in Penn's heirs. The Borough has no title or estate in this property; it has only a right to control and regulate, in the interest of the public, the public use of the 'market place' for the uses and purposes to which it was dedicated. *Hoffman v. City of Pittsburgh,* 365 Pa. 386, 75 A.2d 649.

In my judgment, the ancient documents, plans, letters and other evidence produced, plus the public use of this part of the Square as a market place for over 200 years, were sufficient to establish a dedication for that purpose or use, and consequently the Borough can and should be enjoined from changing or destroying this use."

Routine title searches will not uncover ancient rights unless they are carried forward in the wording of recent documents. Reconnaissance for survey purposes places the searcher in a position to recognize such uses. Research standards requiring a search back to various origins make it a necessity and are for the very reason of discovering and documenting anything affecting the title and uses of a parcel of land.

Parol Dedication The Oregon court addressed the issue of parol dedication in 1873, [217] stating that "It is a well-settled principle of the common law that property may be dedicated to the public use, and that the dedication may be by parol as well as by deed" (citing cases from several jurisdictions).

[217] *Carter v. City of Portland,* 4 Or. 339 (Or., 1873).

It continued,

> to constitute a dedication by parol, there must be some act or acts proved, evincing a clear intention to devote the premises to the public use.
>
> If one owning lands, or having an equitable interest therein (subsequently acquiring the title thereto), lays out a town thereon, and makes and exhibits a map or plan thereof, with spaces marked streets, alleys, public squares, parks, etc., and sells lots with clear reference to said map or plan (though unrecorded), the purchasers of lots in said town acquire, as appurtenant thereto, every easement, privilege, and advantage which the map or plan represents as part of the town. Upon the sale of lots with such reference to the map or plan, the dedication of the spaces marked streets, alleys, public squares, parks, etc., becomes irrevocable.
>
> Formal acceptance by the corporate authorities is not necessary. Where the dedication is irrevocable, it need not be followed by immediate and continued use.

EMINENT DOMAIN

Eminent domain is generally defined as the power of the nation or a sovereign state to take, or to authorize the taking of, private property for a public use without the owner's consent, conditioned upon the payment of just compensation.[218] It is accomplished through the process of condemnation. A well-settled rule is that where land is taken for a public use, unless the fee is necessary for the purposes for which it is taken, the condemnor acquires only an easement in the absence of a statutory provision to the contrary.[219]

In the absence of a statute expressly providing for the acquisition of the fee, or a deed from the owner expressly conveying the fee, when a highway is established by the direct action of the public authorities, the public acquires merely an easement of passage, the fee title remaining in the landowner.[220]

When a railroad company takes by eminent domain a strip of private land upon which to lay its tracks and operate its cars, it ordinarily acquires only an easement in the land so taken. The rule applies to a right of way for main tracks and for necessary sidetracks, chutes, yards, storage places, and the like. Thus, when a railroad corporation acquires land under a general authority to enter and take possession of it, or to appropriate it, or otherwise simply to take it for its own purposes, whether upon payment of compensation, it acquires only an easement in the land, to use for railroad purposes. Sometimes railroads are authorized to acquire the fee in land taken for their tracks, and such statutes have been given effect by a majority of the courts. However, where the statute contains some express provision that is inconsistent with the fee being taken, the court may construe it as granting only an easement although it expressly uses the term "fee-simple."[221]

[218] 26 Am.Jur.2d, § 1.

[219] 26 Am.Jur.2d, § 133.

[220] 26 Am.Jur.2d, § 136.

[221] 26 Am.Jur.2d, § 137.

Under the prevailing rule, a condemnation decree granting a right of way for the purpose of constructing, maintaining, and operating an electric current and power, telephone, or telegraph transmission line creates only an easement in the land and grants the exclusive possession of the property only to the extent that such possession is necessary for the erection, operation, and maintenance of the line, including the right of access.[222]

As a general rule, in the absence of a statute to the contrary, land taken for school purposes does not vest the fee in the school authorities or district, but only in the right to the use and occupation of the land for school purposes.[223]

The uses for which a public park is acquired are continuous and peculiarly exclusive, and it is generally held that either the fee is taken or an exclusive easement that is the equivalent of the fee, except that it leaves in the owner a possibility of reverter in case the land ceases to be used for a park. However, it has been held that the title acquired by condemnation for the purposes of driveways in a park is not a fee but an easement.[224]

Generally, in the case of land condemned for canals, ditches, sewers, pipelines, reservoirs, and the like, unless the statute otherwise expressly provides, it has been held that the fee remains in the landowner, the condemnor acquiring an easement only.[225]

Private Condemnation There exists, in some states, a constitutional provision allowing condemnation for private purposes, depending upon circumstances. For example, Colorado's constitution states, "Private property shall not be taken for private use unless by consent of the owner, except for private ways of necessity, and except for reservoirs, drains, flumes, or ditches on or across the lands of others, for agricultural, mining, milling, domestic, or sanitary purposes. This provision, along with the provisions and procedures in Colorado Revised Statutes, Secs, 38-1-101 and following, allows for private taking of private ways of necessity upon payment of just compensation, the same requirement for the taking of private rights for the public good. In short, Colorado's Constitution provides authorization for private condemnation to acquire a *private way of necessity*. This allows for the acquisition of access to property which would be otherwise "landlocked."

Necessity in this instance is not to be confused with the acquisition of a right of way by necessity through implication, which requires a unity of title of both dominant and servient tenements at some point in time. The leading case explaining the difference is *Crystal Park Co. v. Morton*,[226] which stated that the two "ways of necessity" were very different. "Private ways of necessity, as referred to in the Constitution, are ways which are indispensable to the practical use of the property for which they are claimed" and may also be "ways without which the landowners would have

[222] 26 Am.Jur.2d, § 138.

[223] 26 Am.Jur.2d, § 139.

[224] 26 Am.Jur.2d, § 140.

[225] 26 Am.Jur.2d, § 141.

[226] 27 Colo.App.74, 146 P. 566 (1915).

no access to public roads." There does not have to be a common ownership, or an implied grant.

The right to condemnation of a private way of necessity exists when the need to condemn is reasonably necessary and when the common law or other legal remedies do not provide a present enforceable legal right to an alternate mode of access that is both reasonable and practical.[227]

Recent decisions have held that the burden is on the condemnee to establish the existence of an acceptable alternate route and that the condemnor has a present enforceable right to use that route.[228] The Colorado court has made it clear that the right of private condemnation may not be exercised when the condemnor has a reasonable, common-law way of necessity.[229] In all cases, a thorough investigation must be undertaken to eliminate any other options for access, strict compliance with the statutory requirements must be exercised, and the least burdensome route should be selected.[230]

CUSTOM

Custom may be said to be a poorly understood concept seldom included in treatises devoted to property rights, especially those concerning easements. However, in areas of old settlements, custom, or customary use, can be a valid method of creating property rights, particularly when it comes to easements. Custom is recognized as a valid concept in the case law of several states but has received little recognition throughout the country in general. As the Florida court in *City of Daytona Beach v. Tona-Rama, Inc.*[231] noted,

> occasionally in this country it has been decided that rights to use private land cannot thus be created by custom, for the reason that they would tend so to burden land as to interfere with its improvement and alienation, and also because there can be no usage in this country of an immemorial character. In one state, on the other hand, the existence of such customary rights is affirmed, and in other this is assumed in decision adverse to the existence of the right in the particular case.[232]

An easement, by virtue of custom, is a legal right acquired by the operation of law through continuous use of land for a specific purpose over a long period of time. An easement right acquired by custom lacks a dominant tenement and is generally in favor of the inhabitants of a particular locality.

For the acquisition of a wider right over many tracts of land, or specific types of land regardless of ownership, the right may be asserted by ancient custom of usage.

[227] *Minto v. Lambert*, 870 P.2d 572 (Colo. App., 1993).

[228] *Freeman v. The Rost Family Trust*, 973 P.2d 1281 (Colo.App., 1999).

[229] *Billington v. Yust*, 78p P.2d 196 (Colo.App., 1989).

[230] Kube, *Access at Last: The Use of Private Condemnation.*

[231] 294 So.2d 73 (Fla., 1974).

[232] Referencing *Tiffany Real Property*, (Third Edition), Vol. 3, § 935.

Most of our law is created by legislation, or judicial creation, but custom plays some role in how the law develops. The elements are:[233]

1. Ancient: The use must be long and general in character, so long "that the memory of man runneth not to the contrary."
2. Continuous: The use must have been exercised without interruption. It need not necessarily be exercised continuously, by a person with a paramount right.
3. Undisputed: The use needs be peaceable and free from dispute.
4. Reasonableness: The use must be appropriate in manner for the type of land.
5. Clearly bounded: There must be visible and definable bounds to the right.
6. Obligatory: The right cannot variably applied by individual owners, it must be obligatory upon each and every one.
7. Legal: The use must not be repugnant or inconsistent with any other laws or customs.

Custom differs from prescription, which is personal and is annexed to the person of the owner of a particular estate, while custom is local and relates to a particular district. The distinction has been thus expressed: "While prescription is the making of a right, custom is the making of a law." Lawson, *Usages & Cust.* 15, note 2.[234] *Bouvier's Law Dictionary*[235] defines custom as "such a usage as by common consent and uniform practice has become the law of the place, or of the subject matter to which it relates."

A custom can only exist in favor of the community of a town, village, or hamlet. The difference between custom and prescription is that custom is properly a local usage, not annexed to any one person, but belonging to the community rather than to its individuals. Prescription, on the other hand, is merely a personal use or enjoyment. Custom is an outcome of immemorial usage and will not ordinarily result from adverse enjoyment for the time period required by the statute of limitations.[236]

Quasi-easements founded on custom appertain to many as a class, and not as grantees, nor do such rights require the existence of a dominant tenement. Where, however, rights capable of being the subject of grant as true easements are claimed by custom as belonging to those entitled in respect of their estates, no essential of a true easement is lacking, except that the origin of the right is custom, and not grant or prescription.

At common law, certain rights could be acquired by custom, which, while in the nature of easements, must yet be distinguished from strict easements. Such a right differs from an easement appurtenant because, to constitute an easement of that kind, there must be two tenements, a dominant one to which the right belongs and a

[233] 1 Blackstone, Commentaries, 75-78.

[234] *Black's Law Dictionary*.

[235] *Rawle's*, Third Revision, p. 742

[236] *Gillies v. Orienta Beach Club*, 289 N.Y.S. 733, 159 Misc. 675 (1935).

servient one upon which the obligation is imposed, but in the case of a custom, there is no dominant tenement. It differs from an easement in gross in that the latter is a burden upon the premises during the life of the owner of the easement only, while a right acquired by custom is not subject to such limitation.[237]

> In a Connecticut case,[238] the court defined a right of way by custom and stated that "a right of way by custom in favor of the inhabitants of a particular locality might be set up by the common law of England. It could be proved by immemorial usage. From such proof a presumption was deemed to arise that the usage was founded on a legal right. This right was not assumed to arise from a grant by an owner of land of an easement in it. No grant of that nature can subject the tenement of the grantor to an easement that will outlast the life of the grantee, unless it be made in such a way as to become appurtenant to some other tenement. A right of way by custom appertains to a certain district of territory, but not to any particular tenement forming part of that territory. Nor is it confined to owners of land within that territory. It belongs to the inhabitants of that territory, whether landowners or not. To a fluctuating body of that kind, no estate in lands can be granted. If, therefore, an easement be claimed to exist in their favor, a title cannot be made out by prescription, on the theory of a lost grant. It must have come, if at all, from some public act of a governmental nature".

In this case, the court refused to adopt the law of custom and most cogently stated its reason therefor, saying, "The political and legal institutions of Connecticut have, from the first, differed in essential particulars from those of England. Feudalism never existed here."

One of the classic examples is the common law easement of bathing in ponds or streams in another's land. Such easement may be acquired by prescription or exist by custom. This type of easement is subject to being discontinued or destroyed by the erection of homes in the vicinity of the bathing place, which could render it indecent to bathe there in public.

It is exceedingly important, as well as critical, to ponder the differences between custom, prescription, and implied dedication when it comes to determining the creation of an ancient way. Each carries its own requirements and limitations. Alterations, improvements, and vacations may also be governed accordingly, depending on which category the road falls in, the jurisdiction, and the laws governing actions at a point in time. Since the results of each method of creation are the same in physical appearance, they can be easily confused, and consequently treated incorrectly.

There are major highways in old settled areas, particularly on the East Coast of the United States and in the Southwest, that have existed and have been maintained for a very long time, yet have no formal creation or layout documentation, and perhaps only records of maintenance, if even that. Some of these highways likely were established through customary use, and have long since been forgotten except

[237] Ruling Case Law on Easements (9 R. C. L. 743).

[238] *Graham v. Walker*, 78 Conn. 130; 61 A. 98 (1905).

for modern maintenance and widespread use, whereby they have been taken for granted that they are in favor of the public. This presents a difficult situation for both land surveying and title examination since there is no definition of the rights other than the use itself.

Examples are found with such things as "the old stagecoach road," "by the ancient way," or "the government road." As noted, one may find oral history, description references, maintenance records, and an acknowledgment that "it has always been a road," but no creating documentation exists. The common answer is something like "everyone knows it is a road. Everyone uses it and has used it since day one."

The recent Florida case of *Trepanier, et al. v. County of Volusia*,[239] had to do with public use of the beach and the water beyond a seawall. The court of appeals noted that "the trial court's order in this case is lengthy and understandably complex." They further noted that the right of the public to access and use the beaches of Florida is protected by Florida's public trust doctrine. "The trial court explained that although land held in trust for the people under this doctrine, as set forth in *Article 10, section 11 of the Florida Constitution*, is the expanse of beach below the mean high water line, the public has a right to use and access the beach up to the seawall or line of permanent vegetation in all of Volusia County, by "custom," "prescription," and "dedication."

Under the section on custom, the court stated, "If the public has a right to drive and park on appellant's privately owned platted lots, it most likely will be through application of the law of "custom." Florida's Supreme Court first recognized the public's "customary" right to the use of Florida's privately- owned dry sand beaches in the *Tona-Rama* decision. 294 So.2d at 74.[240] There the court said:

"The beaches of Florida are of such a character as to use and potential development *as to* require separate consideration from other lands with respect to the elements and consequences of title. The sandy portion of the beaches . . . [have] served as a thoroughfare and haven for fishermen and bathers, as well as a place of recreation for the public. The interest and rights of the public to the full use of the beaches should be protected.

The court recognized that the public may acquire a right to use the sandy area adjacent to the mean high tide line by custom when "the recreational use of the sandy area . . . has been ancient, reasonable, without interruption and free from dispute. . . ." The recognition of a right through "custom" means that the owner cannot use his property in a way that is inconsistent with the public's customary use or "calculated to interfere with the exercise of the right of the public to enjoy the dry sand area as a recreational adjunct of the wet sand or foreshore area."

This appeal requires us to confront several issues relating to the law of "custom" applied to Florida's beaches that have not directly been confronted before. Among these are: Did *Tona-Rama* announce, as a matter of law, a right by "custom" for the public to use the entire dry sand beach of the entire coast of Florida? If so, does that right include the right to drive and park on the beach? If *Tona-Rama* did not establish

[239] 965 So.2d 276 (Fla.App., 2007).

[240] See Chapter 13, Case # 18.

a "customary" right, as a matter of law, how is the right established in an individual case such as this one?"

It went on to inquire, "what evidence is required in order to establish entitlement of the public to use of a particular parcel, based on custom? Logically, the place to begin would be with an understanding of how rights are acquired on a theory of "custom" and how, historically, such rights have been legally established. For this purpose, we have been greatly aided by David J. Bederman's fine article, "The Curious Resurrection of Custom; Beach Access and Judicial Takings," published in 1996 in the *Columbia Law Review*. In that article, Bederman explains that, at common law, establishment of a right through custom required proof of several elements. In addition to the temporal requirement of "ancient" use, three other key elements must be proven: peaceableness, certainty, and consistency. Finally, the customary use must be shown to be "reasonable." While some may find it preferable that proof of these elements of custom be established for the entire state by judicial fiat in order to protect the right of public access to Florida's beaches, it appears to us that the acquisition of a right to use private property by custom is intensely local and anything but theoretical. "Custom" is inherently a source of law that emanates from long-term, open, obvious and widely accepted and widely exercised practice. It is accordingly impossible precisely to define the geographic area of the beach for which evidence of a specific customary use must be shown, because it will depend on the particular geography and the particular custom at issue.

The specific customary use of the beach in any particular area may vary, but proof is required to establish the elements of a customary right (see Bederman, *supra*, at 1449). If the only source of a right claimed as "custom" is that a certain thing has been done in a certain way in a certain place for so long that no one can remember when it wasn't done that way, the inability to offer evidence of the custom suggests the weakness of the claim.

In this Court's opinion in *Reynolds,* we interpreted *Tona-Rama* to require the courts to "ascertain in *each case* the degree of customary and ancient use the beach has been subjected to and, in addition, to balance whether the proposed use of the land by the fee owners will interfere with such use enjoyed by the public in the past." Also, as was noted in *Tona-Rama,* some or all of the customary uses by the public may be shown to have been abandoned. We note that the County requested the trial court to take judicial notice of Section 20-82 of the Volusia County Code of Ordinances, the Beach Code, which includes the following language:

It is not the intent of the Charter or of this chapter to affect in any way the title of the owner of land adjacent to the Atlantic Ocean, or to impair *the right of any such owner to contest the existence of the customary right of the public to access and use any particular area of privately owned beach,* or to reduce or limit any rights of public access or use that may exist or arise other than as customary rights.

(Emphasis added). We think this language cogently describes the test: Does evidence establish the existence of the public's right to access and use a particular area of privately owned beach?

To establish a customary right, we do not suggest that the County must prove that cars, horses, or other modes of transportation have customarily traversed and parked on Appellants' specific parcels of property. Rather, we read *Tona-Rama* to require proof that the general area of the beach where Appellants' property is located has customarily been put to such use and that the extent of such customary use on private property is consistent with the public's claim of right.

The County takes the position that the right to drive vehicles onto the beach inherently falls within the public's customary right to use and access the beach. It writes, "[v]ehicles may not be restricted seaward of the mean high water mark any more than the people and their towels." Vehicles or the access they provide are "imperative" to enjoying the beach. On the face of it, this argument is circular because if it is true, it must be so by custom. There are many beaches in Florida where parking and driving are not allowed. Appellants urge that driving and parking on the beach are not properly a customary right because the practice of driving and parking on the beach is not ancient or reasonable and because there is no evidence that driving or parking was ever a public use made of the area of beach where Appellants' property is located. We agree that it is not enough to show that driving and parking are a customary use of some of the County's beach; it must be shown that driving and parking are a customary use of this part of this area of the beach. This could be established by proof that driving and parking are an ancient, peaceable, certain, constant, and reasonable use of the entire coast contained within Volusia County, but that proof has not been made. According to the record, there are parts of the beach in Volusia County where driving and parking are not allowed at all or during certain periods. Also, the record suggests that the County may have abandoned driving and parking on some parts of the beach in order to preserve them in others.

In *Tona-Rama,* the court said that the public may acquire a right to use the sandy area adjacent to the mean high tide line by custom when "the *recreational* use of the sandy area . . . has been ancient, reasonable, without interruption and free from dispute. . . ." Further, the court said that, although the public's customary rights "cannot be revoked by the land owner, it is subject to appropriate governmental regulation and may be abandoned by the public." Id.(emphasis added). Given this language in *Tona-Rama,* it does not appear that the fact that vehicular use may be regulated is necessarily relevant to a determination whether it is a customary use. Driving and parking on the beach may be considered an adjunct to the recreational use of the beach because it is the way to access the beach; it may be viewed as a customary use in its own right based on either a historic custom of using the beach as a thoroughfare; or it may itself be deemed a recreation. If the former, it seems that whether or not driving and parking on the beach fall within the ambit of customary rights is related to the availability of historic alternative access. Otherwise, the availability of alternative access is irrelevant."

The Oregon court in 1969[241] made an important point in its decision: "Because so much of our law is the product of legislation, we sometimes lose sight of the importance of custom as a source of law in our society. It seems particularly appropriate in the case at bar to look to an ancient and accepted custom in this state as the source of a rule of law. The rule in this case, based upon custom, is salutary in confirming a

[241] *State ex rel. Thornton v. Hay,* 254 Or. 584, 462 P.2d 671 (Or., 1969).

public right, and at the same time it takes from no man anything which he has had a legitimate reason to regard as exclusively his."

This decision is included in Chapter 14 as Case #18. Its importance to this discussion is that it analyzes the differences between custom and prescription, discusses in some detail the requirements put forth by Blackstone, and contrasts "custom" with "customary rights."

In addition, and as important, the court offered the following insight to the doctrine of custom:

"Because many elements of prescription are present in this case, the state has relied upon the doctrine in support of the decree below. We believe, however, that there is a better legal basis for affirming the decree. The most cogent basis for the decision in this case is the English doctrine of custom. Strictly construed, prescription applies only to the specific tract of land before the court, and doubtful prescription cases could fill the courts for years with tract-by-tract litigation. An established custom, on the other hand, can be proven with reference to a larger region. Ocean-front lands from the northern to the southern border of the state ought to be treated uniformly.

The other reason which commends the doctrine of custom over that of prescription as the principal basis for the decision in this case is the unique nature of the lands in question. This case deals solely with the dry-sand area along the Pacific shore, and this land has been used by the public as public recreational land according to an unbroken custom running back in time as long as the land has been inhabited."

Practically nowhere, at least in this country, are the customs of a people so entrenched and protected as in the State of Hawaii. Regarding this, the court noted in the case of *Public Access Shoreline Hawaii by Rothstein v. Hawai'i County Planning Com'n by Fujimoto*[242]:

> We perceive the Hawaiian usage exception to the adoption of the common law to represent an attempt on the part of the framers of the statute to avoid results inappropriate to the isles' inhabitants by permitting the continuance of native understandings and practices which did not unreasonably interfere with the spirit of the common law. The statutory exception is thus akin to the English doctrine of custom whereby practices and privileges unique to particular districts continued to apply to the residents of those districts even though in contravention of the common law. This is not to say that we find that all the requisite elements of the doctrine of custom were necessarily incorporated in § 1-1. Rather we believe that the retention of a Hawaiian tradition should in each case be determined by balancing the respective interests and harm once it is established that the application of the custom has continued in a particular area.

Public Trust Easements The ancient laws of the Romans held that the seashore not appropriated for private use was open to all. This principle became the law in England, and these rights were further strengthened by later laws in England and subsequently became part of the common law of the United States.

[242] 79 Hawai'i 425, 903 P.2d 1246 (Hawai'i 1995).

An easement through public trust was one of the claims made in the case of *Akau v. Olohana Corp.*[243] In this case, the plaintiffs had lived or fished in Kawaihae for many years. They represented two subclasses: one containing Hawaii residents who used or were deterred from using the trails, the other containing all persons who own land or reside in the area and used or were deterred from using the trails. The defendants were landowners or tenants who possessed the beachfront land between Spencer Beach Park and Hapuna Beach Park, a span of about 2½ miles along the beach. They had barred all public access across their land to the public beach since acquiring the land in 1954.

Two of the trails in issue run roughly parallel to the beach between the two parks and have existed since before the turn of this century. The Kamehameha Trail is at most points very close to the water and at others about 100 yards away. The Kawaihae-Puako Road is about 150 yards further upland. There are also eleven intersecting trails that run from the main trails to the shore. Plaintiffs allege that these trails had been used by the public until 1954.

The Territory of Hawaii owned the land between the two parks until it was sold to Richard Smart in 1954. The parcel consisting of the Kawaihae-Puako Road was also sold to Smart at that time. Smart conveyed all his land by deed or lease, and all the original defendants were owners or lessees of that land.

Plaintiffs claimed that the trails have been and are public rights of way and asked for declaratory and injunctive relief to that effect. The eight theories plaintiffs rely on are: (1) HRS § 7-1; (2) ancient Hawaiian custom, tradition, practice, and usage; (3) common law custom; (4) easement by implied dedication; (5) easement by prescription; (6) easement by necessity; (7) easement by implied reservation; and (8) easement through public trust. The state was made a nominal defendant to protect the interests of the public. Its position, however, is in support of plaintiffs in the appeal.

The case had to do with class standing as opposed to the state bringing the action, but is illustrative of the theories under which an action of this nature may be brought.

In the case of *Banner Milling Co. v. State*,[244] the New York court explored the differences in land rights held by a state. It said,

> We are of the opinion that the vital question in this case is this: Was the land, the title of which is in question here, owned and held by the State as a sovereign in trust for the People, or as a proprietor only? There is a well-recognized distinction between lands held by the State as sovereign in trust for the public and lands held as proprietor only, for the purpose of "sale or other disposition." (*Weber v. State Harbor Comrs., supra*, 18 Wall. 68). In either circumstance, except a statute (making an agreement on behalf of the People not to sue) authorizes it, lands of a sovereign State cannot be lost to, or taken from, the State by failure to assert her title (2 C. J. 213; *Fulton L., H. & P. Co. v. State,* 200 N.Y. 400; *St. Vincent F. O. Asylum v. City of Troy,* 76 id. 108; *Hays v. U. S.,* 175 U.S. 248.); and, after such a statute has been passed by a State, such lands only as

[243] 65 Haw.383, 652 P.2d 1130 (Hawai'I, 1982).

[244] 117 Misc. 33 (N.Y., 1921).

the State holds as a proprietor may be lost to the State; it cannot lose such lands as it holds for the public, in trust for a public purpose, as highways, public streams, canals, public fair grounds. (*Burbank v. Fay,* 65 N.Y. 57; 2 C. J. 213, 214, 215.)

VOTE OF A GOVERNING BODY

A governing body, in the development of what is known as a "common scheme" may set aside, or create, easements, particularly for, but not limited to, access. This has been known to have been a common practice with proprietorships, and in the early development of the colonial states, corporations known as proprietors held large areas which they developed. Many of these became townships, with lots laid out in ranges and with provisions for access, sometimes known as *rangeways*. While this class of easement creation may fall under the category of express grant, its significance and uniqueness warrants a separate section.

Several state courts have clearly stated in their decisions that the law recognizing the authority of proprietors in common to alienate their lands by vote is well settled, so that the vote is equivalent to a grant.[245] This poses a difficulty for the researcher as the proprietors records themselves must be consulted since no recorded deed will be found as a source of title. Many of these records have become lost, or are housed in obscure places even though accessible using a bit of clever detective-like investigation.

This, however, again underscores the importance of complete research. As the New York court stated in *Dolphin Lane Associates, Ltd. v. Town of Southampton,*[246] "This action, by its very nature, involves the tracing of chains of title going back to the earliest settlement of the Town of Southampton. Thus, the history of the early settlement and development of Southampton must be examined. The past must be explored to understand the present."

The Maryland court in the case of *Ski Roundtop Inc. v. Wagerman,*[247] took it a step further by stating "although the case law in this area is sparse, it appears that a requisite for valid title to real property is an original conveyance of public land by the State. See 3 American Law of Property, § 12:16 (1952); 73B C.J.S., Public Lands, § 188 (1983); 2 Patton on Titles, § 281 (2d ed. 1957). Absent such a conveyance, one purporting to transfer an ownership interest in such property transfers nothing, and no quantity of successive transfers by deed nor the mere passage of time will metamorphose good title from void title."

[245] For example: A grant of land by an act of the legislature vests an actual seizing in the grantee. *The Proprietors of Enfield v. Permit*; 8 N.H. 512 (1837); It has long been the settled law of this commonwealth, that proprietors in common had authority in early times to alienate their lands by vote, which, if duly recorded on the books of the proprietary, passed the title and constituted competent evidence of the transfer. *Campbell v. Nickerson*, 73 Mass.App.Ct. 20 (2008).

[246] 339 N.Y.S.2d 966, 72 Misc.2d 868 (1971).

[247] 79 Md.App. 357, 556 A.2d 1144 (1989).

This case referred to a previous Maryland case, *Maryland Coal and Realty Co. v. Eckhart*,[248] which stated, "An albatross in the wake of every title searcher is the ominous question of whether he has gone back far enough in the chain of title." Despite a chain of conveyances in the Wagerman's chain of record title extending back to 1812, the court found the title void because an original land patent was lacking. (see Case #17 in Chapter 14 for a further example).

Extent and Scope of Easements Some easements are defined according to location and width in the document in which they were created. For others, it depends on the manner in which they were created or even the intended use of the easement. For example, a prescriptive easement is defined by the *use being made*, while an implied easement may be defined by the *use contemplated* by the parties. The location of an easement, by necessity, may be governed by the use made of the servient tenement, while its width may be governed by its intended use.

A way, if not located, must be reasonable and convenient for all parties, in view of the surrounding circumstances. It must have a particular definite line; the grantee does not have the right to go at random over any and all parts of the servient estate. Additionally, a party having a way be necessity is entitled to only one route, and it must be the shortest, most direct passage to the nearest public way as long as the party has a right to a convenient way and has reasonable access to his property.[249]

When a location is not fixed, the owner of the servient estate has the right to designate its location, but if he does not do so, the person entitled to the easement may make his own selection of a location.[250] In either case, the location is determined by the reasonable convenience of both parties, under the circumstances.[251]

In the Massachusetts case of *Campbell v. Nickerson*,[252] the court stated that the following is a long-standing principle:

> Where a right of way, or other easement, is granted by deed without fixed and defined limits, the practical location and use of such way or easement by the grantee under his deed, acquiesced in by the grantor at the time of the grant and for a long time subsequent thereto, operate as an assignment of the right, and are deemed to be that which was intended to be conveyed by the deed, and are the same, in legal effect, as if it had been fully described by the terms of the grant.

[248] 25 Md.App. 605, 337 A.2d 150 (1975).

[249] 25 Am.Jur.2d, § 64.

[250] The right of locating a way by necessity belongs to the owner of the land, but it must be a convenient way. *Russell v. Jackson*, 19 Mass. 574 (1824).

The right of locating a way of necessity is in the owner of the land over which it is to pass. *Heiser v. Martin*, 9 N.J.L.J. 277 (1886).

[251] 25 Am.Jur.2d, § 66.

Where a right of way is expressly granted and its precise location and limits are not fixed or defined by deed, parties may define location and determine limits of right of way by subsequent agreement, use and acquiescence. *Taylor v. Heffner*, 58 A.2d 450, 359 Pa. 155 (1948).

[252] 73 Mass.App.Ct. 20 (2008).

If an easement is in need of relocation, the consent of both parties is needed.[253]

See Chapter 14, Case #4 for an example of defining the width of an easement by reservation.

Use permitted. In ascertaining whether a particular use is permissible under an easement by prescription a comparison must be made between such use and the use by which the easement was created with respect to (a) their physical character, (b) their purpose, (c) the relative burden caused by them upon the servient tenement.[254] This rule recognizes that human behavior and the circumstances that influence it inevitably change. Thus, the beneficiary of an easement established by prescription will be permitted to vary the use of the easement to a reasonable extent. This flexibility of use is limited, however, by concern for the degree to which the variance further burdens the servient estate.

In ascertaining whether a particular use is permissible under an easement appurtenant created by prescription there must be considered, in addition to the foregoing factors enumerated , the needs which result from a normal evolution in the use of the dominant tenement and the extent to which the satisfaction of those needs increases the burden on the servient tenement.[255]

[253] Holder of servient estate cannot have easements relocated, over objection of dominant estate, simply because location and use had become inconvenient to use and enjoyment of servient estate. *Edgell v. Divver*, 402 A.2d 395 (1979).

An easement may not be relocated without consent of owners of both dominant and servient estates. Ibid.

Generally, owner of easement may make changes, not affecting character of servient estate, in manner using easement, as long as use thereof is confined strictly to purposes for which it was created. *Garan v. Bender*, 22 A.2d 353, 357 Pa. 487 (1947).

No alteration can be made to increase easement restriction except by mutual consent of easement owner and owner of servient estate. *Reid v. Washington Gas Light Co.*, 194 A.2d, 232 Md. 545 (1963).

[254] *Restatement of Property*, § 478.

[255] *Wright v. Horse Creek Ranches*, 697 P.2d 384 (Colo., 1985).

CHAPTER 5

TERMINATION OF EASEMENTS

Also known as extinguishment, easements may be terminated (ended) by a variety of methods. Extinguishment of a right to an easement means its total annihilation, and not its suspension. Nonuse of an easement, no matter for how long, will not extinguish an easement, except under specific circumstances as discussed.

1. Expiration
2. Release
3. Merger of Title
4. Abandonment
5. Estoppel
6. Prescription
7. Destruction of the Servient Estate
8. Cessation of necessity
9. Eminent Domain
10. Frustration of purpose
11. Cessation of special purpose
12. Overburden
13. Death: of the owner of an easement in gross

EXPIRATION

Like any other interest in land, an easement may be potentially unlimited in duration or it may be created to last for a limited period of time, in which case the easement will expire according to its own terms.[256] Sometimes known as an *easement for years*,[257] it falls into a category of its own. Where an easement is created for a specified purpose, it expires when the purpose is accomplished. This could be in the form of the happening of a particular event or contingency, or the occurrence, breach, or nonperformance of a condition.[258]

An easement, by necessity, will terminate when the necessity for it ceases or no longer exists. However, most easements have a potentially unlimited duration requiring some special act to terminate them.[259]

Example of a right of way that will expire at the end of a specified period of time:

> The above described premises shall be subject to a conservation easement to be granted to the Hampton Conservation Commission, which easement shall continue for a period of fifty (50) years from the date of this deed and such easement shall be for the purpose of preserving the above described parcel in its present state and condition, it being understood and agreed, however, that the grantee, its successors and assigns shall have a right to construct and maintain a road upon and across the premises.

Example of a right of way that will expire upon the death of parties (contingency):

> There is reserved to Emedia Beaulieu and Blanche Gauthier individually and not to their respective heirs and assigns, a right of way over the granted premises to and from other land now or formerly of said Beaulieu in the rear of the granted premises, said right of way being 20 feet in width throughout its entire length, the Westerly line of said right of way being the Westerly boundary of the granted premises, and this right of way shall terminate on the death of the two said parties and this right of way is to be used for a common right of way.

RELEASE

An easement may be extinguished with a release given by its owner to the owner of the servient land, but only if both parties concur. The easement may be purchased

[256] Easement granted for a particular purpose terminates as soon as such purpose ceases to exist, is abandoned or is rendered impossible of accomplishment. *Trustees of Howard College v. McNabb*, 263 So.2d 664, 288 Ala. 564 (1972).

[257] While all easements may be terminated by an act of the parties or by operation of law, they may initially be created of different durations. Thus, an easement may be created for a specified period of time, such as for a term of years or months, or for the life of either the owner of the servient or dominant tenement, or, if the easement is one in gross, for the life of the person in whose favor the easement exists. *Irvin v. Petitfils*, 44 Cal.App.2d 496, 112 P.2d 688.

[258] An easement does not have to be permanent, but rather an easement may be created which ends upon happening of a condition. *Dotson v. Wolfe*, 391 So.2d 757 (Fla.App., 1980).

[259] 25 Am.Jur.2d, § 99.

An easement may be granted which will terminate on the happing of some particular act or the nonperformance of a condition subsequent. *Akasu v. Power*, 91 N.E.2d 224, 325 Mass. 497 (1950).

by the servient owner as long as the dominant owner is willing to sell, if it would enhance the value of the servient estate by relieving it of its burden.[260]

MERGER OF TITLE

Since one cannot have an easement in their own land, since by definition an easement is a right in land of another, termination is automatic when one entity owns both the dominant and servient estates at the same time. This causes the easement to *merge* into the fee.[261] This is sometimes called *unity of ownership*.

Whether this easement revives when the land is again divided depends on the conditions and circumstances at the time. Most courts have held that once the easement has been terminated it must be created anew if it is later wanted. However, in a few special situations, the courts have allowed the revival of the easement.[262]

This illustrates the necessity of fully researching both dominant and servient chains of title fully from the creation of an easement to date, to determine whether an extinguishment may have occurred through merger of title. As noted, the extinguishment is automatic upon the merger, and no additional documentation or recordation is necessary, or is likely to be found.

ABANDONMENT

Although it is uncommon, an easement may be terminated by *abandonment*. Nonuse does not constitute abandonment,[263] so that the acts claimed to constitute abandonment must be of a character so decisive and conclusive to indicate a *clear intent* to

[260] In order to be effectual in itself, a release must be executed with the same formalities as are generally required in making transfers of interests in land. *Dyer v. Sanford*, 50 Mass. 395.

[261] Unity of ownership and possession of two adjoining lots extinguishes mutual rights of way, as well as other subordinate rights and easements. *First Nat. Bank v. Laperle*, 117 Vt. 144, 86 A.2d 635, 30 A.L.R.2d 958 (1952).

All easements, whether of convenience or necessity, are extinguished by unity of possession, but upon any subsequent severance easements, which, previous to such unity, were easements of necessity, are granted anew in the same manner as any other easement which would be held by law to pass as incident to the grant. *Grant v. Chase*, 17 Mass. 443, 9 Am. Dec. 161 (1821).

In order that unity of titles in the dominant and servient estates should operate to extinguish an easement, the ownership of two estates should be coextensive; and if a person holds one estate in severalty, and only a fractional part of the other, the easement is not extinguished. *Atlanta Mills v. Mason*, 120 Mass. 244 .

[262] 25 Am.Jur.2d, § 114.

Once extinguished, easement does not again come into existence upon separation of the former servient and dominant estates unless proper new grant or reservation is made. *Fitanides v. Holman*, 310 A.2d 65 (Me., 1973).

Easement may remain unaffected by unity of estates, or revive upon separation, if a valid and legitimate purpose will be subserved thereby. *Schwoyer v. Smith*, 131 A.2d 385, 388 Pa. 637 (1957).

[263] 25 Am.Jur.2d, § 105.

abandon the easement.[264] A mere declaration of an intention to abandon an easement does not effect an abandonment,[265] nor does a failure to repair or maintain it.[266]

As a general rule, an easement acquired by grant or reservation cannot be lost by mere nonuser for any length of time no matter how great. The nonuse must be accompanied by an express or implied intention to abandon. However, the nonuse itself, if long continued, is some evidence of intent to abandon. On the other hand, in the case of an easement established by prescription, it is not necessary to show an intent to abandon in order to prove loss by disuse, and it has been held that such an easement is lost by mere nonuse for the same period as was required to establish it.[267]

A right of way, whether acquired by grant or prescription, is not extinguished by the habitual use by its owner of another equally convenient way, unless there is an intentional abandonment of the former way.[268] The use of a substituted way, however, may be evidence of abandonment if necessitated by a denial of the use of, or an obstruction of, the original way.[269]

Abandonment is a complex theory when real property, or real property rights, is involved. One court summarized it this way: A full legal title can be divested by abandonment only when the circumstances thereof are sufficient to raise an estoppel to assert title or when possession is acquired by one in consequence of the abandonment and held under a claim of right for the statutory period of limitations. When there is no estoppel or limitations there can be no abandonment that will affect the right of the holder to the legal title.[270]

As with prescriptions, the burden of proof is on the person making the claim.[271]

[264] 25 Am.Jur.2d, § 103.

[265] 25 Am.Jur.2d, § 104.

[266] Mere neglect of the condition of a way is not enough in addition to nonuser to show an abandonment. *Harington v. Kessler*, 247 Iowa 1106, 77 N.W.2d 633 (1956).

[267] 25 Am.Jur.2d, § 105.

A temporary suspension of the use of an easement is not alone sufficient to show abandonment. *Chitwood v. Whitlow*, 313 Ky. 182, 230 S.W.2d 641 (1950).

Mere nonuse of an easement for a less time than that required by the statute of limitations to acquire a prescriptive right does not raise a conclusive presumption of its abandonment. *Groshean v. Dillmont Realty Co.*, 92 Mont. 227, 12 P.2d 273 (1932).

To constitute abandonment of a right of way created by express grant there must be, in addition to nonuse, circumstances showing an intention of the dominant owner to abandon use of the easement. *Kurz v. Blume*, 407 Ill. 383, 95 N.E.2d 338, 25 A.L.R.2d 1258 (1950).

[268] *Jamaica Pond Aqueduct Corp. v. Chandler*, 121 Mass. 3 (1876).

[269] 25 Am,Jur.2d, § 105.

A right of way, whether acquired by grant or prescription, is not extinguished by habitual use by its owner or another owner equally convenient, unless there is an intentional abandonment of the particular way. *Adams v. Hodgkins*, 84 A. 530, 109 Me. 361, 42 L.R.A.(N.S.)741(1912).

[270] *Southern Coal & Iron Co. et al. v. Schwood et al.; Southern Coal & Iron Co. et al. v. Brush Creek Coal Co. et al.*, 145 Tenn. 191, 239 S.W. 398 (1921). See *Caledonia County School v. Howard*, 84 Vt. 10, 77 A. 877; *McLellan v. McFadden*, 114 Me. 242, 95 A. 1025 (1915); *Houston Oil Co. v. Kimball*, 114 S.W. 662 (Tex.Civ.App., 1908).

[271] The burden of proving abandonment of easement is on the party asserting such abandonment. *Nelson v. Bacon*, 32 A.2d 140, 113 Vt. 161 (1943).

Sabins v. McAllister, 76 A.2d 106, 116 Vt. 302 (1950).

The recent Mississippi case of *Stancil v. Farris*[272] was an issue whether a right of way had been abandoned, and therefore lost, by the holder of the easement. The court stated that "abandonment is a question of fact that requires non-use for an extended period of time as well as the intent to abandon. There are two elements to it. Evidence of abandonment must be full and clear . . . there must be some clear unmistakable affirmative act or series of acts indicating a purpose to repudiate ownership. The Court does not find that that existed here."

The court went on to state that "protracted non-use for an extended period of time manifest[s] a presumption of abandonment, but the Court here does not find any such presumption to apply because there was no protracted non-use for any extended period of time."

The Maine case of *Phinney v. Gardner*,[273] also addressed the issue of abandonment but stated its reasoning. In its decision, the court stated, "The characteristic element of abandonment is the voluntary relinquishment of ownership, whereby the thing so dealt with ceases to be the property of any person and becomes the subject of appropriation by the first taker."[274] It went on to state that "the term is used in connection with personal property, inchoate and equitable rights, and in corporeal hereditaments, but 'at common law a perfect legal title to a corporeal hereditament cannot, it would seem, be lost by abandonment.' 1. C.J. p. 10. Its very essence is inconsistent with the attributes of real estate."

Right to Abandon Location Generally, a public service corporation that has been granted the power of eminent domain and has acquired location by the exercise of this power, may, if it sees fit, surrender its franchise and abandon its location.[275]

One of the fundamental principles of eminent domain is the land taken for a public use shall be devoted to that use within a reasonable time, and this condition to the taking will be implied even though the authorizing statute is silent on the subject. It follows that a failure to devote property taken by eminent domain to any public use whatever may amount to an abandonment unless continued for a sufficient length of time to indicate an actual intention to abandon.[276]

The question as to the extent of the interest acquired by condemnation proceedings has an important bearing on the effect of the cessation of the use for which the property was taken. In that event, if an easement only was acquired by the condemnation of land the right to the possession reverts to the owner of the fee, regardless of whether the taking was by the state or by an individual or a corporation. Thus, when only an easement has been acquired for the public use by condemnation, if the use for which the land was taken is formally discontinued, permanently abandoned in fact, or becomes impossible, or the land is devoted to a different inconsistent use, the

[272] 60 So.3d 817 (Miss.App. 2011).
[273] 121 Me. 44, 115 A. 523 (1921).
[274] 1 R.C.L. p. 2.
[275] 26 Am.Jur.2d, § 145.
[276] 26 Am.Jur.2d, § 146.

easement expires and the owner of the fee holds the land free from encumbrance. In any such event the right to possession does not remain in the condemning party but reverts to the owner of the fee.[277]

ESTOPPEL

An easement may be extinguished by the conduct of the easement holder, even though he had no intention of giving up the easement. To produce an extinguishment in this way, the servient owner must have changed his position in justifiable reliance upon the conduct of the owner of the easement. For instance, the apparent abandonment of an easement could result in its extinguishment: if the servient owner interprets mere nonuse of the easement as an abandonment, and in reliance thereon makes substantial improvements upon his land, to the knowledge of the holder of the easement, the latter may be estopped from asserting his rights.

An estoppel resulting in the termination of an easement can occur whenever three prerequisites for establishing estoppel are present. First, the easement owner must represent that the easement was terminated. This occurs if the easement owner acts in such a way that a reasonable person in the position of the owner of burdened property would be justified in believing that the easement was not longer desired. Second, the owner of the burdened property must perform some acts in reliance on the representations of the easement owner. Third, the owner of burdened property must show that he would suffer significant injury were the easement found to continue to exist despite the detrimental reliance he had made on the easement owner's representation.[278]

PRESCRIPTION OR ADVERSE POSSESSION

An easement may be extinguished as well as created through, or by, prescription. In order for an easement to be extinguished by prescription, there must be a continuous and uninterrupted *interference* with the easement for the same period required by the statute of limitations for adverse possession.

An easement may be lost by possession and use by one other than the easement owner in a manner adverse to the exercise of the easement.[279] The possession and use

[277] 26 Am.Jur.2d, § 147.

[278] A Practical Guide to Disputes Between Adjoining Landowners – Easements, § 1.05(8).

[279] An easement created by express grant cannot be lost by mere nonuser without adverse possession, but the easement may be barred by a complete nonuser for the prescriptive period with possession in another that is inconsistent with or hostile to the right of such easement. *Kurz v. Blume*, 407 Ill. 383, 95 N.E.2d 338, 25 A.L.R.2d 1258 (1950).

An express reservation of easement may be lost by prescription; if servient owner should by adverse acts lasting through the prescriptive period obstruct the dominant owner's enjoyment, intending to deprive him of the easement, he may by prescription acquire the right to use his own land free of the easement. *Russo v. Terek*, 508 A.2d 788, 7 Conn.App. 252.

must be actual, adverse, continuous, visible, notorious, and hostile for the prescriptive period.[280] Similarly, nonpermissive erection and maintenance by the servient owner for the statutory period of permanent structures, such as buildings, that obstruct the use of a right of way, will operate to extinguish the easement. If, however, the act of the servient owner in erecting buildings or other structures on a granted, but unused, easement can be said to have been done with the permission of the dominant owner, the easement is not extinguished by adverse possession.[281] Nor is a right of way extinguished by obstructions, such as fences, gates, or bars, which are removable by the easement owner and that do not preclude its use as a right of way.[282]

As with adverse possession, the burden of proof is on the party so asserting, in the case of an easement, the owner of the burdened property.[283]

DESTRUCTION OF THE SERVIENT ESTATE

A destruction of the servient estate usually extinguishes any easements therein or thereon. For example, an easement a structure, such as a party wall,[284] is extinguished if the structure is destroyed without the fault of the servient owner. Another example is where a stairway through a building on parcel A provided access to the upper floor of a building on parcel B. The termination of the easement may be implied if building A collapses or is otherwise destroyed in a natural disaster.[285]

Similarly, a right of way over a riparian parcel terminates if the servient parcel washes or erodes away. (see the examples in Chapter 6).

Adverse possession by owner of servient estate will extinguish ways-of-necessity provided all other elements of adverse possession are established. *Pencader Associates, Inc. v. Glasgow Trust*, 446 A.2d 1097 (Del., 1982).

Easement acquired by grant may be extinguished through continuous adverse possession for period of 20 years. *Titcomb v. Anthony*, 492 A.2d 1373, 125 N.H. 434 (1985).

Implied private easement may be lost by adverse possession. *700 Lake Ave. Realty Co. v. Dolleman*, 433 A.2d 1261, 121 N.H. 619 (1981).

[280] To sustain the extinguishment of an easement by adverse possession generally, it must be shown that the owner of the easement knew or should have known that an adverse claim was being made. *Philadelphia Electric Co. v. Philadelphia*, 303 Pa. 422, 154 A. 492 .

[281] The erection of a fence and the planting of trees and shrubbery on a right of way, which effectually barred it, constitutes adverse possession which, when continued for the statutory period, extinguishes the easement. *Nauman v. Kopf*, 101 Pa.Super. 262

An obstruction under circumstances that might be termed "permissive" will not extinguish a right of way. *Welsh v. Taylor*, 134 N.Y. 450, 31 N.E. 896 (1892).

[282] 25 Am.Jur.2d, § 110.

[283] The burden of proving extinguishment of easement by open, notorious, hostile, and continuous possession of owner of servient tenement for statutory period is on party asserting such possession. *Nelson v. Bacon*, 32 A.2d 140, 113 Vt. 161 (1943).

[284] A "party wall" is a division between two connecting and mutually supporting structures, usually but not necessarily, standing half on the land of each owner, and maintained at mutual cost and for the common benefit of both parties.

[285] *A Practical Guide to Disputes Between Adjoining Landowners – Easements*, § 1.05(4).

Where on having easement in adjoining wall for support of building removed building, easement no longer existed. *Ansin v. Taylor*, 159 N.E. 513, 262 Mass. 159 (1928).

CESSATION OF NECESSITY

As previously stated, a way of necessity ceases when the necessity ceases,[286] the theory being that it is no longer a necessity. However, an easement created by other methods, does not cease even though the necessity for it does.[287]

This situation can become very tricky and perhaps confusing, as it is sometimes difficult to separate necessity from straight implication, the former arising through the latter. True implied easements, without any consideration of necessity, do not expire even though the necessity for them ceases.

EMINENT DOMAIN

Eminent domain proceedings may terminate an easement. A servient estate may be converted to uses incompatible with its easements. A common situation is where rights of way are blocked or rendered unusable because of a change in the use of a servient estate, such as when a new highway is established, cutting across a right of way.

Major highways such as those in the Interstate System or other "limited access" highways frequently sever access roads and sometimes render parcels or portions of parcels inaccessible. Usually, when this happens it necessitates compensation in the form of payment of damages. Occasionally an alternative access is provided, or created, even to the extreme of constructing a bridge over, or a tunnel under, the newly created highway, as illustrated in Figure 5.1.

FRUSTRATION OF PURPOSE

When there is a total and permanent impossibility of enjoyment of an easement limited by its terms to a particular purpose, the easement may expire. The termination of an easement through frustration of purpose is similar to the rule that an easement for a particular purpose is limited to the purpose stated. If an easement can be used only for a stated purpose, and the purpose disappears, the easement disappears as well.[288]

[286] A way of necessity terminates as soon as the owner of the dominant estate can pass without interruption over his own land to a highway. *Baker v. Crosby*, 75 Mass. 421 (1857).

Way of necessity exists only so long as the necessity itself remains. *Shpak v. Oletsky*, 373 A.2d 1234, 280 Md. 355 (1977).

[287] If an easement is created by grant, it does not cease, although the necessity for it ceases. *Atlanta Mills v. Mason*, 120 Mass. 244 (1876).

[288] In an early Massachusetts decision, the court stated that an easement for "an open dock and common passageway for ships . . ." was eliminated when, "the laying out of a street and filling up of the dock by the city . . . made the enjoyment of the easement impossible, and thereby extinguished it." *Central Wharf and Wet Dock Corp. v. Proprietors of India Wharf*, 123 Mass. 567 (1878).

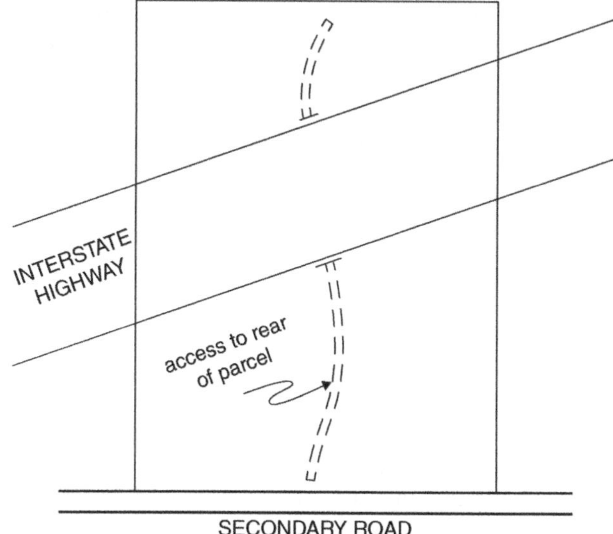

Figure 5.1 Highway cutting across an ownership, resulting in the termination of access to a portion of the property.

"Where a change in land or tenement is of such decisive and conclusive a nature that the easement can no longer be enjoyed, it is extinguished, as, where a piece of land subject to an easement is washed away by the encroachment of a river, the easement ceases."[289]

However, an easement is *not* terminated when the purpose for which it is created is neither *totally* nor *permanently* impossible of enjoyment.

Example: Where a 5-foot passageway created in deed in 1872 for access to a street was blocked off from the street by a railroad station constructed in part over the street's location, the easement was held extinguished by frustration of purpose.[290]

In the case of *Hopkins the Florist, Inc. v. Fleming*,[291] an easement had explicitly reserved the right to a particular view from an existing house. The house was later moved to a location from which the view could not be maintained under any circumstances. The Vermont court held that the easement was extinguished.

Cessation of Purpose There is a well established rule that an easement may be terminated by the completion of the purpose for which it was granted, inasmuch as the reason for, and necessity of, the servitude is at an end. Thus, if an easement is granted for a particular purpose only, the right continues while the dominant tenement

[289] *Atlantic & N.C.R. Co. v. Way*, 172 N.C. 774, 90 S.E. 937.
[290] *New England Mutual Life Insurance Co. v. Agorianitis*, 7 Land Ct. Rptr. 33 (Mass., 1999).
[291] 112 Vt. 389, 26 A.2d 98 (1942).

is used for that purpose but ceases when the specified use ceases. Moreover, a way of necessity is a temporary right in the sense that it continues only so long as the necessity exists.[292]

As with estates in land, an easement may be created which will terminate *ipso facto* upon the happening of a particular event or contingency. Such an easement is known as a *determinable easement*.

The difference between *frustration* and *cessation* of purpose is in the first instance the use inadvertently becomes impossible, whereas in the second, the end of the easement was an intentional act on the part of the creator, intending a termination at some future time, often when a particular event took place.

OVERBURDEN

In general, the dominant estate may not increase the burden of an easement upon the servient estate. Such an increase may, in some circumstances, cause the easement to be extinguished. "If the easement arises by *prescription*, a change in the dominant estate calling for a burden upon the servient land exceeding that devolving upon it by its *customary use* during the prescriptive period, if the increased use is *inseparable* for the former use, will operate an extinguishment of the easement."[293]

An example is the Virginia case of *Ellis v. Simmons*,[294] wherein the court stated, "any increase in the burden of a prescriptive easement upon the servient estate, whether caused by a change of use or by an increase in the area of the dominant estate, is prohibited in Virginia."

Sometimes what appears to be a burden, in reality is not, depending on the nature of the increased use. In another Virginia case, *Shooting Point, LLC v. Wescoat*,[295] overburdening was the main issue. The court stated that a party alleging that a particular use of an easement is unreasonably burdensome has the burden of proving his allegation. Generally, when an easement is created by grant or reservation and the instrument creating the easement *does not limit its use*, the easement may be used for "any purpose to which the dominant estate may then, or in the future, reasonably be devoted." However, this general rule is subject to the qualification that no use may be made of the easement, different from that established when the easement was created, which imposes an additional burden on the servient estate.

The court went on to say, "our decisions in *Hayes* and *Cushman* illustrate the nature of this inquiry. In *Hayes*,[296] an operator of a marina on the dominant estate, a

[292] 17 Am Jur., 1023.

If an easement for a particular purpose is granted, when that purpose no longer exists, there is an end of the easement. *McGiffin v. City of Gatlinburg*, 260 S.W.2d 152, 154 (Tenn. 1953) (quoting Washburn, *Treatise on Easements*, 654 (3d ed.)).

[293] 1 Frederick D.G. Ribble, Minor on Real Property § 110, at 150 (2d ed. 1928) (emphasis in original).

[294] 270 Va. 371, 619 S.E.2d 88 (Va., 2005).

[295] 265 Va. 256, 576 S.E.2d 497 (Va. 2003).

[296] 243 Va. at 256-59, 414 S.E.2d at 820-22.

2.58-acre tract, proposed to expand its marina facility from 84 to 280 boat slips. The easement providing access to the marina was a private roadway about 1,120 feet long and 15 feet wide along its entire course. The agreement creating the easement did not restrict its use.

> We held that the record supported the chancellor's conclusion that the proposed expansion would not unreasonably burden the servient estate, although the "degree of burden" would be increased. We assumed, without deciding, that an expanded use of a dominant estate could be of such degree as to create an additional burden on a servient estate, but concluded that the proposed marina expansion was not shown to create such an additional burden.

Similarly, in *Cushman*,[297] the instrument creating the easement did not contain any language limiting the easement's use. When the easement was established, the dominant estate, a 126.67-acre tract, had two dwelling houses and was used as a farm. The owner of the dominant estate proposed to subdivide his land for a residential and commercial development that would include 34 residential lots.

[297] 204 Va. at 252-53, 129 S.E.2d at 639-40.

CHAPTER 6

EASEMENTS AND DESCRIPTIONS

Because an easement is an interest in land, a *grant* of an easement, whether by deed or otherwise, should be drawn and executed with the same formalities as a deed to real estate. It ordinarily must be in writing. In some jurisdictions, words of inheritance are required to create a perpetual easement. Any writing must contain a description of the land that is subjected to the easement with sufficient clarity to locate it with reasonable certainty.[298]

GENERAL

The standard for sufficiency of a land description is that is must clear enough such that a competent land surveyor can go upon the land, locate it, and mark its boundaries. The Texas court reiterated that in 1958 by stating, "the general rule to be that the certainty required in such descriptions is of the same nature as that required in conveyances of land so that a surveyor can go upon the land and mark it out."[299] The same court stated in 1953, "the sole purpose of a description of land contained in a deed of conveyance is to identify the subject matter of the grant."[300] The Minnesota put it a different way in 1942, by stating "descriptions must be such as to identify the property or afford the means of identification aided by extrinsic evidence."[301] Generally, what is required is a description to identify the area involved and reflect the intention of the parties.[302]

[298] 25 Am.Jur.2d Easements and Licenses, § 20.

[299] 25 Am.Jur.2d Easements and Licenses, § 20; *Compton v. Texas Southeastern Gas Co.*, 315 S.W.2d 345 (1958).

[300] *Kulies v. Reinert*, 256 S.W.2d 435.

[301] *City of North Mankato v. Carlstrom*, 2 N.W.2d 130, 212 Minn. 32 (1942) (1912).

[302] 17 Am.Jur Easements, § 29.

Many jurisdictions have considered the sufficiency of an easement description necessary to burden a subsequent purchaser of property. Like South Dakota, the Supreme Court of Washington requires a conveyance creating an easement to comply with the statute of frauds. *Berg v. Ting*.[303] To comply with the statute of frauds, a conveyance creating an easement must contain either (1) a description of the land sufficient to locate the servient tenement or (2) a reference to another document that contains a description sufficient to locate the servient tenement. *Id.* Although "'a deed [of easement] is not required to establish the actual location of an easement, [it] is required to convey an easement' which encumbrances a *specific* servient estate, the servient estate must be sufficiently described." (emphasis added; quoting *Smith v. King*).[304]

The Supreme Court of New Mexico stated that "the description of real estate in a deed *inter partes* is sufficient if it identifies the property intended to be conveyed by it, or furnishes means or data which point to evidence that will identify it."[305] The court summarized the rules in various decisions, which included a decision by this court, concerning the sufficiency of real estate descriptions. A description is sufficient only if:

> "[T]he description furnish[es] the key to the identification of the land intended to be conveyed,"[306]; or if the description is "either certain in itself, or capable of being reduced to certainty by a reference to something extrinsic to which the deed refers,"[307]; or "if there appears therein enough [in the description] to enable one, by pursuing an inquiry based upon the information contained in the deed, to identify the particular property to the exclusion of [all] others,"[308]; or if the deed itself furnishes" the means of identification,"[309]; or if the description" can be made certain [by] inquiries suggested by the description given in such deed,"[310]; or the description in a deed must be sufficiently certain to identify the land therefrom or furnish the means with which to identify it.[311]

The same sufficiency-of-description rules that apply to descriptions in deeds, apply to descriptions of easements. *See Cummings v. Dosam, Inc.*[312]

The Massachusetts Supreme Court acknowledged the sufficiency-of-description requirement when it struck down a conservation easement on grounds of an insufficiently described servient tenement.[313] That easement prohibited construction on

[303] 125 Wash.2d 544, 886 P.2d 564 (1995).
[304] 27 Wash.App. 869, 620 P.2d 542 (1980).
[305] *Heron v. Ramsey,* 45 N.M. 483, 117 P.2d 242 (1941).
[306] *Smith v. Fed. Land Bank,* 181 Ga. 1, 181 S.E. 149 (1935).
[307] *Buckhorn Land & T. Co. v. Yarbrough,* 179 N.C. 335, 102 S.E. 630 (1920).
[308] *Coppard v. Glasscock,* 46 S.W.2d 298, 300 (Tex.Com.App.1932).
[309] *Ault v. Clark,* 62 Ind.App. 55, 112 N.E. 843, 845 (1916).
[310] *Ford v. Ford,* 24 S.D. 644, 124 N.W. 1108 (1910).
[311] *Hamilton v. Rudeen,* 112 Or. 268, 224 P. 92 (1924).
[312] 273 N.C. 28, 33, 159 S.E.2d 513, 518 (1968).
[313] *Parkinson v. Bd. of Assessors of Medfield,* 395 Mass. 643, 481 N.E.2d 491, (1985).

82 acres of land, yet allowed the use of "[o]ne single-family residence with usual appurtenant outbuildings and structures." The court found that the servient tenement was the 82 acres minus an ambiguous amount of property required for the use of the residence. Because nothing in the instrument creating the easement identified the amount of property required for the residence, the servient tenement was insufficiently described and invalid. The court explained:

> "While no particular words are necessary for the grant of an easement, the instrument must identify with reasonable certainty the easement created and the dominant and servient tenements."[314] The instrument must be sufficiently precise that "a surveyor can go upon the land and locate the easement."[315] If the instrument does not describe the servient land with the precision required to render it capable of identification . . . the conveyance is absolute[ly] nugatory.[316]

The North Carolina Supreme Court has discussed the sufficiency of an easement's description in many cases. In *Allen v. Duvall* it explained: When an easement is created by deed, either by express grant or by reservation, the description thereof "must either be certain in itself or capable of being reduced to a certainty by a recurrence to something extrinsic to which it refers . . . . *There must be language in the deed sufficient to serve as a pointer or a guide to the ascertainment of the location of the land.*"[317]

When an easement's description is patently ambiguous, the language is insufficient to identify the land with certainty, and so the purported easement will be void. "When . . . the ambiguity in the description is not patent but latent—referring to something extrinsic by which identification might be made—the reservation will not be held void for uncertainty."

The North Carolina Supreme Court later struck down a portion of an easement remarkably analogous to the one we consider today.[318] The North Carolina easement purported to burden land identified as "this tract and adjoining tracts being acquired by Grantee." The court upheld the validity of the easement on "this tract" because it was legally described in the deed. But the court invalidated the purported easement on "adjoining tracts being acquired by Grantee" because the adjoining tracts were not otherwise described. The court pointed out that the language "adjoining tracts being acquired by Grantee," was patently ambiguous. "'The description must identify the land, or it must refer to something that will identify it with certainty.' The

[314] *Dunlap Investors, Ltd. v. Hogan,* 133 Ariz. 130, 650 P.2d 432 (1982) (quoting *Oliver v. Ernul,* 277 N.C. 591, 178 S.E.2d 393 (1971); *Hynes v. Lakeland,* 451 So.2d 505, (Fla.Dist.Ct.App.,1984). *Germany v. Murdock,* 99 N.M. 679, 662 P.2d 1346 (1983). *Vrabel v. Donahoe Creek Watershed Auth.,* 545 S.W.2d 53, (Tex.Civ.App.,1976). *See McHale v. Treworgy,* 325 Mass. 381, 90 N.E.2d 908 (1950).

[315] *Vrabel v. Donahoe Creek Watershed Auth.,* 545 S.W.2d 53, (Tex.Civ.App.,1976).

[316] *McHale,* 325 Mass. at 385, 90 N.E.2d 908; *Allen v. Duvall,* 311 N.C. 245, 316 S.E.2d 267 (1984).

[317] 311 N.C. 245, 316 S.E.2d 267, (1984) (quoting *Thompson v. Umberger,* 221 N.C. 178, 19 S.E.2d 484, (1942)).

[318] *Cummings,* 273 N.C. at 34, 159 S.E.2d at 518.

same principle applies to the description of the servient estate in a deed granting an easement."[319]

Also as in the case at bar, a Texas appellate court considered an easement described as "111.0 acres, more or less, out of a 250.5 acre tract of land in the Basil Durbin Survey."[320] The instrument did not describe the location of the 111-acre servient tenement, nor did the instrument reference another writing describing the location of the 111-acre servient tenement. The court concluded that the description rendered the easement void as to third parties. The court explained that for an easement to be sufficiently described, "the description must be so definite and certain upon the face of the instrument itself, or, in some writing referred to, that the land can be identified with reasonable certainty."[321] In *Matney*, the court stated: "Since the description, or the key thereto, must be found in the language of the contract, the whole purpose of the statute of frauds would be frustrated if parol proof were admissible to supply a description of land which the parties have omitted from their writing. So, while a defect in description may be aided by the description shown on a map, in such case the map must be referred to in the contract."[322] The common denominator in these cases is that the conveying instrument must either describe the servient tenement with certainty or make reference to something else that makes the servient tenement identifiable with certainty. South Dakota follows this view.

In *Ford v. Ford*, the South Dakota court stated: "The office of a description in a deed is not to identify the lands, but to furnish the means of identification, and that a description is considered sufficiently certain which can be made certain, and that a description in a deed would be deemed sufficient if a person of ordinary prudence, acting in good faith and making inquiries suggested by the description given in such deed, would be enabled to identify the property."[323]

In *Schlecht v. Hinrich*, this court again required a real estate description to furnish means to identify the property.[324] The court stated, "A description of property in a chattel mortgage is sufficient where it will enable a third person, aided by inquiries which the instrument itself suggests, to *identify the property*." (emphasis added; holding that a misleading property description was sufficient to put third parties on notice of a mortgage).

In DRD's Blanket Easement, the only identifying words in the description are "grantor's land." These two words do not suggest any point of reference by which one could identify the specific property burdened. *See Ford, Schlecht*. "Grantor's land" certainly does not itself, or by reference to an outside aid, identify the burdened land with certainty. *See Cummings, Vrabel*. The broad description "grantor's land" is insufficient to create an easement under the analogous descriptions considered

[319] quoting *Deans v. Deans,* 241 N.C. 1, 84 S.E.2d 321, (1954).

[320] *Vrabel v. Donahoe Creek Watershed Auth.*, 545 S.W.2d 53, 54 (Tex.Civ.App.1977).

[321] citing *Matney v. Odom,* 147 Tex. 26, 210 S.W.2d 980 (1948).

[322] 147 Tex. at 31-32, 210 S.W.2d at 984 (quoting 1 Jones, *Cyclopedia of Real Property Law* 329).

[323] 24 S.D. 644, 648, 124 N.W. 1108, (1910) (holding description of property in a homestead conveyance sufficient to convey the property).

[324] 50 S.D. 360, 363, 210 N.W. 192, 193 (1926).

in *Cummings,* and *Vrabel.* Indeed, "grantor's land" could have included any property that was owned by Dakota Resorts in the vicinity of Emery Nos. 4 and 5 on February 16, 2000. Contrary to the circuit court's opinion, there is certainly nothing in the language of the purported easement suggesting that the burdened land was located between Emery Nos. 4 and 5 and "Terry Peak Summit Road," the latter descriptor being parole [sic] evidence not mentioned in the Blanket Easement.

We conclude that the words "grantor's land" are not by themselves "sufficient to serve as a pointer or a guide to the ascertainment of the location of the land."[325] This nondescriptive language neither describes the land sufficiently to locate the servient tenement nor references another document that does so.[326] The description furnishes no means or data pointing to evidence that identifies the servient tenement.[327] It clearly does not enable a person to identify what lots in Aventure's subdevelopment—to the exclusion of all other lots—are burdened as the servient tenement.[328] The description was inadequate to give notice or be legally effective as to Flickemas.[329]

VOID INSTRUMENTS

There are numerous cases where courts have declared an instrument void for uncertainty of description. Decisions for a particular state should be examined for the court's thinking in that state. Particularly troublesome are so-called blanket easements, as covered in Chapter 7.

The Florida court dispensed with the problem in this fashion:

> The construction of a deed for easement, or a contract as in this case, may be ruled by the principles outlined in *Kotick v. Durrant.*[330] In that case the Supreme Court of Florida noted that the failure to define the boundary of a right of way in an instrument creating an easement does not render the instrument ineffectual since the grantee is entitled to a convenient and suitable way, depending upon the conditions and the purposes for which the easement was intended. The court observed that these determinations invoke consideration of the situation of the parties and the property, the circumstances at the time the instrument was executed, the practical construction of the instrument as evidenced by the conduct and admissions of the parties themselves, and by the contemporary usage with respect to the subject granted.[331]

[325] *See Thompson,* 221 N.C. at 180, 19 S.E.2d at 485.
[326] *See Berg,* 125 Wash.2d at 551, 886 P.2d at 569.
[327] *See Heron,* 117 P.2d at 246.
[328] *See Coppard v. Glasscock,* 46 S.W.2d 298, 300 (Tex.Com.App.1932).
[329] *DRD ENTERPRISES, LLC, Plaintiff and Appellant v. FLICKEMA, and PSC Properties, LLC, Aventure Estates, LLC, Five J Investment Co., LLC, and Pinnacle Holdings, LLC, Defendants.* 791 N.W.2d 180 (S.D. 2010); 2010 S.D. 88 (2010).
[330] 143 Fla. 386, 196 So. 802 (1940).
[331] *Kingdon v. Walker,* 156 So.2d 208 (Fla App, 1963).

Courts look to various factors to establish a reasonable description of an easement. As noted by the Kentucky court, the process "taxes the best resources of judicial ingenuity."[332] Factors of consideration are as follows:

- The purpose of the easement
- The geographic relationship between dominant and servient estates
- The use of each of the estates
- The benefit to the easement holder compared to the burden on the servient estate owner
- Admissions of the parties

North Carolina courts, however, have been known to take a significantly more restrictive approach to the description issue, holding that an easement grant or reservation is void for uncertainty if it does not adequately describe the location of the easement.[333] In this case, the court compared and contrasted two previous cases, which resulted in two different conclusions because of the wording, or lack thereof, in the grants.

The description in *Borders v. Yarbrough*,[334] while indefinite, expressly referred to a preexisting sewer line (for which the easement was created) across the land of the servient estate. The description, therefore, was capable of being rendered to a certainty by a recurrence to something extrinsic (the preexisting sewer line) to which it referred.

In contrast, in the case of *Oliver v. Ernul*,[335] the writing in question failed to create an easement because the description was uncertain in itself and was not capable of being reduced to certainty since it did not refer to anything extrinsic. The grantees in the case attempted to create an easement for a road, and although a road existed prior to the attempted grant, no reference to it was made in the writing. The description was vague and indefinite and, therefore, patently ambiguous and void for uncertainty.

The Supreme Court of Appeals of West Virginia also adopted the approach that an easement is void if it is not adequately described in the instrument seeking to create it.[336]

A few other courts have expressed a comparatively strict standard regarding the designation of an easement's location, requiring that the easement description meet the general conveyancing standard for identifying a parcel of land. Others apply the "surveyor sufficiency" test: certainty such that a surveyor can go upon the land and locate the easement from the description.

[332] *Daniel v. Clarkson*, 338 S.W.2d 691 (Ky. 1960).

[333] see *Allen v. Duvall*, 311 N.C. 245, 316 S.E.2d 267 (1984) and a line of additional cases.

[334] 237 N.C. 540, 75 S.E.2d 541 (1953).

[335] 277 N.C. 591, 178 S.E.2d 393 (1971).

[336] *Highway Properties v. Dollar Sav. Bank*, 189 W.Va. 301, 431 S.E.2d 95 (1993), quoting *Allen v. Duvall* with approval

In addition, a few states, such as Wyoming, have enacted specific statutes relative to the description of easements. In that state, the law requires specific descriptions for easements recorded after May 20, 1981.[337]

INTERPRETATION

Interpreting an instrument describing an easement should be no different from interpreting any instrument describing an estate in land. Since easements are viewed and treated the same as real property, the definitions and description be treated the same way and using the same rules. Mostly it is *rules of construction* that apply when there are problems of interpretation and ambiguities, or conflicts, within descriptions. In interpreting such a description, the Texas court stated it concisely:

> *Since this description is to be tested by the requisites of a deed, we believe that the same rules of construction are applicable. A deed must be construed by reading all its provisions. They must be made to harmonize and to give effect to all of the provisions wherever possible.*[338]

Rules of construction for the interpretation of description and for resolving ambiguities may be summarized as follows. These are the primary ones, and the ones most frequently applied; however, this list is by no means exhaustive.

The basic rules of construction are as follows.

Intent The main object in construing a deed is to ascertain the intention of the parties *from the language used* and to effectuate such intention where it is not inconsistent with any rule of law. Intention, as applied to the construction of a deed, or similar writing, is a term of art, and signifies a meaning of the writing.

Numerous examples have arisen whereby the "intent" of a description has been equated to intention of the parties, or what the parties "meant to do." This problem has been addressed and decided by the courts on several occasions. For illustration, the following decisions are noteworthy:

[The] intent of the grantor as spelled out in the deed itself must be interpreted, not the grantor's intent in general, or even what he may have intended.[339]

[337] Wyo. Stat. § 34-1-141 (easement created without specific location is void unless a description of location is recorded within one year of execution of easement).

[338] *Compton v. Texas Southeastern Gas Co.*, 315 S.W.2d 345 (1958), citing *Benskin v. Barksdale, Tex. Com.App.*, 246 S.W. 360 (1923).

[339] *Wilson v. DeGenaro,* 415 A.2d 1334, 36 Conn. Sup. 200 (1970).

The intention sought in the construction of a deed is that expressed in the deed, and not some secret, unexpressed intention, even though such secret intention be that actually in mind at the time of execution.[340]

It is not what the parties meant to say, but the meaning of what they did say that is controlling.[341]

A court may not substitute what [a] grantor may have intended to say for what was said.[342]

In summary, in ascertaining the intent of the grantor of a deed, it is not the function of the court to attempt to ascertain and declare what he meant to say. Also, when ascertaining the intent, it is the duty of the court to presume that the grantor knew the legal meaning of the terms that he employed in the deed.

The application of that rule may sometimes appear to produce a harsh result. Even so, it is a safer and wiser rule of decision that rests titles to property on the basis of the definite and ascertainable legal meaning of the words that the grantor used rather than upon the always uncertain and imponderable basis of what some court might decide that the grantor meant to say.[343]

In interpreting a boundary line in a deed, the law is clear that the description in the deed, if clear and unambiguous, must be given effect. The inquiry is not the intent of the parties but the intent which is expressed in the deed.[344]

Conditions and Circumstances at the Time It is a self-imposed rule of the court system that writings are to be construed in accordance with the conditions and circumstances of the time of the writing. See Chapter 14, Case #4 for an example.

Description Honored in Its Entirety A writing must be construed as a whole, and a meaning given to every part thereof.

"The primary rule of interpretation . . . is to gather the intention of the parties from their words, by reading, not simply a single clause of the agreement but the entire context. . . . Furthermore, the intention of the parties will be gathered by reading the entire context."[345]

Specific Controls General Description Where a particular and a general description in a writing conflict, the particular will prevail unless the intent of the parties is otherwise manifested on the face of the instrument.

[340] *Sullivan v. Rhode Island Hospital Trust Co.* 185 A. 148, 56 R.I. 253 (1936).
[341] *Urban v. Urban*, 18 Conn. Sup. 83 (1952).
[342] *Phelps v. Sledd,* 239 Ky. 454, 39 S.W.2d 665 (1931).
[343] *Holloway's Unknown Heirs v. Whatley,* 104 S.W.2d 646 (Texas, 1937).
[344] *Har v. Boreiko*, 118 Conn.App. 787, 986 A.2d 1072 (2020).
[345] *Mierzejewski v. Laneri,* 130 Conn.App. 306, 23 A.3d 82 (Conn.App. 2011) citing 23 Am.Jur.2d 205-206, Deeds § 197 (2002).

References Are Part of the Description Any items referenced in a writing are part of the writing, with as much effect and force as if copied into it. This is especially true in the case of plans or plats.

Words and Phrases Ordinary words are to be given ordinary meaning, technical words are to be given technical meaning, and legal words are to be given legal meaning. See Chapter 14, Case #4 for an example.

The case of *Lake Garda Improvement Association v. Battistoi*[346] provides an illustrative example:

> In construing a deed, a court must consider the language and terms of the instrument as a whole; the words are to be given their ordinary popular meaning, unless their context, or the circumstances, show that a special meaning was intended. The inquiry is into the intent as expressed in the instrument, and need not be what the parties actually intended. It is the meaning of what is said that controls not what is meant to be said. The deed [in this case], as a whole, conveys what it terms "roadways." This is a term which easily may be confused with "road," or "traveled surface," or the like. A "roadway" can be a strip of land through which a road is constructed. A "road," on the other hand, usually refers to that portion of a roadway over which vehicular traffic passes. See Webster, *Third New International Dictionary*. Thus, while the terms "roadway" and "road" may, at times, be synonyms they need not be so used. Of the two, "roadway" seems to be generally employed as the broader term. Thus, it is clear that the deed could reasonably be construed to have intended to grant more than the portions of the roadways then in use as passageways. The intent could reasonably be found to include a certain amount of land surrounding the roads. Since the trail court apparently concluded that the broad meaning of "roadway" was intended, the evidence concerning the existence of Beach Road from 1936 to the present, and the portion in use, is of little value in determining the intent of the parties. While the "road" and the "roadway" need not have been synonymous, the existence of the road at least demonstrates the minimum amount of land that could have been contemplated as "the roadway." "To lay out a highway is to locate it and define its limits." There appears to be no reason why such a layout must be on the ground, for this can be accomplished on a map as well, as long as the location is specific. A road, in the narrow sense, once physically laid out on the ground, is visually apparent. Thus, Beach Road is physically laid out and visible. A roadway, in the broad sense, however, is not obvious, as can be seen by looking at any roads surrounded by open fields, beaches, or the like. Thus, the application of the language used in the deed, construed as a whole, supports the reasonableness of the conclusion that the entire area was included in the conveyance.

Construe against Grantor and in Favor of the Grantee While the grantor is in total control of the wording in a deed until delivered to the grantee, and the grantor is the one who signed it, construing in favor of the grantee is generally considered to be a rule of last resort, but useful when all other rules fail.

[346] 160 Conn. 503, 280 A.2d 877 (Conn., 1971).

Exceptions and Reservations The case of *Carr v. Baldwin*[347] not only details the difference between exceptions and reservations through reference to an earlier Kentucky case, but also offers guidance where there is a problem with the exception or the reservation, but not the remainder of the document.

Distinguishing the two, the earlier case[348] stated the following: An *exception* in a deed, as distinguished from a reservation, is some part of the boundary described in the deed that the grantor *retains* title to and does not convey by the deed. A *reservation* is a clause in a deed, whereby the grantor reserves some new thing to himself issuing out of the thing granted and not *in esse* before, but an exception is always a part of the thing granted, or out of the general words or description of the grant. However, a reservation is not always created because the word "reserves," or "reserving," is employed. "Reserving" sometimes has the force of "saving" or "excepting." "Reserving" and "excepting," although strictly distinguishable, are often used interchangeably or indiscriminately, and the use of either term is not conclusive as to the nature of the provision.

What sometimes purports to be a reservation has the force of an exception. A reservation is a thing issuing out of the land granted, whereas an exception is a part of the land granted.[349]

Land embraced in an exception must be described with the same definiteness and certainty that is required when describing the property granted. Where a deed is valid and only the exception is void for uncertainty or vagueness, the title to the whole tract passes; the exception alone is void.

The principle is: An exception should describe the property with sufficient certainty. Uncertainty or vagueness of description renders a reservation void, unless there is something in the exception, deed, or evidence whereby it can be made sufficiently certain. Uncertainty of location can, however, in a proper case, be cured by the grantor's election within a reasonable time, followed by appropriate acts.

COMPILATION

When drafting an express easement, there should be included a description of both the servient tenement and the precise portion of that tract over which the easement runs.[350] However, some servitudes cannot be located with precision. Easements or profits for hunting, fishing, and other recreational purposes. For example, they often allow the user to take advantage of the entire servient estate and, therefore, are unable to be located precisely.

Gurdon Wattles stated in his treatise *Writing Land Descriptions*,[351] "it is important that the purposes and uses of the easement be expressed." There is no doubt that when this is done, the wording can be very helpful in determining the scope of the easement, which is often in question or subject to interpretation.

[347] 301 Ky. 43; 190 S.W.2d 692 (1945).

[348] *Stephan v. Kentucky Valley Distilling Co.*, 275 Ky. 705, 122 S.W.2d 493 (1938).

[349] NcAfee v, Arkuem 83 Ga, 645, 10 S.E. 441 (Ga., 1889).

[350] Hazen, *Easements From the Viewpoint of the Title Insurer*, 15 Cal. St. BJ 28 (1940).

[351] Gurdon H. Wattles, 1976; printed by Parker & Son, Inc.

Items for consideration in drafting an easement, either for grant or reservation, include, but are not limited to, the following.

Using Technical Words to Make the Intent Clear It is suggested that use of the words "grant" and "easement" will assist in dispelling any notion that an estate other than an easement is intended. This is an important consideration, as many arguments arise over the meaning of writings and whether they refer to fee or easement.

Is the Easement to Be Exclusive or Nonexclusive? Use of the term "non-exclusive" is preferable when the easement is not intended to be exclusive. This would be especially true where multiple parties may be involved, such as with rights of way, drainage easements, and the like.

Is the Easement to Be Perpetual or Temporary? When creating a temporary easement, any conditions that will result in its termination should be clearly noted. Also noted should be any language relative to the restoration of the servient estate that is necessary when termination takes place.

Is the Easement Intended to Serve Some Particular Function? If it is intended to be used for "some specific and definite purpose" as the definition suggests, it should be worded as such, so as have limitations on its use. Consideration should be given to related limitations, if they may apply.

Does a Dominant Estate Exist, or Is It an Easement in Gross? If it is appurtenant to a dominant estate, reference should be made to it. If it is an easement in gross, that should be specified.

Does the Easement Create Limitations on the Use of the Dominant Estate? If limitations exist, or are created, they should be specified.

Identify Servient Estate and the Location(s) of the Easement(s) Thereon The servient estate should be carefully and specifically identified. It should also be clear as to whether the entire estate is burdened or merely a specified portion of it.

Location on the Servient Estate The on-the-ground location should be made apparent. Often the best way to accomplish this is through the use of a recorded survey plan. A metes and bounds description may also suffice.

Right of Relocation Sometimes the option to relocate the easement in the future is a consideration. Consideration should also be given to whether the grantor or the grantor's successors in title have that right.

Easement by Reservation It should be established whether the reservation is only to the grantor or to a third party.

Maintenance and Repair Who, specifically, has the right, or responsibility, to maintain the easement? Consideration should be given to having sufficient width to allow for maintenance and repair.

CHAPTER 7

PROBLEM EASEMENTS

Certain easements, in addition to poor descriptions and incomplete definition, pose particular problems.

UNDESCRIBED EASEMENTS, BLANKET EASEMENTS

Unlocated easements are often called *floating easements*. A floating easement is an easement that, when created, is not limited to any specific area on the servient tenement. Floating easements are also sometimes termed *blanket easements* or *roving easements*. Generally, there is no fixed location, route, method, or limit to the right of way.

Floating easements have a significant practical impact on the servient tenement. These easements burden the entire servient estate and therefore tend to hamper development, limit financing possibilities, and impede alienation of the property. They may be public or private, appurtenant, or in gross.

Examples include the following:

> "Together with a right of way thereto"
>
> "excepting a right of way across the property for a telephone line"
>
> "the right of way, or easement and privilege, to lay, repair, maintain, operate and remove pipelines for the transportation of oil or any of its products or other fluids or substances." over and through my lands."
>
> "a reasonable route through the grantor's land that will not cause undue and unreasonable work and engineering."

LOCATING AN UNDEFINED EASEMENT

The Vermont case of *Lafleur v. Zelenko*[352] offers insight as to procedure for locating an undefined easement. First, the court noted, where deeds to lots containing grant of right of way thereto over other land of grantors did not locate right of way, but simply granted right without limiting or defining it, grantees were entitled to a convenient, reasonable, and accessible way, having regard to the interest and convenience of owner of land as well as to their own. Second, in such cases, the owner of the servient estate has, in the first instance, the right to designate the location, but, if he fails to do so, the person entitled to the right of way may select a suitable route, having regard for the interest and convenience of owner of servient estate. Third, the court noted, while an undefined right of way granted by deed may be given definite location by subsequent agreement of parties, in the absence of such an agreement, the practical location and use of such way by grantee under his deed, acquiesced to by grantor at time of grant and for a long time subsequent thereto, operate as an assignment of right, and the way so established is deemed to be what was intended to be conveyed by the deed, and the same in legal effect as if it had been fully described by terms of grant, although grantee is not necessarily entitled to way commonly used prior to the conveyance.

The following legal principles, raised by the complainant, were presented in the decision:

> Where a right of way is granted over certain land without fixing its location, but there is a way already located at the time of the grant, this will be held to be the location of the way granted, and the way which parties had in mind. 19 C. J., p. 972, 211b; *Gerrish v. Shattuck,* 128 Mass. 571; *Peabody v. Chandler,* 40 N.Y. S. 1028; *Stevens v. McRae,* 97 Vt. 76.
>
> Where an easement is granted in general terms, without definite location and description, the location may be subsequently fixed by an implied agreement arising out of the use of a particular way by the grantee and acquiesced in on the part of the grantor. 19 C. J., p. 972, 213; *Peduzzi v. Restulli,* 79 Vt. 349; *Kinney v. Hooker,* 65 Vt. 333; *Stevens v. McRae, supra; Bentley v. Hampton,* 28 Ky. Law, 1083, 91 S.W. 266; *Winslow v. Vallejo,* 148 Cal. 723, 113 A. S. R. 349; *Stetson v. Curtis,* 119 Mass. 266; *Howland v. Harder,* 183 N.Y. S. 129; *Peabody v. Chandler,* 40 N.Y. S. 1028; *Onthauk v. Lake Shore R. R. Co.,* 71 N.Y. 194, 27 A. R. 35.
>
> Where right of way is granted over certain land without fixing its location, the way selected must be a reasonable one as to both parties in view of all the circumstances, and such as will not unreasonably interfere with the enjoyment of the servient estate. 19 C. J. 971; *McKenney v. McKenney,* 261 Mass. 248; *Ballard v. Titus,* 157 Cal. 673; *Rumill v. Robbins,* 77 Me. 193; *McKell v. Collins Colliery Co.,* 46 W. Va. 625; *Holmes v. Selly,* 19 Wend. (N. Y.) 507.
>
> The primary right of selecting the route of an easement of necessity is with the grantor, but must be exercised to respond to the interests of the grantee as well as his

[352] 101 Vt. 64, 141 A. 603 (1928).

own. *Osborne v. Wise,* L. R. 2 Ch. Div. 968 (7 C. & P. 761, 32 E. C. L. 723); *Holmes v. Seely, supra; Russell v. Jackson,* 2 Pick. (Mass.) 574; *Ritchey v. Welsh,* 149 Ind. 214.

A way of necessity exists only in case of necessity, and convenience is not sufficient foundation therefor. *Nichols v. Luce et al.,* 24 Pick. (Mass.) 102, 35 A. D. 302; *Gayetty v. Bethune,* 7 A. D. 188; *Turnbull v. Rivers,* 15 A. D. 622; *Lawton v. Rivers,* 13 A. D. 741; *Oliver v. Pitman,* 98 Mass. 50; 19 C. J. 971; *Kinney v. Hooker,* 65 Vt. 333.

Where parties to a reservation for an easement have failed sufficiently to express their meaning, their intent becomes a question of fact to be ascertained by the court, which in so doing may inquire into and consider all the circumstances. *Callan v. Hause,* 91 Minn. 270, 1 Ann. Cas. 680; *Morrison v. Marquardt,* 24 Iowa, 35, 92 A. D. 453; *Goodall v. Godfray,* 35 Vt. 226.

Inconvenience is not sufficient damage to warrant an injunction. *Nichols v. Luce, supra; Gayette v. Bethune, supra; Kinney v. Hooker, supra; Long v. Gill,* 80 Ala. 408; *Hyde v. Town of Jamaica,* 27 Vt. 443; *Dee v. King,* 73 Vt. 375.

HIDDEN EASEMENTS

There are two features that contribute to (or cause) hidden easements. They are:

1. Nonwritten, thereby not discoverable through ordinary research and title examination
2. Nonobservable, for example, possessing no observable use or being underground and not recognized in an inspection for mortgage purposes

Hidden pipelines account for the majority of hidden easement cases. Similar difficulties arise in cases of minor building encroachments and neglected roadways. Roadways present two hidden easement problems. A nonwritten roadway easement may be created by implication, necessity, estoppel, or prescription. Often such a road has fallen into disrepair and is without evidence of use only to later have the dominant owner reassert his or her rights. Or, a right of way by necessity has been created through the conveyancing and is not visible, since the dominant estate has lain unused for a period of time. In the alternative, if the dominant estate has been used, it may have been accessed through permissive use over lands of a stranger.

Ordinarily, title opinions will list exceptions, two of which are likely to be that the opinion is subject to matters not on the public record, or items that may be disclosed by an accurate survey or by careful inspection of the property. Many hidden easements will evade even the most careful observer, especially if they are, indeed, not observable.

Categories in which persons, surveyors in particular, are likely to have difficulty, either in the location or the width, are any easements that are nonspecific in their description, but, by their very nature, are easements by:

- Implication
- Necessity

- Implied dedication
- Prescription
- Estoppel
- Custom

Many of these are governed by legal principles, some of which will require determination by a court; however, a surveyor can play a key role in the process by locating evidence of use, and extent, and in some cases, in the search for peripheral evidence and documentation, such as tax records, maintenance records, and similar, relevant, information.

ROLLING EASEMENTS

There is a category of easement, in some jurisdictions, that is perpetual regardless of the character of the land. It becomes obvious and of concern when the easement is affected by accretion and/or erosion, which would normally change the character of the easement, even cause it to disappear in extreme cases. With a rolling easement, the location of the easement moves as the character of the land changes, and it retains its usefulness and purpose.

The Texas case of *Severance v. Patterson, et al.*[353] provides an informative study of the category. The court noted that "not only is Texas's coastline expansive, we also have the highest erosion rate in the nation, affecting 'five to six feet of sand annually.'[354] This erosion rate causes coastal property lines to change annually." It went on to emphasize that "property lines on the coast are defined by migratory, dynamic boundaries. In *Luttes v. State* [355], we determined that Anglo-American common law applied to land grants after 1840 and thus affixed the mean high tide as the boundary between state and private ownership of land abutting tidal waters. The beach is commonly known to lie between the mean low tide and vegetation line. For over fifty years, the OPEN BEACHES ACT [OBA][356] has assimilated that common knowledge as a statutory definition as well. All land seaward of the mean high tide[357], known as the wet beach, is held by the state in public trust; see *State v. Balli*[358], recognizing the 'ancient maxim that seashore is common property and never passes to private

[353] 54 Tex.Sup.Ct.J. 172, 345 S.W.3d 18 (2010); see also *Severance v. Patterson*, 485 F.Supp.2d 793 (S.D.Tex. 2007) and *Severance v. Patterson*, 566 F.3d 490 (5th Cir. 2009).

[354] Michael Hofrichter, "Texas's Open Beaches Act: Proposed Reforms Due to Coastal Erosion," *4 ENVTL & ENERGY L. & POL'Y J.* 147 (2009).

[355] 159 Tex. 500, 324 S.W.2d 167 (1958)

[356] Enacted 1959. *TEX. NAT.RES.CODE ANN.* § 61.011(a).

[357] The average of highest daily water computed over or corrected to the regular tidal cycle of 18.6 years. *Luttes* 324 S.W.2d at 187.

[358] 144 Tex. 195, 190 S.W.2d 71 (1945)

hands.'" The land between the mean high tide and the vegetation line is the dry beach and may be privately owned. A landowner may not exclude the public from a beach covered by the ACT.

Following the passage of the OBA, Texas courts have found that the public has acquired easements by prescription or implied dedication or custom on the dry beach along portions of the Texas Gulf Coast. Although the OBA is silent about the effect of erosion on the boundaries of public beachfront easements, Texas courts have also held that, once an easement is established, its boundaries shift with the vegetation line and the line of mean low tide. Under this "rolling easement doctrine," the state is not required to establish a new easement via prescription, dedication, or continuous right as the shoreline migrates landward. Migration of the shoreline frequently accompanies hurricanes or tropical storms. (See Figure 7.1)

The OBA states in part:

> (a) It is declared and affirmed to be the public policy of this state that the public, individually and collectively, shall have the free and unrestricted right of ingress and egress to and from the state-owned beaches bordering on the seaward shore of the Gulf of Mexico, or if the public has acquired a right of use or easement to or over an area by prescription, dedication, or has retained a right by virtue of continuous right in the public, the public shall have the free and unrestricted right of ingress and egress to the larger area extending from the line of mean low tide *to the line of vegetation* bordering on the Gulf of Mexico.

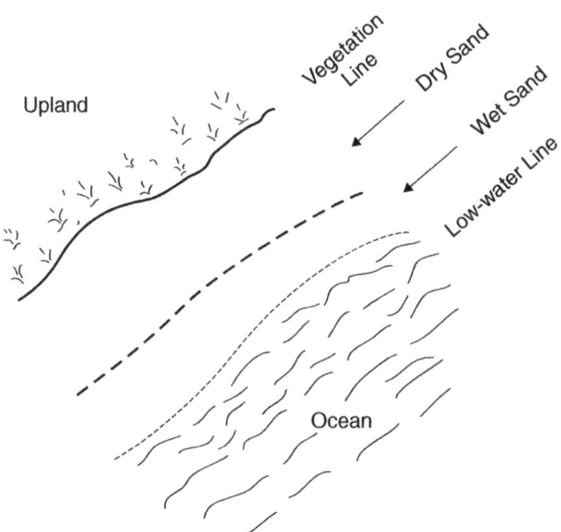

Figure 7.1 Beach title lines in Texas.

The dissenting opinion in this case noted that the language makes clear (and any doubt that might have lingered has been removed by later legislation and the consistent holdings of the Texas intermediate appellate courts) that the landward border of this easement is not and cannot be permanently fixed. That boundary is identified as the "line of vegetation," which, given the vagaries of nature, will always be in a state of intermittent flux.

The practical and legal effect of this shifting boundary is that there is and remains only a single easement of public access to the beach. Shifts in the vegetation line do not create new easements; rather they expand (or, in the case of seaward shifts, reduce) the size and reach of that one dynamic easement. To construe the vegetation line as ever having been a fixed point ignores (1) the plainest interpretation of the unambiguous statutory language, (2) later legislation that reasserts this principle, (3) the ease with which the Texas Legislature could have, had it wanted to, fixed the easement's boundary permanently by specifying a fixed date at which to set the line, (4) interpretations of the Texas courts, and (5) our own prior acknowledgment of this phenomenon.[359] Despite all of this, the majority declares that with each shift of the vegetation line, a new easement is born. There is and always has been one and only one easement, and that easement already encumbered the entirety of Severance's land at the time she bought it, even though the easement's maximum effect and coverage remained inchoate while the vegetation line lay seaward of her property.

The majority's approach perceives the birth of a brand new easement every time the vegetation line shifts, a result I view as a legal impossibility. The line of vegetation, even though relatively stable perhaps, is always subject to flux, as nature is wont to be. Furthermore, the vegetation line can move either landward or seaward from time to time. The majority's analysis, which sees the possibility of infinite mini-takings implicit in the statutory language, appears to assume that the vegetation line moves only landward, for it fails to account for how a property owner may reclaim his land as a result of a seaward shift. Rather than interpreting the plainly worded statute at face value, the majority opts for a more convoluted reading, one that, if accurate, would mire the state in the absurdity of endless litigation over an infinite number of infinitesimal accretions to the dry beach.

Had the Texas Legislature intended such a result, it easily could have fixed the extreme of the landward boundary of the easement. Instead, the state borrowed from the familiar common law approach to property lines expected to fluctuate over time. Were there any remaining doubt as to the result originally intended by the legislature, the lawmakers put it to rest in 1985 when they passed legislation (the "1985 Act"),

[359] We have twice stated this rule. In *Mikeska v. City of Galveston*, we said, with reference to the OBA, "[d]ue to shifts of the vegetation line and the erosion of the shoreline, the natural demarcation lines are not static . . . the legal boundaries of the public easement change with their physical counterparts." 451 F.3d 376 (5th Cir.2006)(citing *Feinman*, 717 S.W.2d at 110-11). And in *Hirtz v. State*, we stated that because of the OBA's definition of the landward boundary of a beach easement as "the line of vegetation" "the location of the easement shifts as the vegetation line shifts." 974 F.2d 663, 664 (5th Cir. 1991).

mandating that sellers of beachfront property inform buyers of the OBA's easement. The 1985 Act states: " Structures erected seaward of the vegetation line (or other applicable easement boundary) *or that become seaward of the vegetation line as a result of natural processes* are subject to a lawsuit by the state of Texas to remove the structures." The 1985 Act makes pellucid that once an easement on the dry beach is established, its landward boundary may therefore "roll," *including over private property*".

A correct determination whether there was one easement or many is the key to the standing analysis in this case. Under the *Palazzolo v. Rhode Island* [360] physical-taking analysis—which the majority purports to employ—Severance lacks standing to bring a Fifth Amendment claim because any taking occurred prior to her 2005 purchase of the property. The majority nevertheless reasons that the rule of *Palazzolo* does not deprive Severance of standing because, when she bought the land, the easement had not attached to the portion of the land that was later affected by Hurricane Rita. As explained previously, this simply is not so: Severance's entire property was encumbered by the easement's very real potentiality—and predictability—that the vegetation line would shift.[361] Simply put, a preexisting restriction such as this cannot be a new taking.

SHORE ROAD ALLOWANCES IN CANADA[362]

Known as a road allowance, the Crown has reserved land in fee-simple along riparian lands. A space that has varying widths depending upon the province, may be held as Crown land, or may be vested in an organized municipality. It is usually 66 feet in Ontario, and 99 feet in Manitoba. Like any other fee-simple parcel adjoining waters, it is subject to loss by erosion or gain by accretion. The use as a road may be added, so that an addition by accretion means an extended road area and erosion means a diminished, or limited, road area.

In New Brunswick, the statute of 1884[363] provided for a reserve, one chain[364] in width, from the banks of specified inland waterways and that the riparian ownership of the streams remained with the Crown. The width of the reserved strip was later increased to three chains, then provision made for disposal of the strip, reserving a public right of passage over a 10 metre width and retaining the bed below the normal high water mark.

[360] 533 U.S. 606, 121 S.Ct. 2448, 150 L.Ed.2d 592 (2001).

[361] The litany of storms that have hammered and at times totally inundated Galveston Island, from "Isaac's Storm" in 1901 through Hurricane Ike in 2008, three years after Rita, leaves no doubt. See e.g., Erik Larson, *Isaac's Storm* (Random House 1999).

[362] Lambden, David W. "Water Boundaries-Inland." In, *Survey Law in Canada*. §§ 6.78-6.81.

[363] *An Act to provide for the survey, reservation and protection of lumber lands,* 1884 47 & 48 Vic., c 7, s. 4.

[364] The [Gunter's] chain is generally defined as 66 feet in length; however, there are a variety of other types of chains of different lengths.

The inner limit of the shore road allowance, or any other shore reservation where the fee-simple title is retained by the Crown, is a different matter from a natural boundary, although it is dependent on the natural boundary at a particular moment in time, the time of the grant, with one exception, in Ontario, for which explanation follows. Generally, the fee-simple owner of an upland parcel receives a parcel of fixed limits and the parcel is not riparian where there is a saving and excepting clause excluding the strip along the shore. The inner limit of such a shore reservation, reserved in fee-simple by the grantor out of the whole parcel described, is fixed in position by the original survey marks, if found as evidence, and as at the date of the grant. The burden is on the surveyor to locate that original natural boundary where sometimes the grant may be one hundred or more years old.

In Ontario, road allowances were defined by the Crown surveys and never formed part of the township lots, they were shown on the township survey plans as extended parcels without a break. They were therefore bounded by an irregular line, distant the specified distance inland from the natural boundary as it existed at the time of the original Crown survey. The Ontario Surveys Act[365] has provided since 1849 that all side lines and limits of lots surveyed are true and unalterable. The courts have consistently held that the inland fee-simple parcel is definite as to extent. This inner limit of the road allowance is not ambulatory as is the water boundary from which it was first defined by dimension. If this were not the case, there would be nonalignment of the road allowance lying in front of the lots granted.

For survey purposes, reference must be made to the patent as the basic title document to know what constitutes the parcel and for the first evidence necessary to locate its bounds. Where a survey plan exists that appears to show the bounds of the parcel clearly, it is still the date of the grant that controls, or where specific items of the Ontario systems apply, the date of the survey.

THE NEW ZEALAND EXAMPLE

To emphasize that the Texas act is not such a unique law, the following example from New Zealand is discussed for comparison and demonstrating that the notion of an ambulatory strip has some fascinating history. The case of *Pipi Te Ngahuru v. The Mercer Road Bd.*[366] held on the facts of erosion into a road reservation along a river that the public were entitled to a road of full width from the new bank. This ruling stood until 1906, when the same court, with the same facts and argument, decided that the owner of the inland parcel had no liability to provide new land for the public road and that the Crown, if it wished to have full width, must acquire the land and pay compensation.[367]

[365] 12 Vict., c. 35, s. 32.

[366] 6 N.Z.L.R. 19 (1887).

[367] See also *Smith v. Renwick*, 3 L.R. (N.S.W.) 398 (Sup.Ct.F.C., 1882); *A.-G. for N.S.W. v. Dickson* (1904) A.C. 273 (P.C.); and *McGrath v. Williams*, 12 S.R. (N.S.W.) 477 (1912) for similar Australian considerations on shore reservations.

The Ontario case of *Herriman v. Pulling & Co.* raised the issue, "It may be upon a matter of nice law that the road reserved . . . would shift . . . over the accretion . . . but this was not argued." However, the earlier decision of *McCormick v. Pelee (Township)*[368] rejected the notion of an ambulatory strip for a shore road allowance. In the case of *Re Monashee Enterprises Ltd. and Minister of Recreation & Conservation for B.C.*,[369] the concepts and the law were neatly presented. Per Seaton, J.A. delivered the judgment of the Court, stating, "It seems to me that the inconvenience of a mobile boundary is such that it should only be found to exist where it is unavoidable. It is unavoidable at the shoreline, but it is not unavoidable at the upland side of the one chain strip." And, "The land gained by accretion is added to and becomes part of the strip."

[368] 20 O.R. 288 (Ch.D., 1890).
[369] 124 D.L.R. (3d) 372 (B.C.C.A., 1981).

CHAPTER 8

THE PROCESS OF REVERSION

To *revert* means to operate by way of a reversion; in a sense to come back to a grantor or lessor.[370]

ESTATE IN REVERSION

An *estate in reversion* is an estate presently vested in interest, to commence in possession in the future after the termination of a preceding particular estate created by the grantor. It is a future possessory estate in either real or personal property arising by operation of law to take effect in possession in favor of a grantor, lessor, or transferor, or his heirs, distributees, devisees, or legatees, after the termination of a prior particular estate granted, demised, or devised. Examples are life estates and estates for years.

A *reversioner* is one in whom an estate in reversion is vested in interest.

POSSIBILITY OF REVERTER

A possibility of reverter is that interest in land remaining in the grantor of a determinable fee-simple estate. It is a reversionary interest, but not an estate in reversion. A distinction is made between a reversion, which is an estate in the present, although to be enjoyed in the future, and capable of being transmitted by descent, devise, or grant, and a mere possibility of reverter. The difference is that reversions are always vested estates corresponding to vested remainders, while the mere possibility of a reverter is always contingent and corresponds to a contingent limitation.[371]

[370] *Glossary of Real Estate Law*.
[371] Ibid.

Example: A, seized in fee-simple, grants to B an improved parcel of land for so long a time as the property shall be used for public school purposes, and no longer. X has remaining, after the grant, a possibility of reverter; that is, a possibility that the property may revert to him and his heirs should the contingency on which the estate is limited occur. Depending on the particular state and its laws, modern usage tends to recognize a possibility of reverter as an estate in land.

Contingencies described in documents need to be carefully considered in accordance with the law in existence at the time. Depending upon the creating documents, and the wording selected, events are not always as might be expected, as illustrated by the following two cases.

In the first, the New Hampshire court stated "where land is taken by a municipality for a public use . . . it is incumbent upon a claimant of damages to establish that at the time of the taking he was an owner of the fee, remainder, or reversion, or a tenant for life or years." It went on to add, "the only practical distinction between a right of entry for breach of a condition subsequent and a possibility of reverter upon a determinable fee is that in the former the estate in fee does not terminate until entry by the party having the right, while in the latter the estate reverts at once upon the occurrence of the event by which it is limited."

The appellant in this case[372] claimed title to the premises as the grandson and only heir of Stephen C. Lyford, who in 1837 gave the Congregational Society a deed of the land in question. The deed conveyed to the society, their successors and assigns forever, a certain parcel of land, giving the boundaries. Immediately after the description of the land are the following clauses: "Said society to hold said premises as long as they occupy the same with a house of public worship and no longer, and when they cease to so occupy said premises then the same shall revert to me and my heirs." The society took immediate possession, and occupied with a house of public worship until the city decided to take the property through their power of eminent domain for a park and library lot. The plaintiff appealed, and must bring himself within the provisions of the statute and establish that he was at the time of the taking an owner of the land taken, that is, an owner of the fee, remainder, or reversion, or tenant for life or years.

The court concluded that "whether the plaintiff's right is a possibility of a reverter upon a determinable fee, or a right of entry for breach of a condition subsequent, he had when the land was taken no right to the land and no possession of it. Wherever the gift is of a fee, there cannot be a remainder, although the fee may be a qualified or determinable one. The fee is the whole estate. When once granted, there is nothing left in the donor by a possibility or right of reverter, which does not constitute an actual estate. All the estate vests in the first grantee, notwithstanding the qualification annexed to it. Whether the event upon which the plaintiff might come into ownership of the land would ever happen was mere speculation. There was no method by which the value of the interest could be assessed that would rise above the dignity of a guess. The plaintiff did not own the land taken. He was not an owner in fee, in

[372] *Lyford v. Laconia*, 75 N.H. 220, 72 A. 1085 (1909).

reversion, or in remainder. He had no subsisting title in the land, but only a possibility that it might revert to him by the happening of the event upon which the estate of the society was determinable.

> As the society owned the use, the taking must have been from the society and not from the plaintiff, who had no right to the use. Whether the plaintiff's possibility of reverter dependent upon a cessation of the public use, should the estate then come to him, is more or less valuable than his similar right upon the cessation of the religious use of the society, is plainly a possibility upon a possibility—a matter too indefinite and vague for pecuniary estimation. Whether upon a cessation of the public use the right to use would revert to the society, or the estate (assuming the validity of its determinable character) would at once vest in the plaintiff, is a question upon which discussion would be useless. If the fee of the society has not been taken, the title must revert to them, and the plaintiff's interest, whatever it is, remains absolutely unimpaired. If the title to the society is gone, then upon expiration of the public use the estate would at once revert to the plaintiff, and it is impossible to say whether he has been damnified or benefited by the taking from the society.

In the second case,[373] the North Hampton (N.H.) Congregational Society conveyed by warranty deed on October 29, 1875, to the North Hampton School District's predecessor in title:

> A certain piece or parcel of land, situate at said North Hampton Centre as the same is now run and marked, and to be fence out for the lot of the new School House, bounded as follows: on the West by the highway, on the North by other land of the grantors, on the line as now marked out, East by other land of the grantors, on the line now marked out and to be fenced, on the South by the old School House lot, upon condition that all the space South of the Southerly side of the new school house, and South of a line running by the Southerly side of said school house to other land of said grantors, on the East shall remain open as a common forever and that no religious services shall ever be held in the Hall above the school room of said school house except by the grantors or by their permission, and that the grantees shall erect and maintain the necessary and suitable fences between the grantors and the grantees forever.

Following the habendum clause were the usual covenants of warranty except that the last of them read as follows: that grantor "will warrant and defend the same to the said grantees, their heirs, successors and assigns against the lawful claims and demands of any person or persons whomsoever, so long as said lot shall be used as a lot for a school house lot as aforesaid."

Centennial Hall was built thereon in 1876 and was used as a school house until 1950, when a new school was constructed on another location. At the 1950 annual

[373] *North Hampton School Dist. v. North Hampton Congregational Soc.*, 97 N.H. 219, 84 A.2d 833 (1951).

school meeting, it was voted to sell Centennial Hall to be moved away or torn down, and to turn over the land to the Town of North Hampton for a public common. The plaintiff petitioned the court to decree it was the owner of said land and building free and clear of any claims of the defendant. The defendant, in answer, sought a decree that said land and building were owned by it in fee-simple free from all claims of the plaintiff or the Town of North Hampton.

In analysis, the court stated, "the issue is the nature of the estate held by the plaintiff (School District) by virtue of their deed." Defendant maintains that the wording "so long as said lot shall be used as a lot for a school house lot as aforesaid" applies to the grant with the result that plaintiff acquired a determinable or qualified fee. Plaintiff, on the other hand, contends that the proviso applies only to the covenant of warranty and that consequently it acquired by said deed a fee-simple, the proviso at most impressing a trust thereon, the purpose of which is set out by it.

The court stated, "although the language used in the deed is not decisive of the issue of what type of estate is created thereby," according to case law, "it is of importance in arriving at the manifest intention of the parties which is the determinative factor. The words 'so long as' used here, as well as 'while,' 'until' and 'during' are the usual and apt words to create a limited estate such as a determinable fee which estate is recognized in this jurisdiction."

The plaintiff took the position that because the proviso came at the end of the warranty clause instead of being in the granting or habendum clauses that it was intended merely to limit the warranty to the school district for use as a school house lot and that this was the only purpose of this limitation. "Viewing the language used 'in light of the surrounding circumstances," and relying on a case decided the year before this deed which held that "a proviso substantially in the same language and place in the identical part of [that] deed was not restricted to the covenant of warranty but must be applied to the grant, thus creating a qualified fee." The court felt this previous decision was of utmost importance in arriving at the intent of the parties. In that case, the discussion centered around the fact that "the limiting proviso although it followed a period after the last covenant of warranty nevertheless began with a small letter instead of a capital as would be the usual custom. The need in this case contains a comma instead of a period after the last covenant of warranty thereby making the proviso which begins with a small letter more in conformity with the usual use of the English language."

It was the court's opinion therefore, that the plaintiff's predecessor (School District) in title by the deed it received from the defendant (Congregational Society) obtained a determinable fee, which terminated when the plaintiff decided not to use said lot as a school house lot. When that event occurred the fee reverted to the defendant, which holds title to it free from any claim of the plaintiff.

Possibly the most important part of this case is the fact that "the building thereon, Centennial Hall, was erected while the plaintiff was the owner of said land. The record reveals nothing which would prevent it from becoming part of the realty. It also became the property of the defendant free from all claims of the plaintiff when it ceased to use said lot as a schoolhouse lot." The principle here is one of longstanding, that buildings on land, unless excepted or exempted in some fashion, are part of the real estate.

These two cases, though the subject matter relates to fee title rather than easements, serve to illustrate the importance of having a full understanding of the estate created at the outset, along with its provisions, qualifications, and the principles of law that affect it and its interpretation.

CHAPTER 9

REVERSION OF EASEMENTS

Any easement may be subject to reversion rights, either specified in the creating document(s) or in accordance with the rules of law or interpretation of the courts. Occasionally, the matter of reversion will be found in an obscure location, such as this nineteenth-century statute:

CHAPTER XLI

An act in addition to and in amendment of chapter eighty of the general statutes relating to school houses

Section 1. If any school district shall neglect or refuse to purchase or procure the land which may be designated in the manner provided by section four, or section five, of chapter eighty of the General Statutes, the selectmen, upon petition, may lay out a lot not exceeding half an acre, and appraise the damages to the owner who shall have like remedy for increase of damages as if the same was laid out for a highway.

Sect, 2. Upon payment or tender of such damages, the land so laid out shall vest in said district, but shall revert to the owner whenever the district shall vote to discontinue the use thereof, or shall cease to use the same for a school house two years successively.

Sect. 3. This act shall take effect from its passage. [Approved July 14, 1871.][374]

[374] New Hampshire General Laws of 1871.

Any estate should be carefully examined for what type of estate it is, and assumptions should never be made that an estate is either fee-simple or an easement, just because it appears that way or that is the *usual* case. It is usually necessary to evaluate the mode of creation or the mode of termination to decide whether reversion takes place, or has taken place, and what its ultimate result might be. Otherwise, there is no understanding whether the servient remains encumbered by the easement.

When easements are terminated they revert [375] [back] to the land from which they were taken.[376] This is true when an estate less than fee-simple absolute terminates, whether it is an easement, a life estate, or other interest. Occasionally, there is a reversionary clause attached to the creation of an easement or other interest, but the usual situation is that reversion is automatic upon termination of the outstanding interest and conditions return to what they were before the creation of the easement, or interest.

The following is an example of a reversionary clause in a mortgage concerning a right-of-way:

> The sole purpose of the granting of this right of way is to insure access by the grantee, her heirs and assigns, to Mohawk Drive in the event of the foreclosure of a certain mortgage recorded in Rockingham County Registry of Deeds in Volume 2069, Page 287. This easement is therefore granted in the condition that the easement, and all rights hereunder, shall terminate and revert to the grantor, its successors or assigns, upon the discharge of said mortgage.

When reversion takes place and boundary lines are involved, one of two sets of conditions arise, neither of which is simple:

1. Boundaries "go back" to what they were before the creation of the easement. This not an accurate description of what takes place, since the original boundaries never changed: an easement does not serve to change original boundaries of land.[377]

[375] The returning of an estate to the grantor, or his heirs, after a particular estate is ended. *Black's Law Dictionary*.

The term "reversion" has two meanings, first as designating the estate left in the grantor during the continuance of a particular estate and also the returning of the land to the grantor or his heirs after the grant is over. 167 S.W.2d 641, 350 Mo. 639.

[376] Abandonment of public highway resulted in reversion thereof to owners of soil, free of all public easements. *Libertini v. Schroeder*, 132 A. 64, 149 Md. 484 (1926).

Owners of property adjoining highway have right to every use and profit that can be derived from it consistent with easement of public, and soil of highway descends to heirs and passes to grantees as an appurtenant t land adjoining, and whenever highway is discontinued, adjoining proprietors hold land discharged of easement. *Antenucci v. Hartford Roman Catholic Diocesan Corp.*, 114 A.2d 216, 142 Conn. 349 (1955).

[377] Rights of owner of fee in land on which was located highway were revived after highway way abandoned and land discharged of public easement. *Burnham v. Burnham*, 167 A. 693, 132 Me. 113 (1933).

This can be a complex situation, since many years may have elapsed since the creation of the easement and conditions may have changed considerably. Also it may be a matter of re-creating property and easement boundaries at the time of creation, which could involve extensive land records research and locating the boundaries that existed at that time.

2. Sometimes boundary lines have changed, or new parcels have been created since the creation of the easement. The chain of events must be determined from the creation of the easement to date, to know what new boundaries have been established.

Such procedure may become quite involved in the case of highway boundaries as will be seen in the examples in the chapters on highways and case studies.

HIGHWAYS

Since highways are a special case, sometimes involving numerous complex issues, they are treated in a separate chapter. In addition to highways, however, any type easement, as well as other estates, may be subject to reversion and demand the applications of appropriate rules to define titles and boundaries.

FLOWAGE

In the case of flowage, frequently there have been camp lot subdivisions created along the shore since the flooding initially took place. These lots now may include some proportionate share of the flooded land that was part of the parent parcel. There have been many cases involving shore boundaries because of this situation, and some states have enacted statutes to solve potential title problems. Basically, however, the usual case is that original titles extend to: (1) the center of a body of water, (2) the low-water mark, or (3) the high-water mark.

By flowage *easement* abutting parcels are partially or wholly submerged and those which were suitable (new or remaining upland), subdivided into smaller lots fronting on the new shoreline. If the flowage easement is terminated, and the water returned to its former level, a determination must be made as to the ownership of that land between the original shoreline and the shoreline after flooding. Determinations are made based on original titles and ownership, original boundaries, subsequent conveyances, and wording in the various deed descriptions. There is never a simple solution, but all ownerships are defined by boundaries in some fashion, and all must be considered. See Figure 9.1.

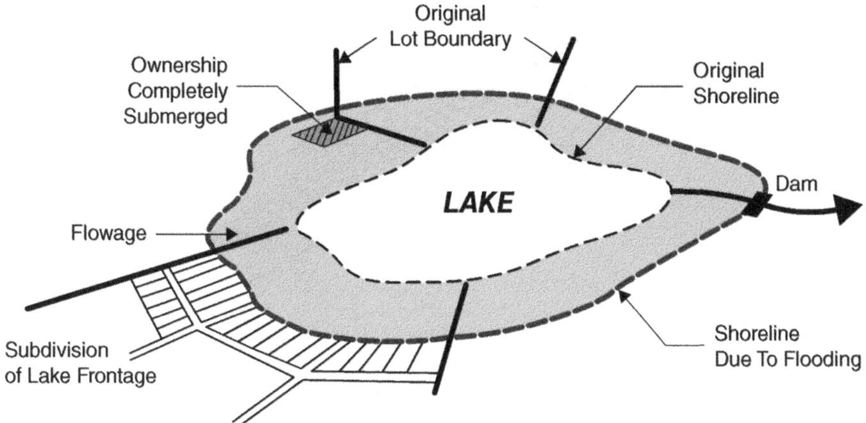

Figure 9.1 Example of flowage easement governing use of upland.

RAILROADS[378]

Sometimes parties conveying easement rights to a railroad company specified what and how reversion will take place upon termination of the easement. Some railroads did this on a regular basis, resulting in reversion occurring in one of three ways, the first two being the most common:

1. Reversion to original grantor or his/her heirs
2. Reversion to heirs and assigns of original grantor
3. Some formal division to abutting landowners, to either centerline or some other defined boundary line(s).

A reversion to the original grantor or *his heirs* can be a very time-consuming endeavor, requiring the determination of whether or not the grantor is still living, and where, or who, his heirs are, or their heirs and so on, and where they reside. Reversion to heirs and assigns of the original grantor demands tracing ownerships forward from the date of the easement to present time in order to identify present successors in title or interest.

[378] A railroad is a "highway" within rule that a grant of land bounded by or abutting a public highway is presumed to carry the fee to the center line of such highway or easement. *Fleck v. Universal-Cyclops Steel Corp.*, 156 A.2d 832, 397 Pa. 648 (1960).

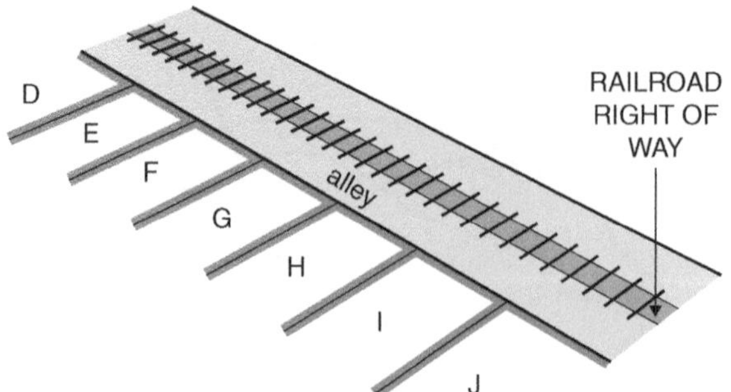

Figure 9.2 Reversion on a railroad.

An interesting case involving reversion of a railroad right of way is as follows (see Figure 9.2):[379]

In this case, an owner deeded to a railroad company a strip of land across his property and included a reversionary clause in case of nonuse for the specified purpose. Subsequently, he sold all the land on one side of the railroad strip to a developer who subdivided it with a marginal alley against the railroad strip. In later years, the railroad gave up its use of the strip, but the grantor had to litigate to regain his physical possession of it. In his petition to the court, he claimed his right to the entire width of the strip, but the court decided that in the sale of the adjoining land, the developer (the grantee) acquired the reversionary rights in and to the half of the railroad strip bounding the subdivided land. Because after-acquired title passes to the current owner, this resulted in each lot along the marginal alley acquiring a portion of the "abandoned" right of way opposite the lot.

As previously noted, any easement or interest in land may be subject to reversionary rights. Since easements, according to law, are viewed and treated the same as real property, they should be treated the same as an abutting property owners. There is a status of ownership, and there are boundaries. Both must be determined. An easement boundary provides no more than a boundary between two separate estates, and is no different from a boundary between two ownerships in fee-simple. (See supporting decisions in Chapter 12).

[379] Wattles, Gurdon. *Writing Legal Descriptions.*

CHAPTER 10

REVERSION RELATING TO HIGHWAYS

(and to Other Types of Rights of Way)

Since highway reversions are the most common, most complex, and most misunderstood, they are treated as a separate topic. Many of the principles discussed here also apply to other reversions.

Many highways are easements. In fact, highways where the governing body owns the underlying fee in a highway are relatively rare, and very much the exception than the rule,[380] contrary to common belief. The average landowner believes the municipality, or the governing agency, "owns the road," and his ownership stops at some definition of sideline. But discontinue public rights, or have an alteration in road location, and this same landowner wants to know how much of the old road he can claim, or has a right to.

While roads that have the entire public right removed are in the minority, they do exist, and someone owns the fee in what was the underlying bed. More common is an alteration of road location, involving a new taking and layout and a discontinuance or vacation of all or part of the old location. Again, someone owns the underlying fee if the road was constructed on an easement. If the underlying soil was owned in fee by the governing body, and it wished to dispose of any portion no longer wanted, it would have to convey it in the same manner as any other parcel of land. Today the conveyance would probably be by deed, whereas in the past there were less formal methods. Additional examples appear in Chapter 10.

[380] The public acquires only an easement in a highway, and the fee in the soil remains in the owners of adjoining lands unless the contrary appears. *Davis v. Girard*, 59 A.2d 366, 74 R.I. 125 (1948).

Example:

> We the Subscribers Selectmen of Poplin have made a little alteration in turning the highway that leads from Jonathan Beedes to Thomas Cases. . . .
> by taking out about Seven Rods of land out of Fellows place now owned by Ezekiel Robinson and for Satisfaction to the Said Robinson we have Given to him the Said Robinson as follows (viz) all the land belonging to the Town in Said highway and range westerly of the following boundaries. . . .
>
> **—April 26, 1806**

DISCONTINUANCE OR ABANDONMENT

The term used to denote the termination of the public easement in a road varies depending on the jurisdiction. The word "abandonment" or "abandoned highway" is frequently used, especially by the public.

"Abandonment" is a poor word to use because in a legal sense it has a particular meaning which does not always apply to the termination of a public easement.[381] As discussed in Chapter 5, a public easement cannot be terminated by nonuse or abandonment,[382] except in special cases, or where the easement was created by prescription. Certain classes of private easements *may* be terminated by abandonment.

Discontinuance is a frequently used term, and is defined by statute in many states. Discontinuance denotes formal action by the governing body, changing the status of the public easement, or terminating public rights altogether.

Vacation is another commonly used term, which means the same as discontinuance. Both serve to achieve the same result, that is, terminate public rights in the easement. When this takes place, reversion is automatic and instantaneous.

ACTUAL HIGHWAY ABANDONMENT

There is a presumption that, once a highway or street is shown to exist, it continues to exist.[383] As discussed in detail previously, to establish abandonment, an intent

[381] Abandonment is the surrender, relinquishment, disclaimer, or cession of property or of rights. *Black's Law Dictionary*.

[382] Intention to abandon is an important factor on the question of abandonment of an easement or other interest in land, and "abandonment" should not be inferred from mere nonuser. *Boston Elevated Ry. Co. v. Commonwealth*, 39 N.E.2d 87, 310 Mass. 528 (1942).

Nonuse is insufficient to establish abandonment of an easement unless an intention to abandon can be shown. *D.C. Transit System, Inc. v. State Roads Commission of Md.*, 290 A.2d 807, 265 Md. 622 (1972).

[383] *Allamakee County v. Collins Trust,* 599 N.W.2d 448, (Iowa 1999).

to abandon must be proven by clear and satisfactory evidence.[384] Actual acts of relinquishment accompanied by an intention to abandon must be shown.[385] Obstructions, encroachments, or the failure to keep a road in repair do not necessarily result in abandonment.[386]

PROCEDURE

In order to determine how reversion rights apply and to identify the rightful owners and what their lines of separation are, it is first necessary to examine the original acquisition.[387] While the Maine court recognized three methods in this case, and the New Hampshire statutes (along with several other states) presently recognize four, there are in fact eight ways by which a *public* highway may be created:

1. Through statutory layout [388]
2. Through dedication and acceptance[389]
3. By prescription
4. By deed
5. By legislative act[390]
6. Through court action
7. By layout over public land[391]
8. Through custom

The creating document and its accompanying description should describe the nature of the easement, the width and survey layout (if it was done by survey), and whose land was encumbered, or "taken." This latter information may only appear as a list of damages paid. It may be necessary to review municipal, county or state records

[384] *Id.; Sterlane v. Fleming,* 236 Iowa 480, 18 N.W.2d 159, (1945).

[385] *Collins Trust,* 599 N.W.2d at 451; *Town of Marne v. Goeken,* 259 Iowa 1375, 147 N.W.2d 218 (1966).

[386] Sterlane, 236 Iowa at 484, 18 N.W.2d at 162; City of Marquette v. Gaede, 672 N.W.2d 829 (Iowa 2003).

[387] A public way can be established by any of three recognized methods: by prescriptive use, by the statutory method of laying out and accepting a way, or by dedication and acceptance. *State v. Beck,* 389 A.2d 844 (Me., 1978).

[388] To lay out a highway is to locate it and define its limits. Such layout need not be on the ground, it can be on a map as well, as long as the location is specific. *Lake Garda Improvement Association v. Battistoni,* 160 Conn. 503, 280 A.2d 877 (Conn., 1971) and cases cited therein.

[389] In order to establish the existence of a public highway, a party must prove that there was dedication as a highway on the part of the private landowner and an acceptance by the pubic authority. *American Trading Real Estate Properties, Inc. v. Town of Trumbull,* 215 Conn. 68, 574 A.2d 796 (Conn., 1990), citing several cases.

[390] Most turnpikes, plank roads, and the like were created by this process.

[391] Extreme caution is suggested when dealing with this category. Early proprietors plans often set aside strips of land for future roads, some of which highways were constructed upon. This area is challenging enough that subsequently a separate section is devoted to it.

to obtain this information. Two typical town highway [statutory] layouts follow, one based on a survey and one not based on a survey:

> We therefore lay out the road: Beginning at a stake standing on the town line on land of Lt. John Haley, thence N 23 deg. W 39 rods, thence N 39 W 33 rods, thence North 40-1/2 rods, thence N 6 and ¾ deg. W 14 rods and 18 links, thence N 17 deg. W 15 rods to land of Joseph S Lawrence Esq., all the above distances through land of said Lt. John Haley, thence N 7 deg. W 51-1/2 rods through land of said Lawrence to the High Road so called near the house of Capt. J.S. Lawrence.
>
> The above described line to be the middle of the highway and the highway to be TWO AND ONE HALF RODS WIDE.
>
> **—June 8, 1830**

> We the Subscribers Selectmen of the Parish for the year 1773: have laid out a Publick Highway for the good of the Parish aforesaid and adjacent Towns and Parishes: and by a Vote of said Parish: Beginning at the old way by the Corner of the road above Mr. Danle Sanborn's house and then to run up as said way is Trod by Mr. Abner Shepards house to the town line: Said road is two rods wide.
>
> **—March ye 12th 1774**

After the original layout has been identified, it is then necessary to determine if any alterations have been made. Doing this may require searching the public records forward to date.

Example:

> We have widened and straightened the highway as follows:-
>
> Beginning at the westerly corner of John Wilson, Jr. Blacksmith shop thence running N 85 deg. W 13 rods on Jeremiah Sawyer's land same course on John Watson's 39 rods, thence N 70 deg. W on said Watson's land 10½ rods, thence N 60 deg. W on said Watson's land 15 rods opposite said Watson's Joiners shop.
>
> Said courses to be the north side of said road and the old bounds for the south side of said road.
>
> **—February 15, 1834**

After all road activity has been accounted for, it is then necessary to search the ownerships under or adjacent to the easement in question. This means the landowners encumbered must be traced forward to determine who the owners presently are. At a minimum, the subject parcel must be searched for the entire period.

Since easements do not cause a change in property boundaries, it is necessary to determine boundary location prior to the creation of the easement and then apply the easement on top of the underlying property or properties. Any parcels created after the easement was created must be treated separately and rules applied depending on

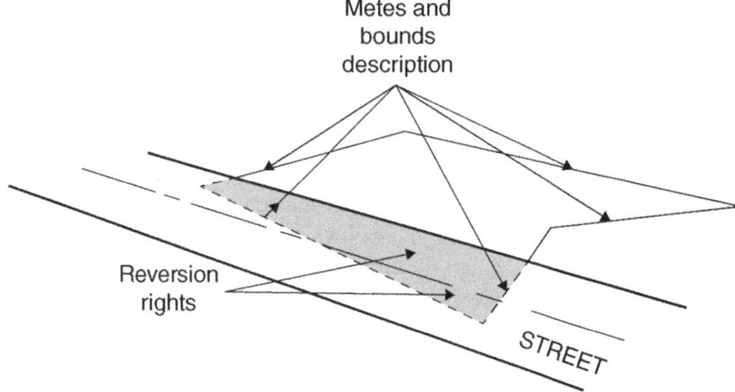

Figure 10.1 Reversion rights are determined by property lines that existed before the creation of the street.

the circumstances and the wording of the conveyance(s). Original boundaries remain as they were; subsequent boundaries are created as new parcels are created. (See Figure 10.1).

PRESUMPTION OF LAW

The principle recognized almost everywhere, either as a general rule of construction or as a rebuttable presumption, is that if land abutting on a highway (meaning public streets, alleys, and roads) is conveyed by a description apt or effective to cover it, and the grantor owns the fee to the center of the highway (or beyond), the grantee takes title to the centerline [392] of the highway as a part and parcel of the grant, unless an intention to the contrary sufficiently appears.[393] The common belief, because of this rule, is that all lands abutting highways, road, and streets, if they are easements, own to the centerline of the easement. Too often the two qualifying statements in the foregoing presumption are overlooked: IF, "the grantor owns the fee to the center of the highway," and "UNLESS an intention [394] to the contrary sufficiently appears."

[392] This means the centerline of the easement, and not the centerline of the traveled portion, as is sometimes believed.

[393] 12 Am.Jur.2d, § 38.

The presumption applies to a judgment on condemnation to the same extent as to a deed of conveyance. *Challis v. Atchison Union D. & R. Co.*, 45 Kan. 398, 25 P. 894; *Michigan C.R. Co. v. Miller*, 172 Mich. 201, 137 N.W. 555 (1912). Annotation: 2 A.L.R. 11, s. 49 A.L.R.2d 984 et seq., § 2.

[394] 12 Am.Jur.2d, § 64. The various rules adopted by the court for construing and interpreting conflicts between calls of description all have for their primary purpose the ascertainment of the intention of the parties.

The case of *Blackwell E. & S.W. Ry. Co. v. Gist,*[395] referenced in Chapter 11, quotes the Oklahoma statute in effect at the time, which allows for those provisions:

> [city councils] shall have the power to annul, vacate or discontinue [any street, avenue, alley or lane] . . . that whenever any street, avenue, alley or lane shall be vacated, the same shall revert to the owners of the real estate thereto adjacent on each side, in proportion to the frontage of such real estate, except in cases where such street, avenue, alley, or lane shall have been take an appropriated to public use in a different proportion, in which case it shall revert to adjacent lots of real estate in proportion as it was taken from them.

It is well settled then, as a general rule, that "a grant of land bounded upon a highway carries the fee in the highway to the center of it, provided the grantor at the time owned to the center, and there be no words or specific description to show a contrary intent.[396] It will not be supposed that a man would care to keep title to the highway in himself, when he had parted with the land bordering thereon.[397]

It makes no difference, in the application of the rule, whether the land abutting upon the highway is a lot that bears a certain number, or is a farm, called Blackacre, or otherwise.[398] Even if the measurement set forth in the deed brings the line only to the side of the highway, the title will still be carried to the center of it, unless such words are used, and such metes and bounds are set forth, as show a contrary intention.[399] Of course, it is understood that the title to the center of the highway, which thus passes by the grant of the adjoining land, is subject to the easement of the public in the highway.[400]

There are a few isolated cases in a few jurisdictions that, under particular circumstances, have held otherwise. The decisions for any particular state, or jurisdiction, should be examined accordingly.

Too often it can be shown that one or the other of these qualifications exist.[401] Few roads are laid out along a boundary line taking one-half the width of the easement from each landowner, even though that is one of the basic theories for the centerline presumption. However, that situation may be found in the public land survey states

[395] 18 Okla. 516, 90 P. 889 (1907).

[396] 3 Kent, Comm. Marg. P. 434; Elliott, Roads & S. p. 549; *Henderson v. Hatterman*, 146 Ill. 555, 34 N.E. 1041 (1893).

[397] *Salter v. Jonas*, 39 N.J. Law, 469; 3 Washb. Real Prop. marg. P. 635.

[398] *Kimball v. City of Kenosha*, 4 Wis. 321; *Berridge v. Ward,* 10 C.B. (N.S.) 400.

[399] 3 Kent, Comm. Marg. P. 434, not 37; Elliott, Roads & S. p. 550; *Paul v. Carver*, 26 Pa. St. 223; *Cox v. Freedley*, 33 Pa. St. 124; *Johnson v. Anderson*, 18 Me. 76; *Cottle v. Young*, 59 Me. 105; *Woodman v. Spencer*, 54 N.H. 507.

[400] *Henderson v. Hatterman*, 146 Ill. 555, 34 N.E. 1041 (1893).

[401] The presumption that the fee of a highway between two adjoining proprietors is owned by them equally to the center fails, where it appears that one of them originally owned the whole or greater part of the land. *Watrous v. Southworth*, 5 Conn. 305 (1824).

There is presumption in New Hampshire that conveyance of land bordered by street conveys title to center of that street, but presumption may be rebutted by showing clear declaration of contrary intent in deed. *Davis v. Lemire*, 449 A.2d 1228, 122 N.H. 749 (1982).

The burden of proof to show that parties conveying land bounded on a highway did not intend to convey to the middle of the street is upon the person so contending. *Buck v. Squires*, 22 Vt. 484.

(PLSS) where highways were created using section lines as centerlines, and the highway extending a certain distance either side of that line.

As an example of the first qualification, consider the following layouts:

"... then from Sd tree northerly 6 rods and at ye end of 6 rods it is 13 feet wide on James Youngs land and 20 feet wide on the heirs of Jona Smith Decd...."

The line above described to be the middle of the highway and the highway to be THREE RODS WIDE except that part bounded on the west by land owned by the heirs of Jeremiah Rowe, which is to be ONE AND ONE HALF RODS WIDE on the easterly side of the above described line."

"...Thence 2° W partly on Simon Robies land and on Josiah lanes land 18 rods to a stake...."

"... the way is to be 12 feet wide on Sd Mardens land and 21 feet wide on said Greens land."

"... Said 2 rods all to be taken out of Sd Martins land and from thence northerly ... one rod out of Kimballs land and one rod out of Butlers land...."

The main basis of the foregoing principle appears to be the unreasonableness, generally speaking, of the supposition that in conveying land abutting on a highway an owner intends to reserve, or by the use of any usual form of conveyance supposes that he is reserving, fee title to any of the highway area, since it is not likely that that area will be of practical importance to him and it is plainly of importance to the grantee.[402] A further ground sometimes assigned for the rule or presumption is found in the recognized principle that when a monument having width is called for as a boundary of premises, the boundary is taken to be the center of the monument. It has also been said that the rule that the abutting owners hold title to the center of a street is found upon the assumption that the owners have each contributed equally to the street.[403]

This presumption is a rather strong principle of law.[404] Some courts have said that the rule applies even though the highway is not mentioned in the conveyance. The omission from the deed of any direct mention of the highway does not render the rule carrying title to the center of the highway inapplicable if the description is by reference to a map or plat which shows the highway.[405] Some of the later decisions hold that a description by metes and bounds which runs along the highway's inner

[402] The intention of the grantor to withhold his interest in a road to the middle of it, after parting with all his right in the title to the adjoin land, is never to be presumed. *Lowe v. Di Filippo*, 12 App. Div.2d 788, 209 N.Y.S.2d 652 (1961).

[403] 12 Am.Jur.2d, § 38.

[404] Sole abutting owner of land on either side of abandoned highway was presumed to own fee thereto. *Nugent v. Vallone*, 161 A.2d 802, 91 R.I. 145 (1960).

[405] Ordinarily a sale of lots by reference to a plan conveys rights in streets and ways shown thereon to both grantee and public. *Callahan v. Ganneston Park Development Corp.*, 245 A.2d 274 (Me., 1968).

side operates to exclude the highway area from the conveyance.[406] The majority, however, hold to the contrary, and it should be noted that among the latter cases are some which take the position that, except where a contrary intention may appear, a conveyance by description calling for a boundary identical with highway's inner edge carries title to its centerline although the instrument makes no mention of a reference to the highway.[407] Some courts have decided that ownership extends to the centerline of a street even if stakes are placed on the sideline and called for in the conveyance.[408] Courts have also ruled that ownership extends to the centerline even if the recited area of the described parcel includes only that portion of the lot outside of the highway and including no acreage under the highway.[409]

The rule that a conveyance of land abutting upon a highway the fee of which is owned by the abutting landowners is presumed to fix the boundary in the center of the highway is *not an absolute rule of law*, but rather a principle of interpretation inferred by law, in the absence of any definite expression of intention, for the purpose of finding the true meaning of the words used.[410] The question of where the boundary is to be fixed in relation to the center of the highway depends on the intention of the grantor, which is to be ascertained from the language used in the conveyance

Deeds sold from maps are construed as passing title to the center of the street, subject to such easement as may have been legitimately obtained. *Wolff v. Veterans of Foreign Wars, Post 4715*, 74 A.2d 253, 5 N.J. 143 (1950).

Where lots are sold with reference to a plat showing a street or way intended for the benefit of adjacent lot owners, each owner acquires title in fee, not only in the lot described by his number and as bounded by the way or street, but also in one-half of the width of the way in front of his lot. *Faulkner v. Rocket*, 80 A. 380, 38 R.I. 152 (1911).

Where a plat shows the boundary of a certain lot to be a line drawn between the lot and a street, a deed describing the lot by its number of the plat does not convey title to the middle of the street. *Tingley v. City of Providence*, 8 R.I. 493 (1867).

Where lots are sold by reference to record plat showing existing or proposed streets which constitute boundaries of lots, a conveyance ordinarily operates to convey to grantee fee-simple to land underlying the adjoining streets and rights-of-way as well as others which do not bound lot conveyed. *Gagnon v. Moreau*, 225 A.2d 924, 107 N.H. 507 (1967).

[406] 11 C.J.S., § 35. A description by metes and bounds, extending to the line of the street or highway, although without express reference thereto, but referring to a plat showing such street or highway, carries title to the center thereof. *Wegge v. Madler* , 109 N.W. 223, 128 Wis. 412, 116 Am. S.R. 953.

[407] 12 Am.Jur.2d, § 39.

[408] A conveyance of a lot, describing the sides as of the length they are outside the street, passes title to the middle of the street. *Oxton v. Groves*, 68 Me. 371, 28 Am.Rep. 75 (1878).

A conveyance to or by the sideline of a public street gives rise to the rebuttable presumption that grantor intended to convey title to the center of the street unless a contrary intent is indicated; however, that presumption does not apply when the land I bounded by a private way not dedicated to public use. *Franklin Property Trust v. Foresite, Inc.*, 438 A.2d 218 (1981).

A reference to sideline of highway as boundary line is not sufficient to reserve an interest in land under the highway. *Grunwaldt v. City of Milwaukee*, 151 N.W.2d 24, 35 Wis.2d 530.

[409] 11 C.J.S., § 35. Where a survey gives the dimensions and quantity of land conveyed exclusive of the public way, it does not operate to destroy the presumption that the fee to the roadbed was conveyed; for the reason that such dimensions and quantity of the usable land is ordinarily deemed by the purchaser of paramount importance in determining its availability for the uses designated by him. *Van Winkle v. Van Winkle*, 77 N.E. 33, 184 N.Y. 193, affirming 89 N.Y.S. 26, 95 App. Div. 605.

[410] 12 Am.Jur.2d, § 39.

construed in the light of the surrounding circumstances.[411] In the case of ambiguity, the construction must favor the grantee.[412]

Application of the rule of construction, which carries title to the centerline of the highway, if the grantor owns such, will result in the grantee's taking title to the whole of the abutting highway area if the conveyance is one of parcels lying opposite each other and on both sides of the highway. But even though the grantor owns the land on both sides of the highway, the grantee ordinarily acquires title no farther than to its centerline if the premises conveyed lie wholly on one side.[413]

If the highway is established on the margin of a tract of land, and in that location lies entirely on such tract, a grantee of the tract, or part thereof bounded on the highway, takes title to the farther edge of the highway unless an intention to the contrary sufficiently appears.[414] (see examples for marginal way, and Case # 8 in Chapter 14).

OVERCOMING THE PRESUMPTION

It is an almost universal rule that if land abutting on a public way is conveyed by a description covering only the lot itself, nevertheless, the grantee takes title to the center line of the public way if the grantor owned the underlying fee, unless the contrary intention sufficiently appears from the granting instrument itself, or the circumstances surrounding the conveyance. This is where many people are misled, in that they apply the centerline presumption without investigating the title, the highway documents, or the surrounding circumstances.

The presumption that the owner of the adjoining land intended to convey his interest in the highway may be overcome, either by the use of express terms excluding it, [415] or by such facts and circumstances as show an intention to exclude it. The intent to exclude the highway must appear from the language of the deed, as explained by surrounding circumstances.[416] The New Hampshire court stated that in order to rebut the presumption, a "clear and unequivocal declaration" of a contrary intent in the deed is required.[417]

[411] 12 Am.Jur.2d, § 40.

Where it clearly appears from the whole instrument, in conjunction with surrounding circumstances, that the grantor intends to reserve the fee in the highway or street, such intent will be given effect. *Nashville v. Lawrence*, 153 Tenn. 606, 284 S.W. 882, 47 A.L.R. 1266 (1926).

[412] 11 C.J.S., § 35.

Matter of Ladue, 23 N.E. 465, 118 N.Y. 215; *Van Winkle v. Van Winkle*, supra.

Slight evidence of intention to exclude the street would have the effect of doing so. *Palmer v. Mann*, 201 N.Y.S. 525, 206 Apps. Div. 484, reversing 198 N.Y.S. 548, 120 Misc. 396, affirmed 143 N.E. 765, 237 N.Y. 616 1924).

[413] 12 Am.Jur.2d, § 42.

[414] Ibid.

[415] See Case # 6 in Chapter 14 for examples.

[416] *Mott v. Mott*, 68 N.Y. 246; Elliott, Roads & S. p. 550.

[417] *Luneau v. MacDonald,* 103 N.H. at 276, 173 A.2d at 46 (1961); *Duchesnaye v. Silva*, 118 N.H. 728, 394 A.2d 59 (1978).

It is possible to convey a parcel abutting a highway without conveying any of the fee under the highway; however, the language in the conveyance must be clear enough to leave no question about the intent. Being bounded by a road or running "along said road," does not exclude the road. "Along the east side of said road," and "side line of said road" may or may not exclude the road, depending on other circumstances and court decisions of the particular state. "Excepting the road" and "excluding the road" would seem to be clear, but may not necessarily be.[418] "Excepting the road" may have meant "excepting the rights of others to use it."

In order to except the underlying fee in a road or street there must be definite statements showing a clear intention to exclude the road.[419] Unless the deed clearly states that the road is excluded, or clearly indicates so by its language, the conveyance is to the center line of the road, provided the grantor owns the bed, or if not, to whatever extent he or she does own.

Grantor Doesn't Own It, or Part of It Clearly, if the grantor does not own the land under the roadway, he or she cannot convey it. For whatever their own reasons are, modern requirements in some towns require that, when land subdivisions are approved, they require a deed for the streets. One must be very cautious about this as to timing. If the lots are conveyed first, and the centerline presumption applies, the grantor no longer owns the underlying fee to the streets, so can convey nothing. If the grantor conveys the streets first to the governing body, then conveys the lots, the grantor does not have title to the underlying fee therefore the centerline presumption cannot apply, except where in some cases the deed is interpreted as merely an easement and not the fee.

In one instance, a grantor gave a deed to the governing municipality, who failed to record it. Subsequently, he sold all the lots, all of which deeds were recorded. Since that created overlapping titles to the street, it was a matter of law as to who had the better title to the land under the streets. Since the state in which this took place is a notice state, the underlying fee to the streets belonged to the abutting lot owners, since they were first on the public record.

Intent of the Parties to the Transaction The intent of the parties is always the controlling consideration in the interpretation of title documents. As elaborated earlier, the intent is to be determined from the language used in the document, not one of the parties' intent in general, what they may have intended,[420] or some secret, unexpressed intention.[421]

[418] A reservation "excepting the roads laid out over the land" does not exclude the bed of the street from the operation of the presumption, as the phrase is construed to a reservation of the public easement only. *Peck v. Smith*, 1 Conn. 103.

Where a farm was divided into two parts by a town road, which ran through it, a deed to the farm "reserving the town road" did not reserve the fee of the road, but only its use. *Day v. Philbrook*, 26 A. 999, 85 Me. 90 (1892).

[419] *Nashville v. Lawrence*, supra.

[420] *Wilson v. DeGenaro*, 415 A.2d 1334, 36 Conn.Sup. 200 (1979). See Chapter 12, Case # 4.

[421] *Sullivan v. Rhode Island Hospital Trust Co.*, 185 A. 148, 56 R.I. 253 (1936).

Language Excludes Street In some instances, the language used by the parties in the transaction is interpreted to exclude the underlying fee in a street or roadway. Depending on the language used, the intent of the parties and the jurisdiction, such language may or may not be sufficient to exclude the underlying fee to the street. See Chapter 14, Case #6 for an extensive analysis by the New York court as to language interpretation.

ABANDONMENT, STRICTLY SPEAKING

As previously discussed, the term "abandonment" may be misleading, even though it is commonly used. Strictly speaking, "abandonment" means the surrender, relinquishment, disclaimer, or cession of property or of rights. It is the relinquishing of all title, possession, or claim, or a virtual, intentional throwing away of property.[422]

Long-term disuse and lack of maintenance, as when a road eventually becomes impassible, may give the impression that it is abandoned and therefore vacated. In fact, it may have fallen into a different category from that when it was normally used, and it holds its status as a way. This is why it is imperative that the underlying fee title be understood.

Generally, real property and real property rights cannot be abandoned except in rare cases under special circumstances. There are, however, exceptions in that certain types of easements may be abandoned when the right circumstances exist.[423]

Railroads Railroads have been created in a variety of ways, which has resulted in a variety of estates held by railroad companies: deed (of various types), condemnation, lease. Again, examination of the acquisition document(s) is a necessity, to determine exactly what was created, and its characteristics. Estates created may be fee or easement, or variations of either. Reversion will likely depend on what was created and how. Sometimes reversionary language is included within the creating document.

In the case of *Swisher v. United States of America*,[424] the court found that the plaintiff "failed to satisfy the typicality requirement of Rule 23(a)(3)" governing class action suits, because, it stated, "class members . . . [have] cases with different conveyance language, different governing state laws and different factual bases to establish the railroad's abandonment." There are too many variations in railroad estates to establish a "class," which provides guidance for the necessity of determining exactly what a railroad acquired, and what are the associated features.

Road Reserves In some states, particularly within the original colonies but not limited thereto, proprietorships and others set aside strips of land for future

[422] *Black's Law Dictionary*.
[423] See *Phinney v. Gardner*, supra.
[424] 189 F. 638 (1999).

roads (if wanted) so that most if not all parcels would in theory have access and not be landlocked. Some of these strips were built upon, others, especially where obstacles such as rock outcrops, wetlands and steep slopes, have not. In lieu of paying monetary damages, at times there existed a provision to exchange a section

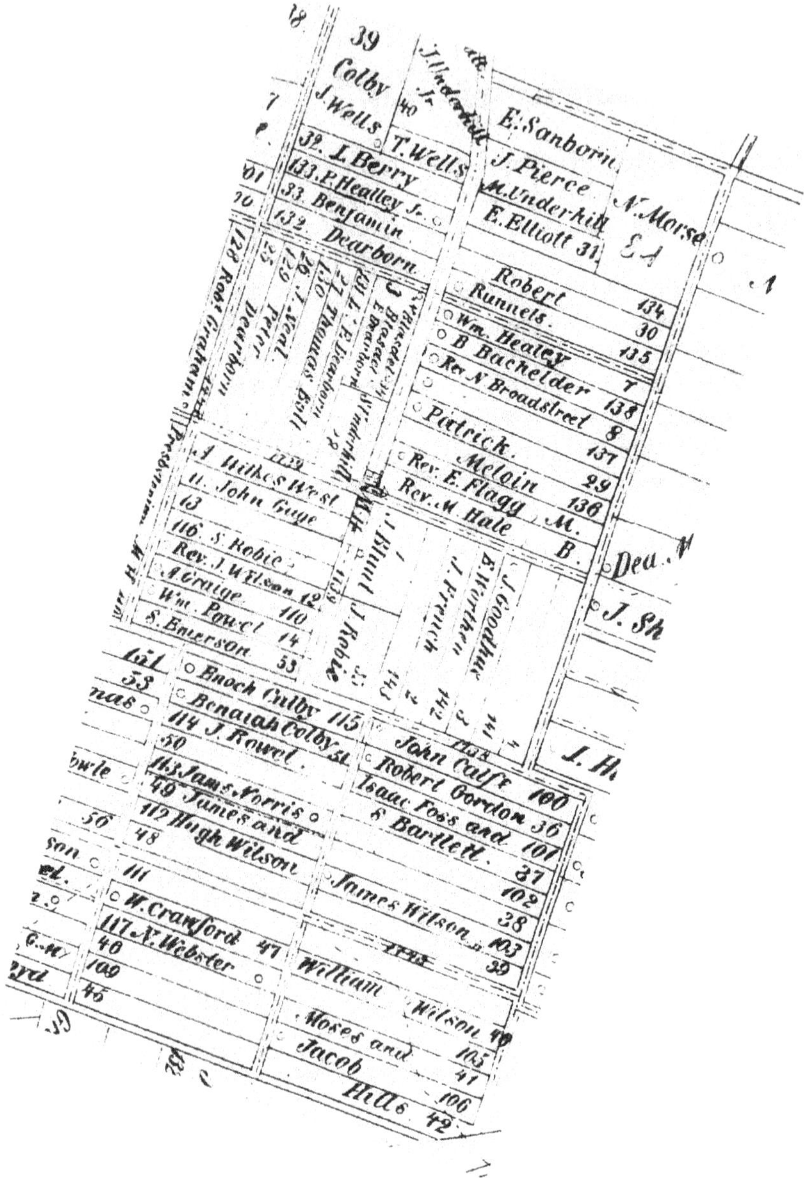

Figure 10.2 Diagram of original lotting showing road reserves, some of which have highways constructed on them.

of strip for an easement across someone's land. Satisfaction of the compensation requirement there may come as the receipt of a parcel of land rather than money. Usually this is noted in the road documents, and only occasionally might one find a recorded deed.

When a public way is located in this fashion, and subsequently vacated, it is important once again to determine the owner of the underlying fee, since in may still be the governing body, and therefore no private reversion exists. In other words, abutters may not take to the centerline, since the strip is a separate title to a parcel of land in which they have no claim. (See Figure 10.2).

CHAPTER 11

RULES OF LOCATING AND DEFINING REVERSIONS

The ownership lines in a street or highway are determined by the ownership lines as they existed before the dedication or creation of the road easement. *Brown's Boundary Control and Legal Principles* states that the problem of apportioning the limits of ownership based on the presumption that fee to the center line of the street is conveyed is sparsely found in the court records.[425] Within subdivisions, where the limits of private ownership are seldom defined, the problem becomes complex on curved and irregular streets. The following rules, partially based on customs of surveyors and a few court decisions, represent the accepted practice.

First and foremost, basic surveying rules govern the location(s) of boundary lines. The actual survey (on the ground) controls all paperwork, with very few exceptions: plats, deeds, highway layouts, and a host of additional documentary sources.

The Texas case of *Outlaw v. Gulf Oil Corporation*[426] puts the different elements embodied in a survey in perspective. It states "A map is a picture of a survey; field notes a description thereof. The survey is the substance and consists of the actual acts of the surveyor. If existing established monuments are on the ground evidencing these acts, such monuments control. Such established monuments control because same are the best evidence of what the surveyor actually did in making the survey. More, they are part at least of what the surveyor did. 9 Corpus Juris, 210." It went on to add "Insofar as the field notes govern an area conveyed, that inquiry is not what the surveyor intended to do, but, if it can be determined, what he actually did, and cited *Blackwell v. Coleman County*, and *Blake v. Pure Oil Co.,* 128 Tex. 536, 100 S.W.2d 1009.

[425] Robillard, Wilson, and Brown, 6th edition, 2009, Chapter 8.

[426] 137 S.W.2d 787 (Tex.Civ.App., 1940).

The Maine case of *Stetson v. Adams*[427] also summarizes this nicely. The judge in this case instructed the jury, "A man may have a lot of land and lay it out with a surveyor, running lines, laying out streets, putting down stakes, making roads and parks, and then afterwards endeavor to make a plan of them from what has been done. It is a picture of what has been done. Now, the plan is the picture; it is the guide; it is the finger post. But what controls is what was done really upon the surface of the earth—the lines as they were run; the stakes as they were put down. The bounds as they were made are the controlling bounds, even though they may vary from the plan, and even though the plan may [not] have been accurate in locating them. Where the boundaries are first—especially where the lines are first run—there they stay."

BASIC RULE

The basic rule is that, unless a deed or map indicates otherwise, reversion rights extend from the street termini of the property lines to the center line of the street in a direction that is at right angles to the center line of the street (see Figure 11.1). Direct prolongation of the lot line to the center line of the street is not a generally accepted method, although some states adhere to that rule. In Kansas[428] and Oklahoma[429] division is made the same as for accretion to rivers and is by proportion. Figure 11.2 illustrates the problems that arise when straight prolongation of lot lines is attempted (see Figure 11.2).

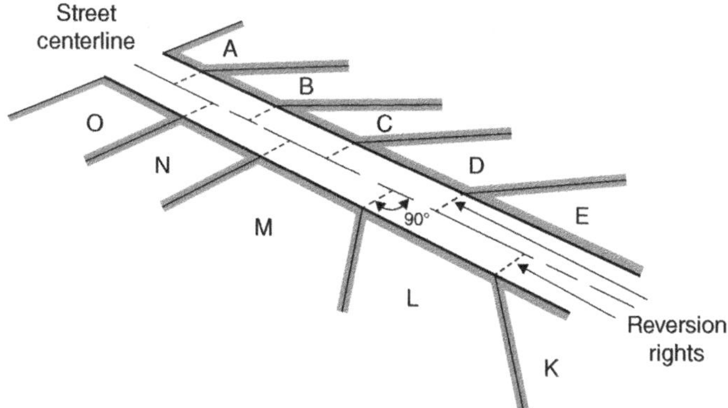

Figure 11.1 Reversion rights, unless otherwise indicated, are at right angles to the center line of the street.

[427] 91 Me. 178, 39A. 575 (1898).

[428] *Showalter v. Southern Kansas Ry. Co.*, 49 Kan. 421, 32 P. 42.

[429] *Blackwell, SW Ry. Co. v. Gist*, 18 Okla. 516, 90 P. 889 (1907).

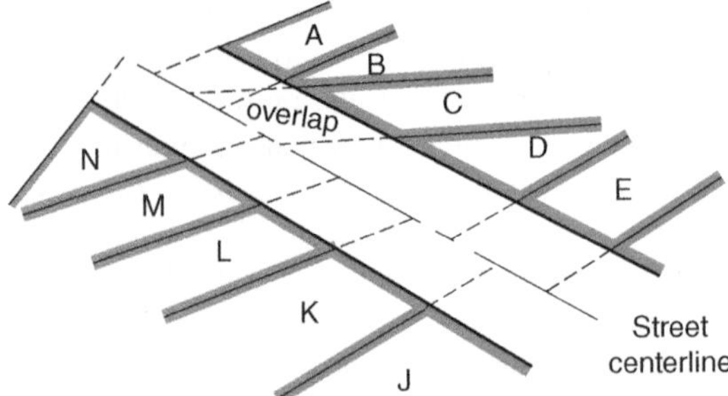

Figure 11.2 The alternative solution, of prolonging property lines to define reversion rights. Problems of overlaps and gaps are likely to occur.

CURVED STREET

Reversion rights on a curve extend radially to the center line of the street. Radial lines are at right angles to a tangent to a curve (see Figure 11.3). This principle may be considered a special application of the former principle. The following example illustrates the more common application of the principle.

STREET INTERSECTION

Extending lot lines at right angles to the street with ownership to the center of the street results in following illustration of reversion rights at a street intersection (see Figure 11.4).

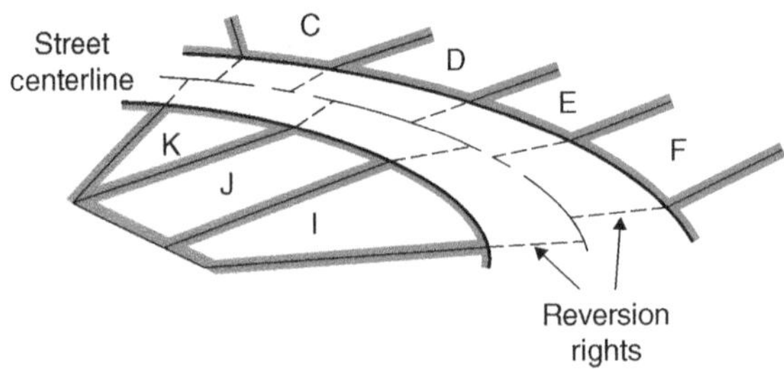

Figure 11.3 Reversion rights on a curved street extend along radial lines to the center line of the street.

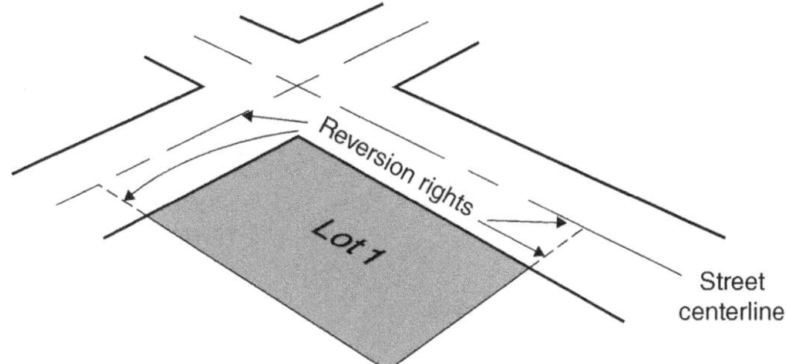

Figure 11.4 Reversion rights at a street intersection.

OWNERSHIP AT INTERSECTION WITH REVERSION ONLY AT ONE STREET

The foregoing example assumes that there are reversion rights in *both* streets, which would likely be the usual case. The following example illustrates the procedure when only one of the two streets is subject to reversion rights (see Figure 11.5).

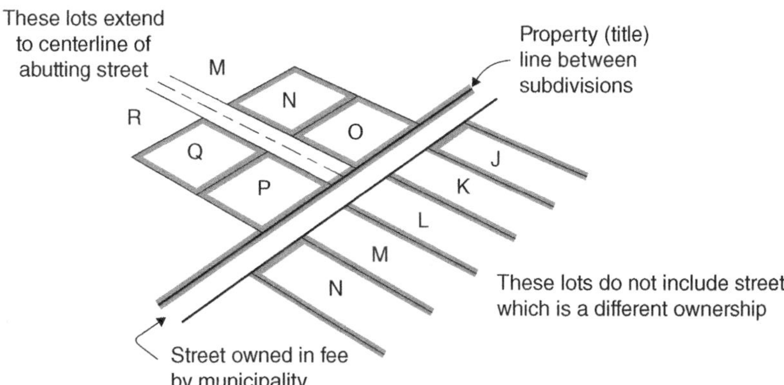

Figure 11.5 Reversion rights where one street is subject to the rights, but not the other.

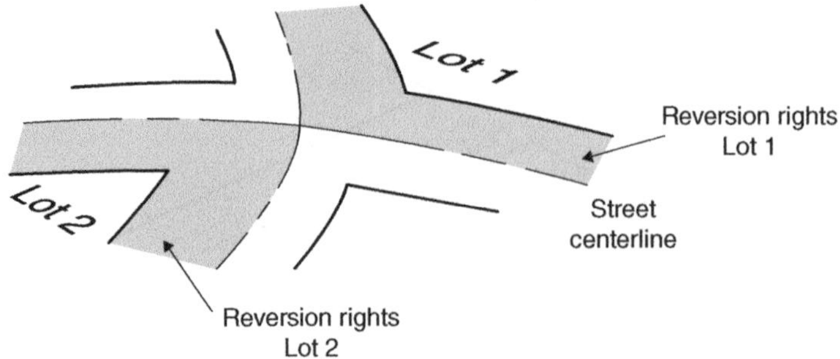

Figure 11.6 Reversion rights at a curved intersection.

CURVED STREET INTERSECTION

At a curved street intersection reversion rights also extend to the center line as determined by a curved line. This curve may be simple, compound, reverse, or spiral. In any case, ownership extends to the center line of the street or highway (See Figure 11.6).

LOTS AT AN ANGLE POINT IN THE ROAD

Reversion rights are defined by the bisecting line in the following example, where there is an angle point in the road. Lot H will receive a greater amount of land than Lot D, even though both lots have equal amounts of road frontage (See Figure 11.7).

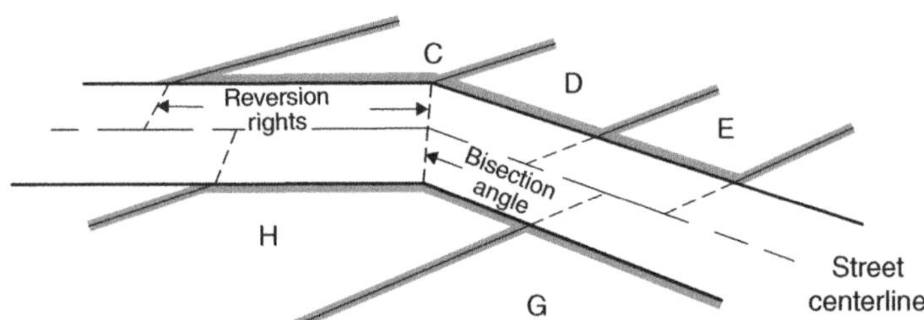

Figure 11.7 Reversion rights at an angle point in a road.

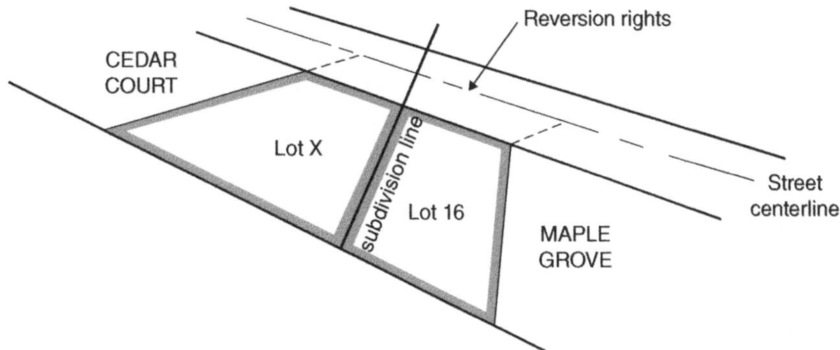

Figure 11.8 Example of reversion rights involving a subdivision boundary.

LOTS ADJOINING A SUBDIVISION BOUNDARY

Ownership to the center of the street takes place *when the grantor owned that far*. The following two examples are based on situations where such was *not* the case.

In the first example, Lot X in Cedar Court and Lot 16 of Maple Grove are in different subdivisions, the heavy line representing the boundary between the parent parcels. Both parent parcels were created after the road, so therefore ownership of them extends only to the center line (See Figure 11.8).

In the next example, the road cut across the parent parcel and therefore lines of lots created by later subdivision could only extend as far as the grantor's ownership. That ownership did not extend as far as the center line for some lots but extended beyond the center line for others (See Figure 11.9).

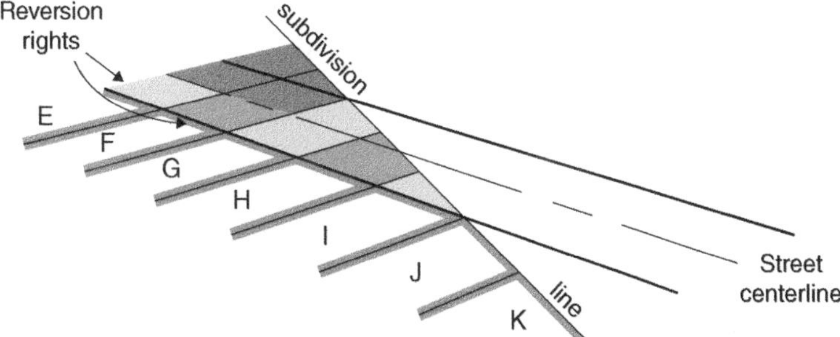

Figure 11.9 Reversion rights of lots along a subdivision boundary which is intersected diagonally by a highway.

MARGINAL ROAD

In the case of a marginal highway, ownership is presumed to extend across the entire highway (See Figure 11.10).[430] Also see example 8 in Chapter 14.

SPECIAL CASES

Courts have decided a number of cases dealing with special circumstances, and state decisions must be reviewed to determine the courts feeling in these cases. For example, in New Hampshire, the court decided the following (See Figure 11.11):

> N.H. 1971. Defendants in quiet title action who held deeds to property bounded by the edge of road which ran between property and the ocean, and not be the edge of the ocean, gained title to the land extending to the ocean when town officially discontinued the public road, and master's report concluding that plaintiff, an adjacent landowner who produced recent quitclaim deed containing vague language which might have referred to the property, had no title to any land between the ocean and former boundary of defendants' property was not inconsistent or against the weight of the evidence. [431]

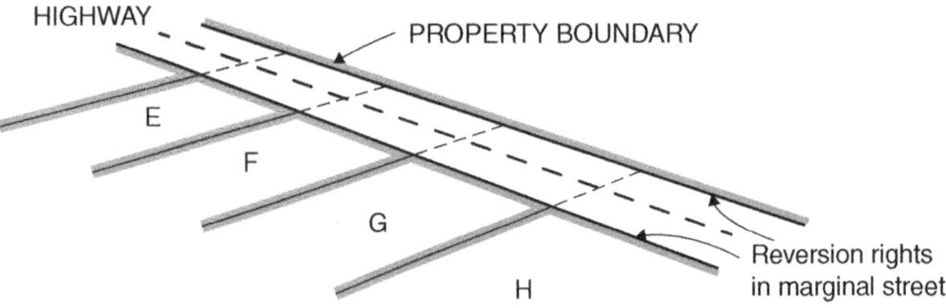

Figure 11.10 Reversion rights on a marginal highway extend across the entire highway.

[430] Owner of lot adjacent to marginal highway is presumed to take a naked fee to entire bed of the highway. *Brighton Const., Inc. v. L & J Enterprises, Inc.*, 296 A.2d 335, 121 N.J. Super. 152 (1972), affirmed 313 A.2d 617, 126 N.J. Super. 186 (1974).

12 Am.Jur.2d, § 42.

[431] *Sheris v. Morton*, 276 A.2d 813, 111 N.H. 66, certiorari denied 92 S. Ct. 727, 404 U.S. 1046, 30 L. Ed.2d 735.

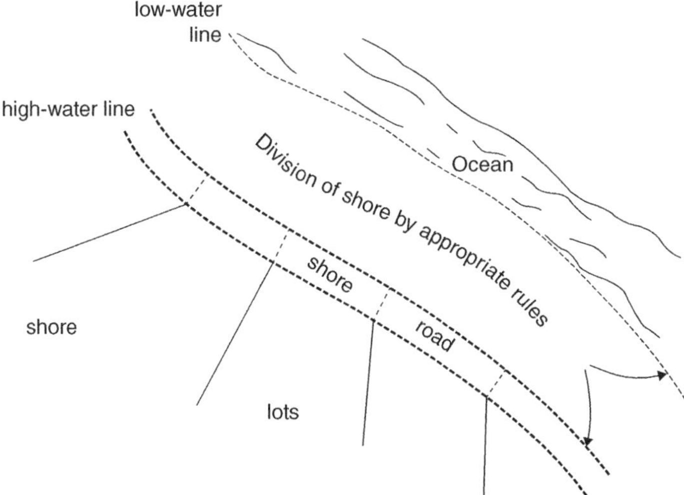

Figure 11.11 Marginal road on a shoreline illustrating division of private ownership.

This case results in two, perhaps even three, sets of conditions in the case of reversion. The road itself would need to be divided, either as a whole strip, or one set of rules applied to the near side of the center line and another to the area on the far side. Then an additional set of rules is needed for the division of the shore when ownership extends to low-water mark.

The case of *Taylor v. Haffner*[432] presents another twist to the presumption wherein the court noted that since a railroad was a special entity, it would not be the beneficiary of the one-half of vacated highway. See Case Study #14 in Chapter 14.

The marginal highway rule was apparently not considered in this case.

PROBLEM CASES

Sometimes odd circumstances lead to problematical situations. Following are two examples presented in *Land Survey Descriptions*:[433]

A single right angle street can also be a problem depending on the circumstances (See Figure 11.12).

A dead-end street is a worse situation. Ordinarily, Lots B and D would share the land under the street, each to the center line. If that were applied here,

[432] 114 P.3d 191 (Kan.App. 2005).

[433] William C. Wattles, Rev. and Published by Gurdon H. Wattles. 10th Edition. 1974.

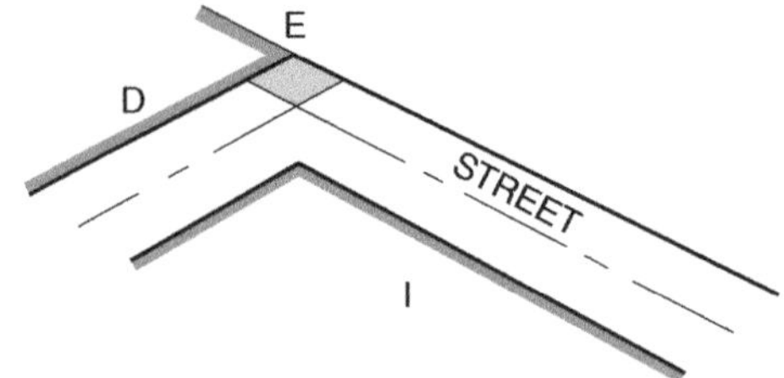

Figure 11.12 Area of reversion on a right angle street.

Lot C would not get any part of the street and may have a problem of access (See Figure 11.13).[434]

Road not in right of way. Not at all uncommon is the situation where the actual traveled way is found to be either wholly or partially outside of the easement (See Figure 11.14). Some courts have addressed this problem,[435] and it may be necessary to seek court determination in each case.[436] If sufficient time has passed as required by statute for the creation of a prescriptive easement, then the public may have acquired rights to the traveled portion through use, thereby having title to all of it, the unused portion being part of the easement through the original establishment. In the case where sufficient time has not passed, the public may not acquired title to the traveled portion; therefore, full interest would remain in private parties. If this is the case, the governing body would need to be approached about paying damages, a land exchange, or some other means whereby the discrepancy is resolved.

Some courts have stated that it is the road as actually laid out that controls, rather than its record layout.[437] See the referenced decisions, and Case Study #9.

[434] The Massachusetts courts have addressed this problem. Landowner at the end of a way cannot acquire any fee interest in the way without encroaching on the property rights, if any, of the abutting side owners. *Emery v. Crowley*, 359 N.E.2d 1256, 371 Mass. 489 (1976).

Term "abutting," in context of fee ownership of ways after conveyance of property bounded on a way, refers to property with frontage along the length of a way. M.C.L.A. c. 183, §§ 58, 58(a)(i,ii), *Emery v. Crowley*, Id.

[435] A deed bounding the land conveyed as on the west line of a highway, where the highway was built four rods outside of the recorded location, conveys land to the limit of the road as laid out and actually used for travel. *Brooks v. Morrill*, 42 A. 357, 92 Me. 172 (1898).

[436] 11 C.J.S., § 36. When a highway or street is referred to in a grant or other conveyance as a boundary, the way as opened and actually used is construed or understood to be the boundary, rather than the way as formerly existing, or as existing of record, or as surveyed for a proposed change, or, in the absence of any evidence to show a different intention, as platted. *Johnson v. Palmetto*, 77 S.E. 807, 139 Ga. 556 (1913); *Rosenblath v. Marabella*, 3 La.App. 584; *Davis v. Monroe*, Tex. Civ. App. 289 S.W. 460 (1926); *Southern Iron Works v. Central of Georgia R. Co.*, 31 So. 723, 131 Ala. 649 (1901); *Winchester v. Payne*, 102 P. 531, 10 Cal. App. 501; *Hill v. Taylor*, 4 N.E.2d 1008.

[437] When a street is referred to in a deed as a boundary, the street as opened and actually, rather than its record lay-out, is the boundary intended. *Falls Village Water Power Co. v. Tibbetts*, 31 Conn. 165; *Sproul v. Foye*, 55 Me. 162 (1867); *Brown v. Heard*, 27 A. 182, 85 Me. 294 (1893); *O'Brien v. King*, 7 A. 34, 49 N.J.Law, 79; *Aldrich v. Billings*, 14 R.I. 233.

Brooks v. Morrill, 42 A. 357, 92 Me. 172, supra.

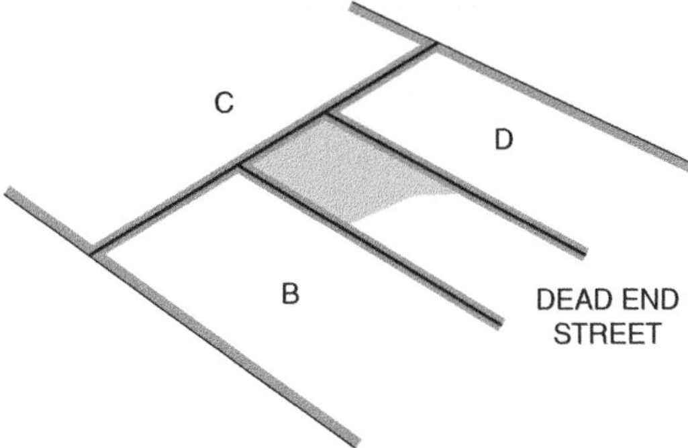

Figure 11.13 Area of reversion on a dead-end street.

The Maine case of *Sproul v. Foye*[438] illustrates how the court decided the location of a highway constructed well outside of its intended layout. The argument was between the two owners on opposite sides of a road, both of which had descriptions calling for the road. One wished to honor the layout, the other the constructed way. The court stated that a layout for a road is not a road, and therefore it is the constructed location that governs. (Refer to Case Study # 9 for the diagram).

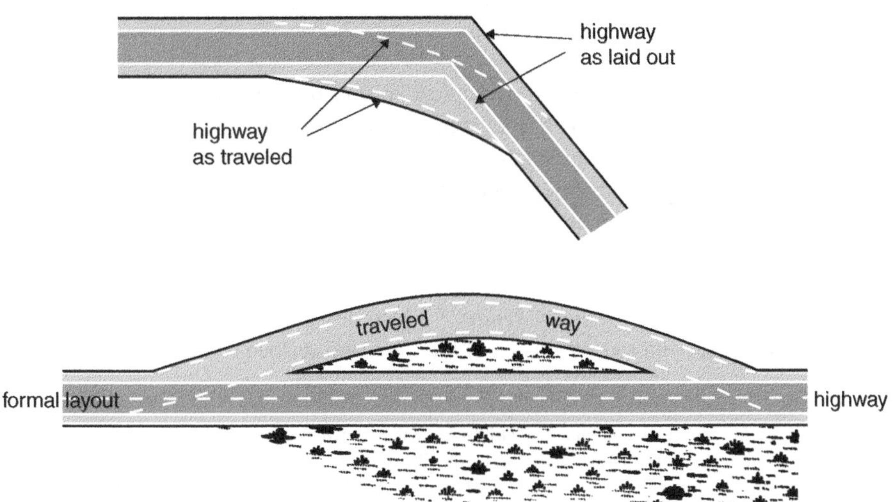

Figure 11.14 Examples of traveled way falling outside of the easement.

[438] 55 Me. 162 (1867)

DOCUMENTS INDEFINITE OR NOT AVAILABLE

Many documents are not definite or detailed in easement location or in definition. Frequently, roads lack a good description of layout or fail to define the width. This particularly true of the older layouts:

> "laid out a highway, beginning at a white oak marked H.I.B., and 4 rods in breadth along by the head of Joseph Bunker his land from thence to the King's thorofair Road. Laid out this 9th of April 1703 by us.
>
> **—Jno Tuttle, Jere Busnam, James Davis, of the Committee"**

> "We have this day Laid out a highway on Joseph Halls land, Beginning at Thomas Merrills southwest Corner and so Running Westerly as the fence now Stands as far as Hall's house it contains about twenty square rods of land."
>
> **—Dec. 15, 1788."**

The State of Maine recognized this problem, and in 1971 enacted the following statute:

> *§2103. Lost or unrecorded boundaries.*
>
> When a highway survey has not been properly recorded, preserved or the termination and boundaries cannot be ascertained, the board of selectmen or municipal officers of any municipality may use and control for highway purposes 1 ½ rods on each side of the center of the traveled portion of such way.

It would seem that this is the most logical approach to this type of problem. One cannot do more than act on the best evidence that is available.[439]

However, professional judgment must be exercised in research of any kind, as records can sometimes exist but be located in unconventional places. How many places to search and for how long are often questions that must be addressed. Experience tells us that most people give up to quickly or too easily; still others are lacking in the necessary expertise to be successful at difficult research projects.

The Illinois case of *Klose v. Mende*,[440] illustrates the type of problem that can sometimes be encountered. In this case, the township wished to improve a road and

[439] In absence of evidence to contrary, the earliest known center line of traveled highway is considered center line of highway right-of-way, and when original survey of such road runs single line such line is presumed to be center line of highway. *Clark v. State*, 246 N.Y.S.2d 53, 41 Misc.2d 714 (1963).

Where the width of a way is not specifically defined in the grant or reservation, the width is such as is reasonably convenient and necessary for the purposes for which the way was created. In such a case, the determination of width becomes basically a matter of the construction of the instrument in the light of the facts and circumstances existing at its date and affecting the property, the intention of the parties being the object of the inquiry. 25 Am.Jur.2d, § 78.

[440] 378 Ill.App.3d 942, 882 N.E.2d 703 (Ill.App. 3 Dist. 2008)

informed one of the abutting owners, who stated that they owned the road in fee by warranty deed. The road commissioner produced a statutory dedication from the municipal records, whereupon the parties brought an action. The court found that the dedication failed because the Illinois law required additional records, specifically a petition, a surveyor's report, a survey and plat, as well as the road commission's order granting the dedication. Although they had searched for these items, they failed to find them among the available records.

Two years later, during a cleaning of the municipal offices, records were found in a closet that was previously inaccessible. Therein were the necessary documents to comply with the statutory requirements. This time, the court found that there was indeed a highway, whereupon the property owners brought an action for negligence, stating that the persons in charge had not acted properly in fulfilling their duties. The court found otherwise, but the dissenting opinion in the case stated that municipal officers have a duty to maintain careful records and make them available upon request.

There are many examples of the failure to produce records, and sometimes persistence and perseverance are what is necessary to acquire the necessary information. Too many people, meeting with resistance, or letting time or monetary constraints dictate, fail, and resort to guesswork. As pointed out earlier, it is far too easy to be wrong, and therefore do more harm than good, or something contrary to the law. This case demonstrates needless long-term frustration and expense, because proper care was not taken by the responsible parties.

This is a situation, not unfamiliar to many researchers, where the authoritative body cannot, or will not, produce the necessary or required documents. It is particularly frustrating for a title examiner or a surveyor, since everything may be dependent on the creating documentation and without it one cannot know the character of highway, its extent, its location, its width, or the ownerships burdened by it.

SUMMARY OF PROCEDURE FOR DETERMINING REVERSION RIGHTS IN VACATED HIGHWAYS

1. First, procure the original documents creating the highway and determine if layout is fee, easement, or some other lesser estate. These may or may not contain language as to what reversion, if any, is to take place in the case of termination of an easement. The information should locate the easement and specify its dimensions, at least its width, although sometimes all of this information is lacking. The description should also specify the landowners burdened at the time, at least a list of owners to whom damages were paid should be included. The document should also spell out any particular characteristics, such as term of years, restrictions on use, and the like.

2. Once the creation document(s) have been obtained, next obtain a copy of the law providing for such creation at that point in time, whether by statutory layout, prescription, dedication, or whatever means. Compare the requirements of the statute with the creation of the highway to ensure completeness and correctness.

Some courts have ruled that if statutory requirements have not been met, carefully and completely, the road procedure fails, that is, there is not a "legal" road.[441]

3. Next, determine the boundaries and ownerships affected by the layout at the time. Deeds and related title documents must be assembled for this point in time or earlier to know which properties are affected and what their boundaries are. A title document for each affected ownership should be obtained as documentation of the existence of the affected party.

4. Research the records to determine if the layout has been altered in any way. Alterations to be on alert for are relocation of the right of way, particularly if obstacles were encountered at the time of actual layout, survey or construction, or in the form of a widening or straightening, or combination thereof.

5. Ownerships must then be traced forward to date and compared with present ownerships for accuracy and continuity. It is necessary to determine if any parcels have been created through the subdivision of original parcels or if there have been any changes in original property lines. There is one set of conditions for the time of creation, another for present day. Even though they are seldom the same, they must be related to one another.

6. Review the discontinuance/vacating document(s). There may be special considerations, a partial or temporary discontinuance, or particular wording that would affect the ultimate result. Also be certain that the statutory requirements were satisfied; otherwise, the discontinuance may fail.

7. Apply the appropriate rules. In the case of problem situations, it may be necessary to involve all potentially affected owners, or failing that, a petition to the court for a determination, such as an action to quiet title to a particular parcel or land, may be necessary. Remember that the placing of an easement on, under, over, or through a parcel of land, by itself, does not change the parcel boundaries. It only burdens the property it affects.

[441] One must keep in mind that even if the formal creation of the road fails, and there physically a road in existence, over time it may have become public but created in a different manner, for example, dedication, prescription, etc.

CHAPTER 12

EASEMENTS AND THE LAND SURVEYOR

Boundaries of easements are boundaries separating estates in land, the dominant estate from the servient estate. Since easements are viewed, and treated, like real property, from a surveying standpoint, the same rules apply to easements as apply to the survey of a parcel of land. Consequently, the same standards apply. The land surveyor would be well advised to carefully read and understand the *minimum*[442] *standards for surveying land* in whatever state they are practicing in.

ALTA/ACSM STANDARDS[443]

The minimum standards for Land Title Surveys according to the American Land Title Association and the American Congress on Surveying and Mapping, jointly, include the following items relative to easements:

- *Adhering to the normal standard of care.* "Surveyors should recognize that there may be unwritten local, state, and/or regional standards of care defined by the practice of the 'prudent surveyor' in those locales." (Item 3C)
- *Following appropriate boundary law principles* for the determination of boundary lines and corners. (Item 3D)
- Corners referenced to the *nearest right of way line.* (Item 5Bi)
- *Determining width or location* of any abutting, street, road or highway right of way. (Item 5Bvi)

[442] The key word here is *minimum.* That is the least that should ever be done; in many cases the nature of the project, or problem, requires more, sometimes considerably more, effort than what would be involved as a minimum. The standard is what the ordinary, prudent individual would do in like circumstances.
[443] Minimum Standard Detail Requirements For ALTA/ACSM Land Title Surveys. Adopted 2010, Effective February 23, 2011.

- Disclosing evidence of "any *easements or servitudes* burdening the surveyed property." (Item 5Ei)
- Locating *cemeteries, gravesites, and burial grounds.* (Item 5F)
- Determining *width of abutting ways* and the source of such information. (Item 6Civ)

RIGHT OF WAY AS BOUNDARY LINE

A valuable lesson may be learned from the 1992 Ohio case of *Manufacturer's National Bank of Detroit v. Erie County Road Commission.*[444] This case had to do with the location of a highway right-of-way line wherein there was a trespass into the right of way by a cornfield, thereby obscuring the line of sight down an intersecting highway. As a result, there was a collision of two vehicles resulting in death. The landowner claimed that requiring a land user to determine exactly where right-of-way lines are located places too great a burden on the land user. The court stated that a "border of right-of-way is boundary line" like any other. The court continued, quoting a long-established standard, "a landowner or occupier is under an obligation to know the boundaries of the property." Consequently, when easements are involved in a survey, they should be approached the same as any other survey or boundary location. That is, they are either equivalent to the survey of a subject parcel or of an abutter, depending on their character or relationship to the land in question.

RETRACEMENT OF RIGHT OF WAY LINE

The retracement standard of "following the footsteps of the original surveyor" was applied in the case of *Johnson v. Westrick.*[445] In this case, the city informed a landowner that his fence was obstructing another street running through the plat. The landowner testified that he had erected the fence according to the original stakes marking the right of way line. The court held "that the east line of the street was where the original surveyor placed it, not where it should be according to resurveys or subsequent surveys; that subsequent surveys are worse than useless; they only serve to confuse, unless they agree with the original survey."

This decision equates the easement surveying standard with the land surveying standard and emphasizes what a correct location should be. This is an important decision with wide import, as often city officials and others, including surveyors, attempt to make an easement its reported width, regardless of where it was originally located. Doing so ignores the very basic principles of retracement as well as well-established rules of deed construction. Original surveys are the prime consideration, and when a conflict is encountered, such as a fence versus a stated width, the basic rule of monuments taking priority over distances should, in most cases, be an obvious choice.

[444] 63 Ohio St.3d 318; 587 N.E.2d 819 (1992).
[445] 85 Wis. 80, 55 N.W. 177 (1893).

RETRACTMENT OF ORIGINAL SURVEY OF HIGHWAY

In the California case of *People v. Covell*,[446] the court dealt with a highway survey of 1872 done by the county surveyor and approved by the Board of Supervisors. A later survey was done by a deputy county surveyor, who concluded that one of the abutting parcels encroached anywhere from 6 to 10 feet along the line of the highway. The court stated, in a previous case, "it was sought by a later survey to move the lines established by a prior survey. This court held that whether accurate or inaccurate, the original survey granting and establishing certain rights, fixed the rights not only of the government, but of the landowners, and that the government, after establishing such a line and granting and conveying certain rights, possessed no power thereafter to change the course of that line."

The court continued, "That principle is applicable here. When the [first] survey was made and the report thereof giving all the data necessary to establish the survey showing the lines followed, the government corners found, and such data as ordinarily appears in report of a surveyor in laying out a public highway, and that report being adopted by the board of supervisors, and the road laid out by [the first surveyor], that highway so surveyed by [the first surveyor] became fixed, and so far as rights of the county are concerned and the rights of landowners adjacent thereto are affected, no change thereafter could be made. The rights of the respective parties, and the lines of the highway were limited and fixed by the [first] survey."

> Practically all of the cases herein referred to support the principle that in making a resurvey to determine the boundaries of lands which have been granted, the tracks of the original survey should be followed, so far as it is possible to discover the same, and locate upon the grounds the lines of the lands patented as nearly as may be according to the original survey thereof and in reference to which lands have been sold.

EASEMENT PLANS ARE LAND SURVEYS

A short time ago the Connecticut Board of Professional Engineers and Land Surveyors examined a case involving an architect who prepared some "easement plans" required by a planning and zoning commission when the architect's client was applying for site approval. The easement plan prepared by the architect was based on a survey previously prepared by a licensed land surveyor, in which the architect "utilized (the surveyor's) survey to locate perimeter boundaries, information, and three easements." The Connecticut Board found against the architect in that (1) preparation of an easement plan constitutes land surveying, and (2) a licensed architect may not prepare an easement plan without having a license to practice surveying. Although Connecticut statutes allow architects to perform work incidental to their profession, the board's opinion was that the map was not a site plan within the scope of the practice of architecture . . . The work product he produced was solely within the practice of land surveying and not incidental to some architectural work he was performing on the development.

[446] 62 P.2d 602; 17 Cal.App.2d 627 (1936).

LIABILITY OF THE LAND SURVEYOR

In 1986, the Louisiana Appellate court decided that a surveyor and a lawyer were both liable to their client for failing to discover a right of way adjoining an existing highway.[447] The lawyer was retained to examine the title for the client who wished to build a convenience store. He reviewed a book published by a local abstract company which listed property conveyances, and issued a written title opinion that the client had "good and valid title" based on his examination of local public records for "at least the past sixty years." The lawyer then hired a survey firm to survey the property and mark the boundaries. The survey plat did not indicate any easements for rights of way.

On the basis of the foregoing opinions, the client then began construction of his store according to the boundaries shown on the survey plat. Before completion of the building, the state Department of Highways informed him that the State had a 50 foot right of way on each side of the highway. The building encroached on the easement, and a 9 foot by 12 foot section of the building had to be removed.

The client sued both the lawyer and the surveyor for his costs, claiming negligence for failing to discover the existence of the right of way. The lower court dismissed the claim against the lawyer, but found the surveyor negligent, and stated that the client himself was partially liable. On appeal, both the lawyer and the surveyor were found liable. They stated that the title opinion letter was false, and that if the lawyer had actually examined the records, he would have discovered the right of way. The court also noted that "professional services" upon which others rely must be rendered in accord with reasonable professional standards. Less than reasonable skill will result in liability.

EASEMENTS ARE SIMILAR TO OTHER LAND

The only differences between an easement on, over, or under a fee-simple parcel of land is that the easement rests in another person, persons, or entity, and is a lesser estate. It exists as a burden on the underlying fee or servient parcel, and in many respects is treated exactly like fee-simple ownership: it has a title, boundaries, description, measurements, location, and extent. Since *boundaries* and *location* are involved, easements fall under the purview and responsibility of the land surveyor.[448]

Therefore any easement, of whatever character, and however situated, should be treated like any other parcel of land. Researchers would be well advised to be familiar with the professional standards governing their work and take care to act in accordance therewith.

[447] *Crawford v. Gray and Associates*, 493 So.2d 734 (1986).
[448] *Rivers v. Lozeau*, 539 So.2d 1147 (Fla.App. 5 Dist. 1989). "Although title attorneys and others who regularly work with them develop expertise as to land descriptions, the only professional authorized to locate land lines on the ground is a registered land surveyor." Citing § 472.005(3), Fla.Stat.

CHAPTER 13

EASEMENTS AND THE TITLE EXAMINER (OR RECORDS RESEARCHER)

While the foregoing ACSM/ALTA title standards apply to examinations for title insurance purposes, there are (1) additional considerations than aforementioned, and (2) standards that are state specific. Some examples follow.

Most title policies and title opinions carry standard disclaimers in their form that the policy, or opinion, is subject to the following items.

ITEMS OUTSIDE THE PERIOD OF SEARCH

State standards generally require a search of the records for a limited period. Referring back to the discussion in the case of *Ski Roundtop v. Wagerman*, the Maryland court noted that the 60-year period as practiced in Maryland may not be sufficient, and held to a standard of searching back to an original conveyance from the sovereign.

Most title companies in the western portions of the United States routinely take title examinations back to the patent from the U.S. government.

ITEMS NOT ON THE PUBLIC RECORD AT THE COURT HOUSE

Items not of public record are usually not searched, nor are they generally considered. However, some state standards have a provision for searching these items if one is on notice of them. Many items are a matter of public record although not at the court house. Such items may be found at state or federal agencies, or among municipal records. Most of these are available for inspection during regular hours.

ITEMS TO BE SHOWN BY AN ACCURATE SURVEY

This presumes there is, or will be, a survey, in order to afford adequate protection. A survey, in theory, will uncover such things as encroachments, undisclosed burdens on the property, evidence of possible easement use, and so forth, is visual items not available to the searcher of the records.

IMPLIED DEDICATION AND/OR ACCEPTANCE

Some policies will include this item. Since there is no direct record of such activity, there is no way for a searcher of the records to determine either one. Only with a search of other records at other locations would any researcher uncover evidence of such activity, or possibility.

WHAT INSURANCE DOES NOT COVER

Both the standard owner's policy and the standard lender's policy are based on an examination of public records of the recording district in which the land is located. Neither policy ensures against matters that could be disclosed by actual inspection or survey of the property. These policies do not ensure against certain matters not shown by the public records such as unrecorded easements, liens or money obligations, unrecorded utility rights of way, public or private roads, community driveways and other types of encumbrances, or against the rights or claims of persons in possession of the property that are not shown by the public records. However, some of these may be covered under an "extended coverage" policy.

The suggestion is made, and sometimes at least implied by the court system, that the title examiner, or researcher, be *intimately* familiar with the state standards for title examination. Keep in mind that *standards* are only a minimum; in many cases, the prudent individual would expand upon the minimum set forth. There are many instances where considerably more investigation is necessary, depending on the circumstances. As examples the following states have addressed certain aspects. Knowledge of what is contained in the applicable Marketable Title Act for the particular jurisdiction is an absolute necessity.

The Ohio standards make note of the length of the period of search:

> A period of examination made pursuant to the Ohio Marketable Title Act, R.C. §5301.47, et seq., shall be sufficient. [This standard was suspended by the Council of Delegates, Ohio State Bar Association, effective November 15, 1986.] (*Ohio State Bar Association Report*, Vol. 59, No. 48, Section One, December 15, 1986, p. 1968.) The Council authorized the following Comment in lieu of the Standard: "There is nothing in the Ohio Marketable Title Act that entitles a title examiner to rely upon a simple forty year search period. He or she must be aware of the several exemptions in the Act that are not barred by the mere passage of 40 years. [Also, see *Heifler v. Bradford*, 4 O.S. 3d 49, 446 N.E. 2d 440 (1983)]

[The Governors of the Real Property Section, Ohio State Bar Association, issued the following observation in its *Report of the Real Property Section (Ohio State Bar Association Report*, Vol. 59, No. 41, October 27, 1986, p. 1668):

> This standard is commonly misread by title examiners. In fact the Marketable Title Act does not present any length of time for a search period. Since the Marketable Act does have several provisions which become operative over a period of 40 years, however, it has been commonly misconstrued as providing for such a search period. This misconstruction has caused the standard to be more misleading than helpful, and its suspension, while seeking to develop a more accurate standard for this purpose, is recommended.

The title standards for South Dakota include this note:

9-01. EASEMENTS AS ENCUMBRANCES.

Easements, including those reflected on plats, servitudes or non-appurtenant restrictions on the use of real property should be noted as encumbrances.

The Vermont title standards include the following:

The "Chain of Title' concept is a principle of common law, developed to protect subsequent parties from being charged with constructive notice of the contents of those recorded instruments which a title searcher would not be expected to discover by the customary search of the general grantor-grantee indices and other appropriate indices and diligent inquiry of the Town Clerk[449] as to matters left for recording, but not indexed. Notwithstanding the holding of *Haner v. Bruce* (146 Vt. 262), it is not reasonable or customary to examine the indices of the individual record books, where a general index is maintained. This concept limits the duties and liability of a title examiner, to a search of those documents which appear not only in the appropriate land record

[449] Vermont is one of several states where records are kept by the Town Clerk for the town in which the property lies, rather than at the County Clerk's Office, as in most states.

indices, but also in the chain of title to the particular parcel being searched. A subsequent party in the chain of title will not be charged with notice of an instrument which is outside the chain of title.

Comment 1.

The term 'recorded instruments' includes, but is not limited to, deeds, leases, decrees, liens, judgments, maps, documents imposing covenants, restrictions or *easements* on property, agreements adjusting boundaries and *all other documents by which an interest in real property may be transferred or claimed.* [emphasis by author].

Comment 8.

There is an additional circumstance which the title examiner must consider. It is derived from the rule of law announced in the line of cases that includes *Clearwater Realty Company v. Bouchard,* 146 Vt. 359 (1985), *Crabbe and Sweeney v. Veve Associates,* 150 Vt. 53 (1988), and *Lalonde v. Renaud,* 157 Vt. 281 (1989) and the applicable provisions of the Vermont Marketable Title Act. The rule of law in the *Clearwater* line of cases may be stated concisely as—rights of way, easements, and the designation of areas as common space on a recorded plan used as the basis of the description in connection with the conveyance of one or more of the lots shown on the plan vests rights in the grantee and the grantee's successors in title rights in those areas designated on the plan as rights of way, easements, and common space. In deciding the Clearwater line of cases, the issue of the provisions of the Marketable Title Act has not arisen. The provisions of 27 V.S.A. 604 exempt easements granted, reserved or retained in a deed from the provisions of the Marketable Title Act that would otherwise extinguish such rights, and therefore the *rights of way shown on very old plans that are outside the chain of title may still be encumbrances on the title.* [emphasis by author].

Notice. Notice is knowledge of facts which would naturally lead an honest and prudent person to make inquiry of everything which such inquiry pursued in good faith would disclose. Notice is *actual* or *constructive*. Actual notice is notice expressly and actually given, and brought home to the party directly. It may be either *express* or *implied*. Express notice is considered to be that which includes all knowledge of a degree above that which depends upon collateral inference, or which imposes upon the party the further duty of inquiry. Implied notice imputes knowledge to the party because he or she is shown to be conscious of having the means of knowledge. Constructive notice is information or knowledge of a fact imputed by law to a person (although he or she may not actually have it) because they could have discovered the fact by proper diligence, and his situation was such as to cast upon them the duty of inquiring into it.

Constructive notice is a presumption of law, making it impossible for one to deny the matter concerning which notice is given, while implied notice is a presumption of fact, relating to what one can learn by reasonable inquiry, and arises from actual notice of circumstances, and not from constructive notice.[450]

[450] Black's Law Dictionary.

As an obvious example, documents are placed on the public record for the world to see, thereby offering everyone constructive notice of their existence as well as their contents.

Duty of an examiner. Ordinarily, an abstracter has no duty, in his or her investigation, to go outside the records unless such is specified in their contract; but if, in the search or examination of the records, he or she is put on notice of anything outside thereof which may affect the title, he or she should either investigate the same, or at least call attention thereto in the abstract or report so that interested persons may make further investigations.[451]

Notice in title examinations. The Colorado title standards[452] discuss notice to great length. Any examiner or researcher should consider this section:

1.2 NOTICE IN TITLE EXAMINATIONS

1.2.1 Types of Notice

There are three types of notice of concern to title examiners—actual notice, constructive notice and inquiry notice.

1.2.2 Actual Notice

Actual notice may be defined as knowledge of the contents of a document or of other facts which may affect title to an interest in real property. If a title examiner has actual notice of a recorded document, it is immaterial whether the document appears in the record chain of title. If a title examiner has actual notice of an unrecorded document, it is immaterial that the document is not recorded. A title examiner must consider the effect of any document of which he or she has actual notice in the preparation of his or her title opinion.

1.2.3 Constructive Notice

Constructive notice may be defined as being charged by law with notice of the effect on title to an interest in real property of the contents of a document or of other facts without knowledge of the document itself or the facts themselves. A document recorded in the real property records in the office of the county clerk and recorder is constructive notice of its existence and of its contents to all persons subsequently acquiring an interest in the real property affected by that document even if the document is not properly indexed or copied in the records by the clerk and recorder. While the recording of a document is constructive notice to the persons subsequently acquiring an interest in the real property affected by that document, a title examiner is only responsible for analyzing the effect on title of those recorded documents which would be revealed by a properly conducted search of the real property records by the title examiner (*See* Title Standard 1.1.3) or which are contained in the abstract of title examined by the title examiner (*See* Title Standard 1.1.4).

[451] 1 C.J.S., Abstracts of Title, § 10a; *Keuthan v. St. Louis Trust Co.*, 73 S.W. 334, 101 Mo.App.1.

[452] *Colorado Real Estate Title Standards* (revised 2003).

1.2.4 Inquiry Notice

Inquiry notice may be defined as being charged by law with notice of the effect on title of facts that would have been revealed by an inquiry if known facts would cause a reasonable person to inquire. If a person acquiring an interest in real property has knowledge of facts which, in the exercise of common reason and prudence, ought to put him or her upon particular inquiry as to the effect of such facts on the title to such real property, he or she will be presumed to have made the inquiry and will be charged with notice of every fact which would in all probability have been revealed had a reasonably diligent inquiry been undertaken. Whether the known facts are sufficient to charge such person with inquiry notice will depend upon the circumstances of each case. It should be noted that, because Section 38-35-108, C.R.S. provides that a reference to an unrecorded document in a recorded document is not notice to any person other than the parties to the recorded document, a person acquiring an interest in real property is not charged with notice of the effect of the unrecorded document on the title to such real property. If, in the course of a title examination, a title examiner discovers a document which is not in the record chain of title but which sets forth or refers to facts (other than the existence of an unrecorded document) that would cause a reasonable person to inquire about the effect of such facts on the title being examined, the title examiner should disclose such facts in his or her title opinion so that the person for the whom the title opinion is written may determine whether to undertake an inquiry.

LIABILITY OF THE TITLE EXAMINER

Liability results from lack of responsibility, or failing to do what you purport to do. Doing a title examination, or undertaking title research, demands following certain procedures and being thorough, adhering to certain standards of practice.

The New Jersey court summarized in the case of *Bayerl v. Smyth*[453]as follows: "An attorney employed to examine the title to real estate must exercise reasonable care in the matter, and failure to do so is negligence for which he will be liable to his client in damages"[454] The case had to do with the failure to report a judgment against the property.

In the Maryland case of *Ryan v. Brady v. Hawkins*,[455] a problem with a boundary in that the purchaser's belief (Ryan) was that it was a straight line, when in fact it was not, as shown on the recorded plat. While the case mostly had to do with the question of right of rescission of the purchase of the property, the court, in its decision, provided helpful discussion concerning attorney responsibility. "The primary reason a title examiner must carefully examine the most recent plat of the property, the title

[453] 189 A. 93, 117 N.J.L. 412 (N.J.Err. & App., 1937).

[454] Negligence, in most jurisdictions, requires proof of a duty, a breach of the duty, the breach being the proximate cause of the damages, and actual damages. It is often defined as the failure to do what the prudent individual would do under like circumstances.

[455] 366 A.2d 745, 34 Md.App. 41 (1976).

to which he is searching, is to make sure that all deeds thereto subsequent to the date the plat was recorded embody the same description of the property as that shown on the plat. If they don't and nothing has been sold off or added to the property since it was surveyed and platted, then it would be necessary to obtain a confirmatory deed from one or more of the sellers' predecessors in title to reconcile the recorded deed descriptions of the property with its configuration as shown upon the recorded plat thereof and thereby remove that cloud upon the title. Furthermore, as a general rule, the most recently recorded plat is the best source for the description of the property."

CHAPTER 14

CASE STUDIES

The following are examples of some fairly routine problems. Typical of situations encountered with older discontinued highways and elusive or unclear records, they present a challenge to the landowner, title attorney, and land surveyor. All these examples demanded expertise beyond that of the average treatment. Several were decided by the courts.

These situations and their analyses demonstrate the importance of assembling all of the facts before making a decision, and a working knowledge of the relevant law is essential. Many problems can be avoided through adequate research of the facts and law important to the individual case.

CASE #1 WHO OWNS THE ROAD?

In 1982, a survey was conducted for parcel B. The surveyor located the remains of an old road between the parcels and, from sideline evidence in the form of ancient fences, determined that the road was two rods wide. In addition, relying on the presumption that abutting owners have title to the centerline of an "abandoned" (discontinued/vacated) road, he produced a plan showing the centerline of the road as the boundary between lots A and B.

A dispute arose and the owner of Lot A blocked off the road, preventing its use by the owners of Lot B. This was critical to Lot B for access to the barn and to the back land. Late in 1982 the owners of Lot B instituted an action against the owners of Lot A to quiet title in one-half of the old road. As a result of the research done in preparing for court testimony, the following facts were uncovered (see Figure 14.1).

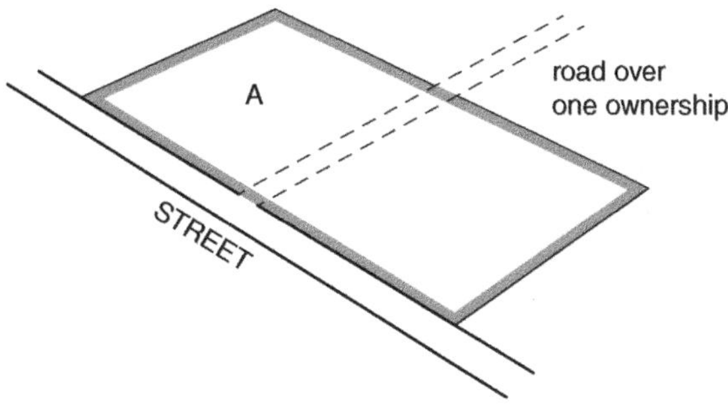

Figure 14.1

The road in question was laid out in 1742 under the requirements of the statute in existence at the time:

> August 13, 1742. A return of an open highway laid out by Selectmen and a committee chosen, beginning at a highway that leads from Salisbury through South Hampton to Kingston near ye Meeting house in said town, beginning first at ye passway 2 rods westerly of Timothy Johnsons line at a stake in said Johnsons line at ye foot of ye hill 2 rods in width westerly across said Timothys land to a stake near the brow of the hill from thence across Thomas Morrills land 2 rods in width & across lands of Joseph Maxfield, John Ordway, Moses Bartlett, Samuel Flanders, Jacob Fowler to back River so called and from Said River on Asa Browns land and land of Abraham Brown to the highway That leads from Salisbury through a part of South Hampton called ye Poock.

The road at the area in question was actually laid out entirely on land of Timothy Johnson (see Figure 14.2).

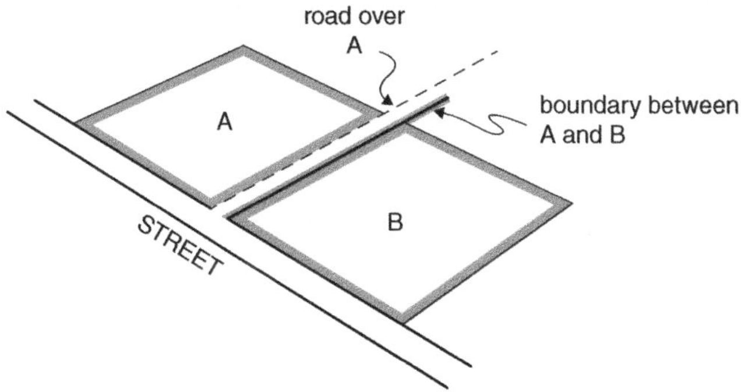

Figure 14.2

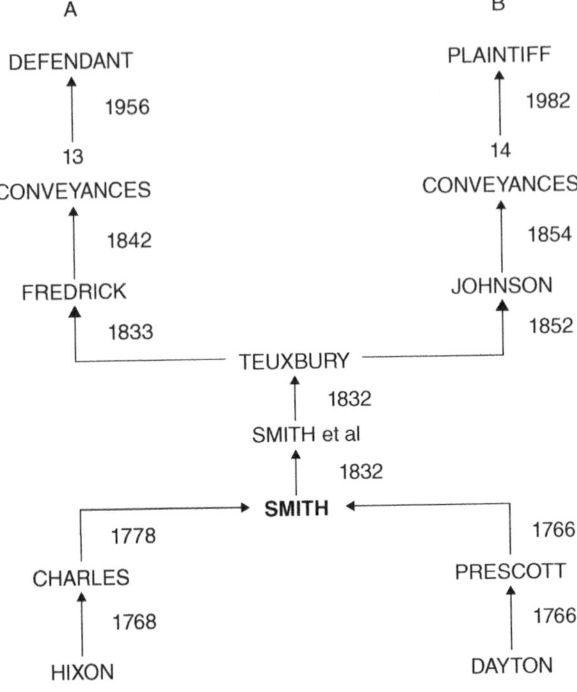

Figure 14.3

In 1842, the Town voted to close the road.

To understand all the events relating to the road, comprehensive deed research was done on both chains of title. This resulted in the following (see Figure 14.3).

A merger of title took place as follows (see Figure 14.4).

Since parcels A and B are in one ownership, there are no longer two parcels separated by a road, but one parcel with a road crossing it.

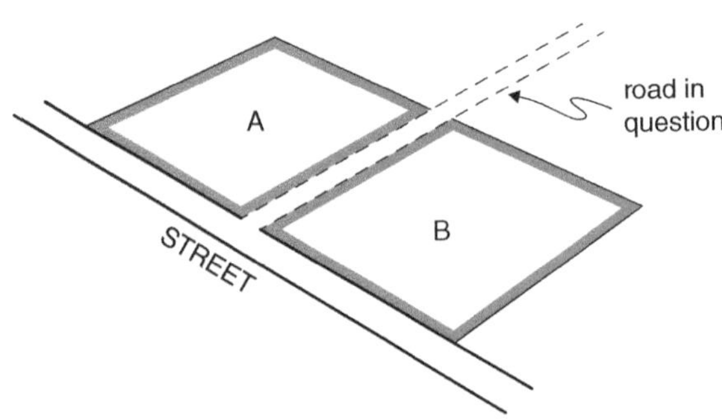

Figure 14.4

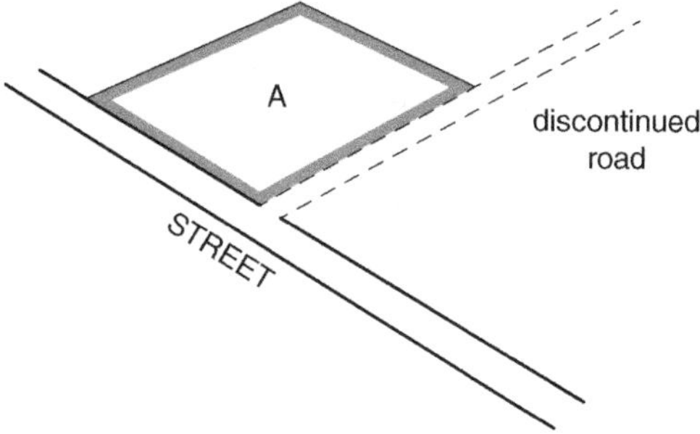

Figure 14.5

The next important step was to examine the descriptions of the outsales from this one, parent, parcel. The following occurred:

Westerly parcel (see Figure 14.5).

"... bounded easterly by the road from the meetinghouse to the peak."

Easterly parcel (see Figure 14.6).

"... bounded westerly by the road from the meetinghouse to Kensington."

The road was in existence at the time the aforementioned two parcels were created, not being discontinued until some years later, in 1842. Examination of the chains of title did not disclose any alterations in the road; review of town records did not reveal any changes in road status except the 1842 discontinuance.

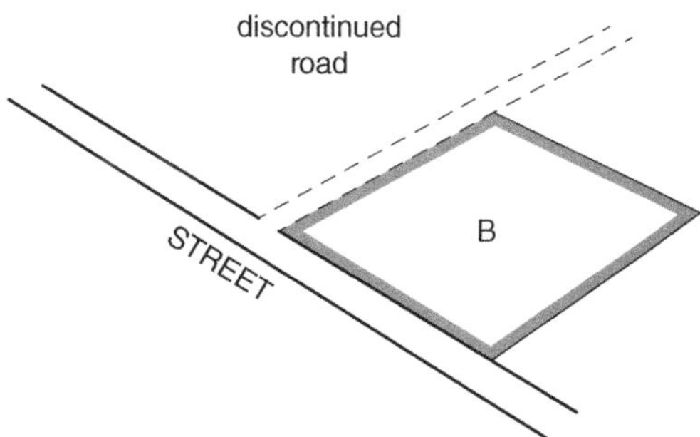

Figure 14.6

Since the parcels in question originated from a common source parcel with a road easement over it and used the road as a common boundary between them, it remains to review where the actual boundary lies in relation to that road.

The presumption is that abutting owners take to the centerline of a highway (easement) upon discontinuance *as long as the grantor owns that far and unless the contrary appears.* In this case, the grantor did own the land and the contrary did not appear. Since there was no language in any of the deeds to limit the boundary to the sideline of the road, or otherwise, the centerline of the easement should be the boundary between the two parcels.

The case was tried in 1986 and the court found for the plaintiff (Lot B). It stated that the boundary between the parcels was indeed the centerline of the old road and not the sideline as was the position of the owners of Lot A.

This case is somewhat interesting in that knowledge of the entire history of the title is necessary to realize that a merger of title took place when the road was an open, public easement. If that had not been the case, all other facts being the same, the boundary would not have been the centerline, but would have been the easterly sideline, as originally laid out over Johnson's land. The easement would revert and would not alter the boundary. The change in boundaries took place because of a merger of tile and the creation of two new parcels.

It should be kept in mind in a case like this that the physical characteristics and evidence on the ground remained the same, but changes were found within the records.

—Unreported case of *Nichols et al. v. Imbrescia* (1986)
Rockingham County (N.H.) Superior Court

CASE #2 WHO OWNS THE LAND?

In 1971, a major subdivision was done abutting a state highway, which was a primary road in the state. In 1988, a survey was done for Lots 1, 2, and 3, and in doing so it was necessary to determine the ownership of that portion of land between the lots and the highway. (See Figure 14.7.)

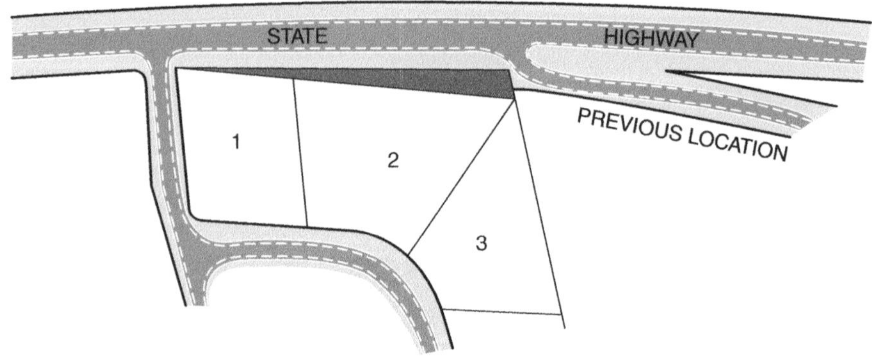

Figure 14.7

It seemed a simple matter to consult the state highway department and inquire about the ownership, or determine whom they identified as owners when they changed the location of the road. Before doing that, however, background research was undertaken to get an idea of the details of the situation.

1. Early maps of the town were located in an attempt to determine how long the original road had been in existence. Maps dating back to 1774 showed the road in question.
2. Highway layouts plans for the town were researched in the town records and the following was found:

> Return of the Publick Highway laid out in the year 1792
>
> First beginning at the town line the south side of the river where the way is now open and passable and running in the same path or way as it is now opened and passable to Andrew Folsoms House and then running on the Line between 37 & 42 to the corner tree marked 30 & 37 and then into the Road that is cut and make passable and then by Nathl Ambrose Barn by Peter Warrens House and to Road and to be 3 rods wide.

This layout, while not very definitive, at least gave width of 3 rods (49.5 feet).

3. A copy of the present layout by the state shows the road as being 83 or 100 feet wide, depending on location, and its position at the area in question is the result of an alteration. Discussion with the state right-of-way department resulted in there having been an alteration in 1954.
4. Deed research was undertaken to determine what land was conveyed to the state for highway purposes.

"Parcel #1 – All the land belonging to the Grantor that comes within a distance of fifty (50') feet measured Easterly of the center line as shown on a plan on file in the records of the Department of Public Works & Highways; and to be filed in the Registry of Deeds; between land now or formerly of Leslie Moody on the South near Station 57+00, and other land of the Grantor on the North near Station 65+00."

"Parcel #2 – All the land belonging to the Grantor that comes within a distance of fifty (50') feet measured Easterly and fifty (50') feet measured Westerly from said center line, between land now or formerly of Earl A. and Lela H. Lawrence on the South near Station 62+50, and land now or formerly of Almond O. Bryant on the North near Station 75+50.

It was determined that the State obtained fee title to a 50-foot strip from Station 57+00 to Station 75+50, along with other land.

5. In the discussion of the state right-of-way department, it was found that the state released all its rights to the town for all interest between Station 65+50 and Station 81+50.
6. Contact with the town disclosed that it had never discontinued the "old highway," nor had it relinquished any of its rights.

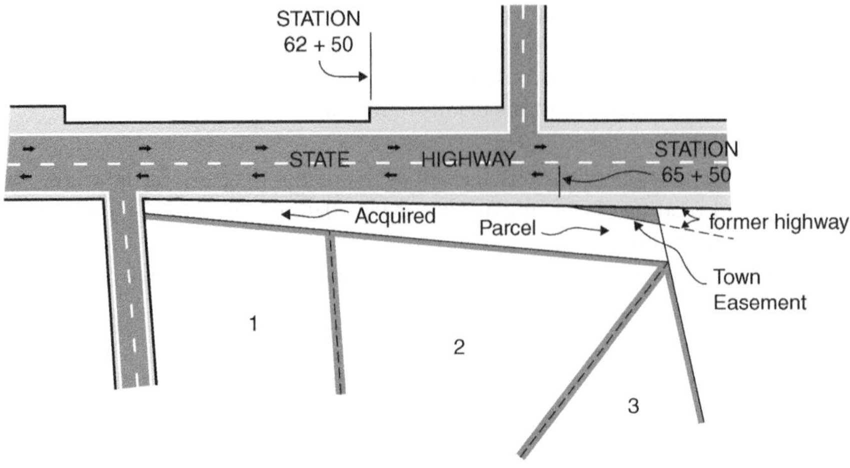

Figure 14.8

The application of the rule of presumption of conveyance to centerline would give the present owners title to the area in question. To be certain, the attorney for the previous grantors (a corporation) was contacted, and it was found that (1) the corporation had been dissolved, (2) the officers had intended to convey everything they owned, and (3) they were available and willing to execute a deed to all right, title, and interest, if any, in any land that remained. Once that was accomplished, the survey was completed (see Figure 14.8).

CASE #3 HOW MUCH RESEARCH IS NECESSARY?

In 1990, a subdivision was done of a tract of land which was a remainder parcel due to the alteration of a public highway (see Figure 14.9).

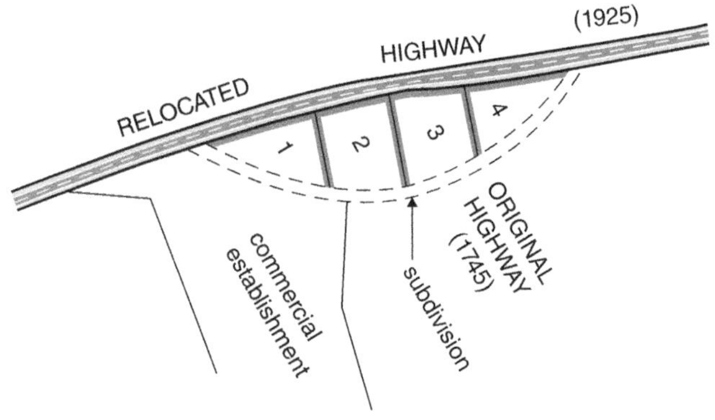

Figure 14.9

In doing the subdivision, the surveyor obtained a copy of the current highway plans and inquired of the town road agent as to the status of the old highway. He was informed that the road was "abandoned," and as a result he assumed title to the centerline. Sideline location and width were determined from the existence of the parallel stone fences, which were obviously very old and were approximately two rods apart (see Figure 14.10).

The owner of Lot 1, having been deeded one-half of the old road, decided to exercise his right in preventing any one else from using that portion. He had no use for

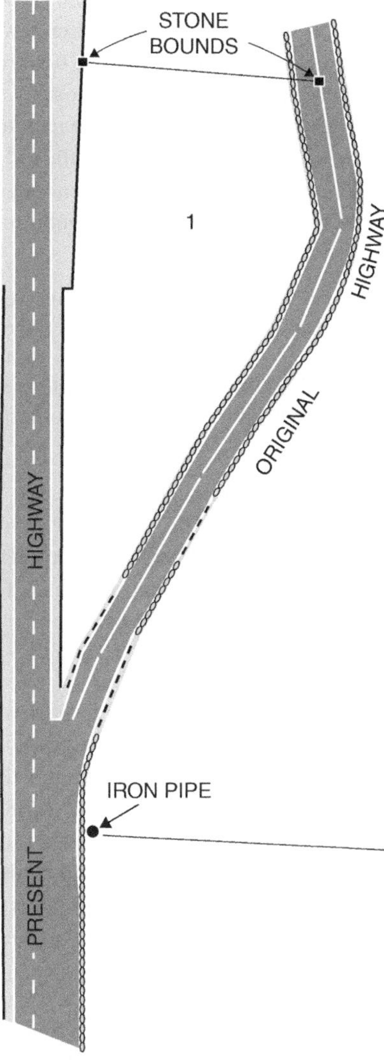

Figure 14.10

it since his access was from the new road location. The problem that arose, however, was that, in blocking off part of the old road, he interfered with the access to the commercial establishment on the opposite side of the old road. This owner decided to investigate his rights, and instructed his surveyor to "do what was necessary" to determine all the facts available. Research into the problem produced the following:

1. State highway records indicated that the road alteration took place in 1926 when the state widened and straightened the route, leaving a portion of the old road outside of the limits of the new right of way. The parcel of land later subdivided was that which was severed from the owner on the westerly side, who still owned land on the westerly side of the new road. Discussion with state officials resulted in the conclusion that the town was the only entity having an interest in, or jurisdiction over, the old road.
2. Town records were examined from 1925 to date to determine if there had ever been a discontinuance or other action affecting the old road. A discontinuance was found in 1937 for a portion of the old road about one-half mile northerly of the location in question, but nothing was found relative to the section of road in question.

> "warrant to discontinue the portion of old dirt road now on westerly side of Route # 108 and connected at both ends to Route 108 and lying South of residence of E.G.W. Currier."
>
> "voted to close the portion of old dirt road now on westerly side of Route No. 108 and connected at both ends to Route 108 and lying south of residence of E.G.W. Currier. March 9, 1937"

3. Town highway layouts were researched, and it was found that the original road had been laid out in 1745. To do this, since no one knew how long the road had existed, except that it was "very old," required that the chains of title for both easterly and westerly parcels be traced back for reference to the road and to determine "when the road was *not* there."

The parcels were traced back to 1712 and 1718, respectively.

After the "window of time" was identified, town highway layouts were reviewed for the layout. The point in time was a very important item in this case for two reasons. First, the town was formerly part of another town until 1738. If the layout occurred prior to that year, it would be found in the parent town's records. Second, the road runs through eight towns, and the layout could be found in all or any one of the eight towns' records, their parent towns, or state or county records, since several towns are involved. As it turned out, the layout in question was found in the present town records in 1725:

> August the 6th day 1745 then laid out by us the Subscribers Select Men a high way on the South Side of Trickling falls bridge & Running from thence Southerly to a heape of Stones & from thence Southerly to within three rods due west from the west ende of

Theophelus griffins house two Rods wide on each side of the Senter of the now trodden path and so Running as far as the town Ship Shall Extend & this is the return we make of Sd way Elisha Sweet Elisha winslo John honton Selectmen"

4. The deed descriptions presented a new fact that was an important consideration. The source parcels of the two lots in question were conveyed in 1712 and 1718, calling for the front lot lines being "2 rods easterly and westerly of the old path." This explains why the road was laid out in 1745 as 4 rods wide. If the reserved roadway and the "old path" were not made part of the conveyances, then the fee to this 4-rod strip may still rest with the town. If that is true, and the 1745 road was laid out on that strip, then neither lot may be entitled to any part of it through reversion, since the road would not be an easement, but a road laid out over a fee strip, belonging to the town.

—Unreported case of *MSK Lumber Co., Inc. v. Murphy* (1994). Rockingham County (N.H.) Superior Court

CASE #4 HOW WIDE IS THE RIGHT OF WAY?

WILSON v. DEGENARO
36 CONN. SUP. 200; 415 A.2D 1334 (1979)

On January 27, 1927, Millard K. Palmer deeded land to Helen Louis Cornish, reserving for himself a right-of-way, as follows: "Reserving to me, the said grantor, my heirs and assigns, the right to use, for travel, telephone, electric service, water and other public utilities, the driftway *as now laid out* extending from said first mentioned driftway, through the tract of land hereby conveyed to the tract of land still owned by me, bounding the tract of land hereby conveyed to the south." (Emphasis added.) At issue is the width of that right-of-way. The plaintiffs claim the evidence has established that width as it was "laid out" in 1927, and that evidence of post 1927 occurrences is irrelevant. The defendant, who counterclaimed, argues that the words "as now laid out" are ambiguous and that the court must determine the intent of the grantor. [See Figure 14.11.]

The court, upon stipulation of the parties, agreed to hear evidence as to the claims of both parties, subject to the reservation by the plaintiff of an objection to evidence of any post 1927 occurrences. By agreement, it was left to this court to determine later what evidence should be considered. This and other generous stipulations by counsel have allowed the court to shorten what could have been an extremely lengthy trial. Extensive, incisive argument and briefs have supplemented the evidence. Both sides were ably served by counsel.

The rule in Connecticut is stated in *Peck v. Mackowsky*, 85 Conn. 190, 194, 82 A. 199, 201: "The words 'as now used,' with which the reservation concludes, were employed to designate the location of the way, and not to define or restrict the use to which the way might be put." The defendant concedes in his brief that the words "as now laid out" are not materially different from the phrase "as now used," but appears to argue that the "location" does not include the width of the way.

Figure 14.11 Shell road.

[1] Courts look to the usual meaning of words in construing a deed. "Lay out" is defined by The American Heritage Dictionary as "[t]he laying out of something; The arrangement, plan or structuring of something laid out; overall picture or form." Ballentine's Law Dictionary defines "laying out" as "Locating and establishing a new highway." The American Heritage Dictionary defines "locate" as "to determine or specify the position and boundaries of" and "location" as "the fact of being located or settled." Similarly, Ballentine's Law Dictionary defines "locate" as "to set in a particular spot or position" and "to set in a particular spot or position" and "to designate a particular portion of land by limits." It further defines "location" as "in the case of land, defining the boundaries or otherwise describing it so that another may know where it lies and its extent."

Our Supreme Court, in adopting these meanings, has declared that "to lay out a highway is to locate it [or] define its limits." *Crawford v. Bridgeport*, 92 Conn. 431, 436, 103 A. 125, 126; *Lake Garda v. Battistoni*, 160 Conn. 503, 280 A.2d 877. It refers to the creation or establishment of a road. *Wolcott v. Pond*, 19 Conn. 597, 601; *Woodbridge v. Merwin*, 27 Conn.Sup. 469, 473, 244 A.2d 57.

[2] The words "as now laid out" are not only analogous to the phrase "as presently used" interpreted in *Peck v. Mackowsky*, supra; they speak even more clearly to the subject of location. Giving the words their ordinary meaning, "as now laid out" clearly referred to the "location" of the right-of-way when it was reserved in Palmer's deed in 1927.

[3] Since the Supreme Court found that the words "as now used" clearly defined the location of the way, this court has no difficulty in finding that the words "as now laid out" similarly define the location of this right-of-way.

> The defendant's claim that "location" does not relate to width denies the word its common meaning. Location, by definition, signifies and implies the determining of applicable boundaries. The width of the right-of-way in 1927 was one of those boundaries. The grantor referred to something actually in existence. It was there. It would be illogical to assume he intended to specify its location (boundaries)—but without so stating, not specify its width—which was inevitably part of those boundaries. Perhaps he could have done so, but he did not. His referral to the right-of-way "as now laid out" clearly tied it down to the dimensions then and there existing. *Peck v. Mackowsky*, supra. His intention was not in doubt.
>
> *[4]* The plaintiffs, in final argument, claimed the way did not exceed twelve feet. The court, having heard the evidence and determined its credibility, is in agreement. While precise mathematical certainty is difficult to determine in cases such as this, the court is satisfied that credible evidence supports the existence in 1927 of a right-of-way with a traveled portion not exceeding eight feet. There was some evidence that the road consisted of more than the traveled portion. The court agrees. "Shell Road" consisted basically of oyster shells, combined with other materials. The evidence supports the reasonable conclusion that it could not rise perpendicularly from its base in the wetlands but was somewhat sloped. It is logical to conclude that the two sides of that road accounted for some additional footage. This accords with concessions made by the plaintiffs in argument, surveys in evidence and evidentiary admissions by the defendant that the road did not exceed twelve feet. Twelve feet is also the present width of the road. It leads to the ultimate conclusion that the defendant should not be allowed to encroach on any more than twelve feet of that road.

The court agrees with the holding in *Burroughs v. Milligan*, 199 Md. 78, 85 A.2d 775, where grantor's reservation of a right-of-way "by the existing road," was held not a way of necessity but a definite reserved way. The court held a way of necessity or a general right-of-way without location have no place where a definite right-of-way exists as evidenced by an "existing road." Such rights do not grow with increased traffic; "We are dealing with rights, and not questions of desirability, and we cannot give the appellees more than they are legally entitled to because it would be convenient for them to have it." Id., 89, 85 A.2d 781.

As early as 1886, the Massachusetts court was in accord. In *Dickinson v. Whiting*, 141 Mass. 414, 6 N.E. 92, a conveyance allowed the use of a specific road ("The lane on the south side of said premises"). Since a defined way was in existence at that time, the court held it was not a general right-of-way convenient for the grantee of such a width as might thereafter be determined by various circumstances, but was of the land so existing. Similarly, in *Burgas v. Stoutz*, 174 La. 586, 141 So. 67, the court held that a "paved runway" being a physical object on the surface of the ground, ad its length and width being ascertainable, the defendant's claims that the language failed to state sufficiently the length or width of the passage was without merit. See also *Lattimer v. Sokolowski*, 31 N.Y.S.2d 880 (sup.)

> *[5]* The annotation "Width of way created by express grant, reservation, or exception not specifying width" in 28 A.L.R.2d 253, sums it up: "Ordinarily, a grant or reservation of a right of way by instrument referring to an existing way

> at the place contemplated, and not otherwise indicating the width of the passage, operates to limit the width to that of the existing way . . . ." Id., 267. Such a reference leaves "little ground for the contention that the intended width was other than that of the existing way." Id., 268.

Despite this conclusion, the court has examined the defendant's counterclaims and the evidence advanced to support them in order to ensure that he receive a full hearing of his claims of law and fact. He argues that the deed is ambiguous as to the extent of the easement, positing that since it was not limited to travel alone, but included public utilities, was must find specific intent to create a much broader way than the existing road referred to; that it must include what is necessary to the reasonable enjoyment of the road, and should confer a "practical right" to allow the full and complete "future development and use of 5 acres of undeveloped land."

> *[6]* The short answer is that the deed is not ambiguous as to the reserved way. But assuming for the sake of argument that an ambiguity exists, the defendant failed to produce sufficient credible evidence to meet the burden of proof necessary to sustain the premise of his claims.

When a deed is ambiguous: "'There should be considered, when necessary and proper, the force of the language used, the ordinary meaning of words, the meaning of specific words, the context, the recitals, the subject matter, the object, purpose, and nature of the reservation . . . and the attendant facts and surrounding circumstances before the parties at the time of making the deed.'" Andrews *v. Connecticut Light & Power Co.*, 23 Conn.,Sup. 486, 490, 185 A.2d 78, 80. The court has determined from the evidence that none of those considerations support the defendant's claims.

> *[7]* The intent of the grantor as spelled out in the deed itself must be interpreted, not the grantor's intent in general, or even what he may have intended. *Lake Garda v. Battistoni*, 160 Conn. 503, 280 A.2d 877. The credible evidence does not sustain the burden of proving that said intent requires a right-of-way in excess of twelve feet. No satisfactory evidence indicates that he intended the land to be fully developed. It is just as reasonable that this oysterman wished his "Shell Road" to remain basically as it was to preserve the ecology of the area so near the water. To find such ambiguity, that the "reasonable uses" of the property by 1979 development standards must be determined, would replace factual interpretation with pure speculation designed to suit the defendant's convenience. This the court cannot do absent credible evidence.

Evidence received from representatives of public utilities was tentative, unsure and probative of little but speculation. Nor could the court accept testimony, modified on cross-examination, that twenty to twenty-five feet was required to introduce a sewer line to the area. Interestingly, it was conceded that any sewer line used to service the defendant's property would be connected to the municipal sewer system. There is no evidence that the grantor intended that sewage was to be handled by a public utility; if he intended future connection to the municipal system, he was not, by definition, including swage removal within his concept of the public utilities that were to serve the property. Similarly, the court finds that the grant to the Greenwich

Gas Company was given by a third party, Walter B. Palmer, at a later date, 1933, and is not probative of the intent of Millard K. Palmer in 1927. Neither can an easement granted by Walter B. Palmer in 1949 shed any light on the grantor's intent in 1927. Even with respect to utilities contemplated to run above ground, the testimony regarding telephone poles was entirely inconclusive.

The defendant made an ingenious comparison of maps in attempting to show that the right-of-way has meandered or changed its location, although, significantly, not its width. But a review of the maps and of all of the accompanying testimony left the court without confidence in the accuracy and validity of those comparisons, especially since the defendant's survey followed two widenings of the way to the defendant.

Giving the defendant's evidence its strongest interpretation, the court does not find that the defendant sustained his burden of establishing a right-of-way in excess of twelve feet.

> *[8, 9]* Starting in 1949, deeds in the defendant's chain of title refer to a recorded map description of the right-of-way. While the defendant appears now to question the accuracy of some of the surveys, they portray a road which clearly is not shown to exceed twelve feet. The defendant had notice of both the reference to the existing road in 1927, and the recorded survey description. He knew from the records what he was getting. See *Stanio v. Berner Lohne Co.*, 7 Conn.Sup. 452, 455. It is the policy of our law that all land titles should appear upon the land records. "[T]he real nature of the transaction, so far as it can be disclosed, must appear upon the record, with reasonable certainty; and if the deed does not actually give notice of any condition or other circumstance, which might be important, it should at least point to a track, which the enquirer may pursue to obtain it." *North v. Belden*, 13 Conn. 376, 379-80. Unquestionably, the deed in question did so. It is those land records upon which Connecticut's property rights are based. It is elementary that nothing should be allowed to weaken the effectiveness of that system of records. *Ashley Realty Co. v. The Metropolitan District*, 13 Conn.Sup. 91, 95. To act otherwise would invite chaos.

The evidence also revealed a recognition by the defendant of a twelve foot right-of-way. In a 29 September 1977 letter to the Greenwich town attorney, he stated: "A copy of the actual survey indicates, the right of way across Mr. Wilson's property is approximately 12 feet wide throughout its entire length . . . . The fact that I may have improved the right of way is completely irrelevant and in any event, since the right of way is twelve feet wide, I had a perfect right to do so."

It was stipulated in lieu of testimony, that the defendant would testify that he was not stating his own opinion, but merely what the map showed. Yet he clearly set forth and relied upon the effect of that map in 1977.

In a variance request filed 23 August 1976, the defendant included a 1976 survey. It was stipulated at trial that he would testify that it was a map of his own property and not of the right-of-way which he did not fee was accurately portrayed. But the right-of-way was pictured as being approximately one-third the width of the thirty foot road with which it connects. It may be that the right-of-way was not accurately portrayed. But against the background of the evidence, including the other surveys, it is doubtful that any inaccuracy was substantial enough to materially aid the defendant's claims.

Nor was the zoning board informed of any inaccuracy in the map. While it was a map of his own property, this was the accessway to that property, and it is doubtful that the defendant would submit a survey to the board with dimensions of that way which he knew to be radically wrong, without so stating. After using the documents before an official body and taking those documents in conjunction with his 1977 letter, it is difficult to permit the defendant to attempt to negate, in an equitable proceeding, the accuracy of the very documents upon which he relied at another time.

> *[10]* Even the acceptance of the defendant's premise that the reservation was general as to its extent would not avail him; he then falls within the purview of another doctrine damaging to his position. Connecticut courts have held that a right-of-way stated in general terms is limited to a way actually taken and used by the grantee of that right. *Richardson v. Tumbridge*, 111 Conn. 90, 96, 149 A. 241; *Pudim v. Moses*, 20 Conn.Sup. 311, 313-14, 134 A.2d 478. See also *Lake Garda v. Battistoni*, 160 Conn. 503, 280 A.2d 877. The evidence satisfactorily determined the way taken and used in 1927 and subsequent years. Other states are in agreement. In *Winslow v. Vallejo*, 148 Cal. 723, 84 P. 191, the court held that the use of a general grant, with consent, in a certain course fixes its boundaries; and that what was once indefinite becomes fixed by use and cannot be changed at the pleasure of the grantee after it is established. See also *Dudgeon v. Bronson*, 159 Ind. 562, 64 N.E. 910, holding that once a way is fixed by acceptance and use, the user may not change it to a right-of-way by necessity.
>
> *[11, 12]* Generally, land rights are to be determined at the time they are granted. *Drummond v. Foster*, 107 Me. 401, 78 A. 470; *Lattimer v. Sokolowski*, 31 N.Y.S.2d 880 (sup.) Where a way is created without specifying any particular width, even though the road was in existence, the right-of-way is limited to the width as it existed at the time of the grant. *Good v. Pettcrew*, 165 Va. 526, 183 S.E. 217; *Palmer v. Newman*, 91 W.Va. 13, 112 S.E. 194; *Kotick v. Durrant*, 143 Fla. 386, 196 So. 802.

Therefore, even if the contested phrase in the deed was considered to be ambiguous, the facts of this case and the prevailing case law would not help the defendant; the credible evidence still does not permit finding a right-of-way in excess of twelve feet.

The parties stipulated that either party in whose favor an injunction is entered will, if necessary, amend the pleadings and prayers for relief to conform to the court's determination of the width of the right-of-way.

Since the court finds that the right-of-way is not in excess of twelve feet, an injunction shall issue in favor of the plaintiffs as follows: The defendant,* his agents, servants and employees are enjoined from performing any and all acts outside an area six feet from either side of the center line of the right-of-way as presently located on the property of the plaintiffs.

* To eliminate the need for adding a party, the defendant with the plaintiffs' agreement, gratuitously stipulated that any injunction granted against him would remain in effect against his wife, Mary Jane DeGenaro, as well.

CASE #5 WHEN DOES A ROAD BECOME NOT A ROAD?

AVERY v. RANCLOES
459 A.2D 622, 123 N.H. 233 (1983)

March 31, 1983, *Upton, Sanders & Smith,* of Concord (*Ernest T. Smith, III,* on the brief and orally), for the plaintiff.

Law Offices of Philip R. Waystack, Jr., of Colebrook (Vickie M. Bunnell on the brief, and Mr. Waystack orally), for the defendants.

King, C. J. The Plaintiff, Gloria R. Avery, brought a bill in equity to establish her title to certain land in Clarksville. The defendants, Frank Rancloes and his wife, Glenna Rancloes, are the owners of property, which is contiguous to the plaintiff's property. The parties' dispute concerns the proper boundary line between their property. The defendants argue that the Trial Court (Dunn, J.) correctly found it to be the westerly side of Hurlburt Farm Road. The plaintiff contends that either the centerline of Hurlburt Farm Road or the barbed wire/cedar-post fence on the easterly side of Hurlburt Farm Road is the proper boundary. (See diagram attached to this opinion [Figure 14.12].) We affirm.

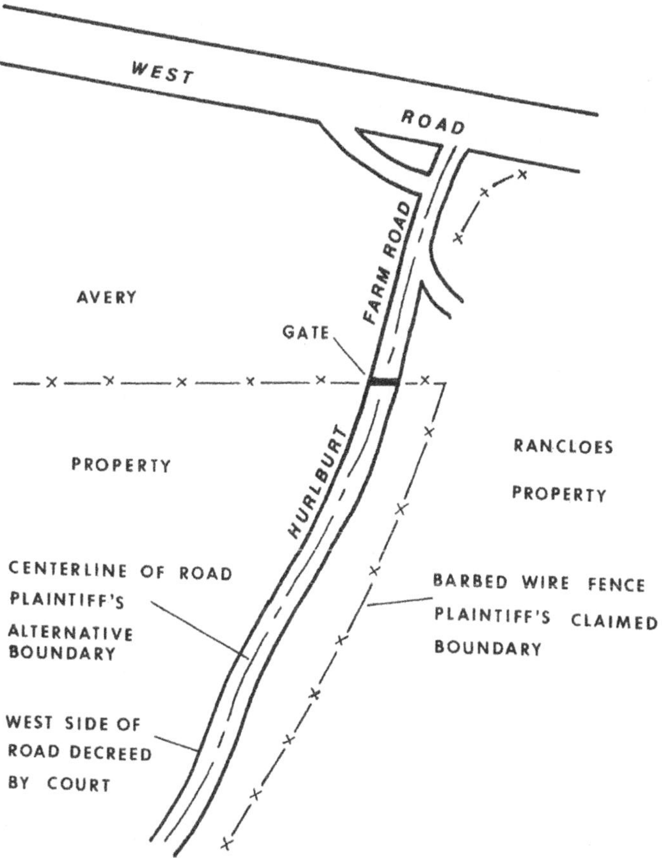

Figure 14.12 Court diagram for foregoing case.

The parties share a common predecessor in title, George W. Anderson, with respect to the land in dispute. On May 15, 1934, Anderson conveyed to Perley Chappel a triangular piece of land described as follows:

> "A certain tract or parcel of land situate in the Town of Clarksville, in the County of Coos in the State of New Hampshire, on the southerly side of the highway [West Road] leading from Keazer Corner, so-called in said Clarksville to Beecher Falls Village, bounded and described as follows:
>
> Northerly by the aforesaid highway [West Road]; *Easterly by the highway leading to the Hurlburt Farm,* so-called, and now owned by Perley Chappel; Southerly and Westerly by a stone wall. Said tract being triangular in form and containing two (2) acres, more or less . . . ." (Emphasis added.)
>
> The following year, this triangular piece of land was conveyed by Perley Chappel to Merle J. Young and Bessie I. Young, the plaintiff's predecessors in title, as part of a larger tract. The physical description of the triangular piece of land included it he larger tract was almost identical to the description in the deed from Anderson to Chappel. Finally, in May 1974, the Youngs conveyed a portion of their land, including the triangular piece originally convened by Anderson to Chappel, to the plaintiff. All of these deeds in the plaintiff's chain of title were recorded.

The defendants acquired their property, which lies both to the north and the east of the plaintiff's property, directly from George W. Anderson and Annie E. Anderson in July 1958. Their deed is also recorded.

The trial court determined that some time between 1935 and 1958, Anderson and the Youngs probably erected the barbed wire/cedar post fence along the east side of Hurlburt Farm Road. It also determined that the defendants maintained the fence after they acquired their property in 1958. The court found that the fence was built to keep cattle on the property, and was not intended to mark the boundary between the properties.. The court also found that Anderson permitted the Youngs to walk their cattle along Hurlburt Farm Road, and also permitted them to install a gate across the road to control cattle and keep trespassers out.

In January 1980, Mrs. Avery, who had purchased her property from the Youngs, closed and locked this gate. She then filed a bill in equity to establish title to the road alleging either that the Youngs' deed to her transferred title to Hurlburt Farm Road, or that she held title to the land by virtue of adverse possession. She claims, in the alternative, that even if she does not have title to the road, that she acquired an easement to use the road because of adverse use.

The trial court rejected these arguments. It held that the boundary of the plaintiff's property was the west side of the road. The court stated that Chappel's deed to the Youngs established the boundary at the west side of the road and that the Youngs had owned no part of the road. Consequently, when the Youngs conveyed the land to the plaintiff, they could convey only what they owned, and, thus, the plaintiff's property reached only to the west side of the road. Finally, the court held that the plaintiff failed to prove her claim of adverse possession. The plaintiff appealed.

The plaintiff argues that the court erred in holding that she acquired no part of the road by deed. She claims that, in construing Anderson's deed to Chappel, the court should have applied the general rule that a conveyance of property bounded by a street or highway conveys title to the center of the boundary street unless clearly contrary language appears in the deed. See *Duchesnaye v. Silva*, 118 N.H. 728, 732, 394 A.2d 59, 61 (1978). We hold that the general rule was inapplicable, and therefore find no error in the trial court's holding.

The rule that a conveyance of property bounded by a street is intended to convey title to the center of the street is based on two presumptions. The first is that the owners of property adjoining the street originally furnished the land for the right of way in equal proportions. The second presumption is that an owner selling land bounded by the highway did not intend to retain the narrow strip of land which constituted the road, and therefore intended to sell to the center line of the street. 6 G. Thompson, *Commentaries on the Modern Law of Real Property* § 3068, at 669-70 (1962 Replacement).

It is clear that the general rule is inapplicable when we examine the facts of this case. In 1927, Charles Felton, Anderson's predecessor in title, owned the land on both sides of the disputed portion of Hurlburt Farm Road. Until that year, Hurlburt Farm Road was a public road. In that year, the Town of Clarksville voted to discontinue the road. When it did so, full ownership of the road reverted to Felton See *Sheris v. Morton,* 111 N.H. 66, 71-72, 276 A.2d 813, 816-17 (1971), *cert. denied,* 404 U.S. 1046 (1972). As the Minnesota Supreme Court stated in describing a similar situation, "[w]hat had been a street would be mere land. . . . The land which had been a street assumed exactly the same legal status as any other land which had not been impressed with a public easement." *White v. Jefferson*, 110 Minn. 276, 284, 124 N.W. 373, 375 (1910). *See Sanchez v. Grace M.E. Church*, 114 Cal. 295, 298-99, 46 P. 2, 3 1896).

After the Town of Clarksville discontinued the road, Felton had title to the land which had constituted the road and to the land on both sides of it. There was nothing legally to distinguish the land which had constituted the road from the surrounding land. Thus, in 1934, when Felton transferred the entire tract to Anderson, he conveyed one large tract of land which was not encumbered by any highway or easement or a private road.

Nor did the rule apply when Chappel later conveyed the same property to the Youngs, using language virtually identical to that used in Anderson's deed, because Chappel could convey only what he had received from Anderson. Similarly, when the Youngs conveyed the property to the plaintiff, they could convey only what they received from Chappel. For these reasons, we hold that the trial court was correct in not applying the rule suggested by the plaintiff.

The plaintiff's next claim is that she either holds title to the land west of the fence which runs along the east side of the Hurlburt Farm Road based on adverse possession or that she has an easement to use the road based on adverse use for a period of twenty years. She claims that the barbed wire/cedar post fence was built and maintained by her predecessor, Merle Young, and his son from 1933 to 1974; that the fence was understood by Merle Young and George W. Anderson to be the true boundary between their contiguous land; that a gate on the Hurlburt Farm Road was built and maintained by the Youngs from 1935 to 1974; that the maintenance of the fence was of a character calculated to give notice to Anderson, the Rancloes' predecessor in title, of an adverse claim to the land by Avery's predecessors in title; and finally, that Gloria R. Avery and her predecessors in title, Merle Young and Bessie Young, have claimed all of the land westerly of the barbed wire/cedar post fence, open and notoriously, without interruption for twenty years continuously.

Although the testimony on the claim of adverse use was conflicting, the court found that the fence had probably been built by Anderson and the Youngs together not to mark the boundary between the properties but as a fence to keep cattle in. The trial judge also found that, after 1958, the fence was maintained by the defendants and that the gate was probably built by the Youngs to control cattle and keep trespassers out, with the permission of Anderson. He further found that Anderson and the Youngs were friends and that Anderson allowed the Youngs to walk their cattle along Hurlburt Farm Road, and Anderson also walked his cattle along that road. He held, therefore, that there was nothing adverse or hostile about the Youngs' use of the road or installation of the gate. The Court held that the first time Gloria Avery took action which put the Rancloes on notice of her claim of adverse possession was on January 3, 1980, when she closed and locked the gate.

Both of these claims required the plaintiff to show that the nature of her use, and that of her predecessors, was sufficient to put the owner on notice that an adverse claim was being made to the property. *Town of Weare v. Paquette*, 121 N.H. 653, 657, 434 A.2d 591, 594 (1981); *Ucietowski v. Novak*, 102 N.H. 140, 144, 152 A.2d 614, 618 (1959). It is well established that permissive use can never ripen into an adverse claim. Id. At 145, 152 A.2d at 618. Whether a use is adverse or permissive is an issue of fact to be determined by the trial court. *Ellison v. Fellows*, 121 N.H. 978, 981, 437 A.2d 278, 280 (1981). There was sufficient evidence before the trial court upon which it could reasonably find that the Youngs' use was not adverse, and, consequently, we will not overturn its findings. Id., 437 A.2d at 280; see *Zivic v. Place*, 122 N.H., 451 A.2d 960, 964 (1982).

Affirmed.

All concurred.

CASE #6: PRESUMPTION OF OWNERSHIP TO CENTERLINE OVERCOME

IN RE PARKWAY IN THE CITY OF NEW YORK; 209 N.Y. 344 (1913)

James R. Deering and James A. Deering for James A. Woolf et al., appellants. The deed from Thomas O. Woolf, Joseph A. Woolf and John A.Woolf to Claus Young, dated July 19, 1855, did not include any part of the land within the boundaries of Walnut street on the Mount Eden map. The grant was limited to the exterior lines of Walnut street and Second avenue. (*English v. Brennan*, 60 N.Y. 609; *White's Bank v. Nichols*, 64 N.Y. 65; *Tag v. Keteltas*, 92 N.Y. 625; 16 J. & S. 241; *Dexter v. R. &*

O. Mills, 39 N.Y. S.R. 933; 133 N.Y. 684; *Matter of St. Nicholas Terrace*, 76 Hun, 209; 143 N.Y. 621; *Deering v. Reilly*, 38 App.Div. 164; 167 N.Y. 184; *Tietjen v. Palmer*, 121 App.Div. 233; *Trowbridge v. Ehrich*, 191 N.Y. 361; *Bartow v. Draper*, 5 Duer, 130; *Morison v. N.Y. El. R.R. Co.*, 74 Hun, 398.)

Tallmadge W. Foster and Lawrence S. Coit for Mary J. Woolf et al., appellants. The point of beginning in the description is the point of intersection of the northerly line of Walnut street with the easterly line of Second avenue. (*English v. Brennan*, 60 N.Y. 609; *Tietjen v. Palmer*, 121 App.Div. 233.) The location of the point of beginning at the intersection of the exterior lines of the two streets necessarily excludes the land in the streets. (*White's Bank v. Nichols*, 64 N.Y. 65.)

Harold Swain for respondent. Assuming that the question is reviewable here, the deed, Woolf to Young, dated July 19, 1855, included the land in Walnut street to the center line thereof. (*Van Winkle v. Van Winkle*, 184 N.Y. 193; *White's Bank v. Nichols*, 64 N.Y. 65; *Hennessy v. Murdock*, 137 N.Y. 317; *Bissell v. N.Y.C.R.R.Co.*, 23 N.Y. 61; *Matter of Ladue*, 118 N.Y. 213; *Trowbridge v. Ehrich*, 191 N.Y. 361; *Woolf v. Woolf*, 131 App.Div. 751.)

Hiscock, J.

This appeal involves a controversy over the right to an award in proceedings to acquire title for street improvements. Without going into the details of the proceeding it is sufficient to state that the underlying and decisive question is whether under the description used in one or more conveyances, the grantee took title to the middle or only to the exterior line of a public highway known as Walnut street, now in the city of New York.

The description thus to be interpreted ran as follows: 'All that certain lot, piece or parcel of land, situate, lying and being in the Town of West Farms, aforesaid, and which is known and designated by the number 267 (two hundred sixty-seven) on a certain map * * * , the premises hereby conveyed being bounded and described as follows: Beginning at corner of Walnut Street and Second Avenue as laid down on said map; thence running westerly along said street fifty (50) feet; thence northerly parallel with said avenue one hundred (100) feet; thence easterly parallel with said street fifty (50) feet to said avenue; thence southerly along said avenue one hundred (100) feet to the corner aforesaid and place of beginning.'

The map in question showed lot 267 as abutting on said highway, and if this description carried title to the center of the street, then the award has been properly distributed; if it carried title only to the exterior line of Walnut street, then the orders thus far made were erroneous and should be reversed.

There is no dispute concerning the general principles which govern the interpretation of a conveyance which bounds and describes lands by reference to a public highway. The language is to be interpreted most favorably in favor of the grantee, and under ordinary circumstances, where the conveyance is of a lot abutting on a highway or where the descriptive lines run to or along such highway, the presumption is of an intent on the part of the grantor to convey title to the center of the highway. This presumption, however, must yield to language which shows an intent on the part of the grantor to exclude from his conveyance title to the bed of an abutting street, and to limit such title to the exterior lines thereof. Notwithstanding the views of the learned courts below, we think that the effect of the language used in the conveyances before us for consideration was to thus exclude title to the bed of Walnut street.

The description of the premises to be conveyed places the starting point at the 'Northwesterly corner of Walnut street and Second Avenue, 'as laid down on the map referred to in the description. It is so well settled as to be practically conceded by counsel for respondent that the terms thus used meant the corner formed by the exterior lines of Walnut street and Second avenue, and did not mean the intersection of their center lines. Thus we have the starting point of the description in the exterior line and on the side of Walnut street rather than in its center. It is again well settled that the starting point in such a description is of great importance, and while it has been said that its effect is not conclusive but must yield where it is inconsistent with other lines in the description (*Van Winkle v. Van Winkle*, 184 N.Y. 193, 204), it must control the other parts of the description at least in the absence of some irreconcilable inconsistency. (*White's Bank of Buffalo v. Nichols*, 64 N.Y. 65, 71.)

I not only do not find the lines of description employed in the conveyances inconsistent with, but rather, as it seems to me, entirely consistent with this point of commencement. We find that the description thus commencing in the exterior line of Walnut street runs along said street fifty feet; thence northerly from it one hundred feet; thence easterly parallel with the street fifty feet, and thence southerly one hundred feet to the starting point. In other words, we have a piece of land of rectangular shape, and the course which runs away from Walnut street is of the same length as that which runs back to said place of beginning in the exterior line of the street. The length and terminus of the final course in the exterior line of the street being thus definitely fixed, it would make an unusual description to run the parallel line of equal length to the center rather than the side of the street.

While it is impossible to make any complete review of the authorities even in this state which have dealt with this general question, reference may be made to a few of them which seem especially pertinent.

English v. Brennan (60 N.Y. 609) involved the interpretation of a description in a deed which ran as follows: 'Beginning at the southwesterly corner of Flushing and Clermont Avenues, running thence westerly along Flushing Avenue twenty-five feet; thence southerly at right angles to Flushing Avenue seventy-nine feet nine inches to a point distant forty feet seven and a half inches westerly from the westerly side of Clermont Avenue, and thence northerly along Clermont Avenue seventy-five feet to the place of beginning.' It will be noted that this description was very similar to the one now before us in that it commenced in the exterior line of a street and then ran 'along' the highway.

It was held that the description did not include title to the bed of the street, and judge ANDREWS writing the opinion of the court after recognizing the general rule that in a grant of land lying adjacent to a highway it is to be presumed that the grantor intended to convey his interest in the street, said: 'But this presumption is rebutted if it appears by the description in the deed that he intended to exclude it from the conveyance. It is in all cases a question of intent to be determined by the description in the grant construed in view of the presumptions referred to. * * * We are to express our judgment on the case before us, and determine whether Clermont avenue is or is not excluded from the grant to the defendant by the description in his deed. We are of opinion that it is, and that his lot is bounded by the sides, and not be the thread or center of Clermont and Flushing avenues. This point is made necessarily by the intersection of the side and not of the center line of the street. * * * The second course terminates at a point 40 feet 7 ½ inches westerly from the westerly side of Clermont avenue, showing that in making the measurements the grantor had the side of the street in view. The next course is defined to be at right angles to Clermont avenue 40 feet 7 ½ inches to that avenue. If it had run to the avenue generally a different question might have been presented, but the precise length of the line being stated, this, in connection with the other parts of the description, seems to render it quite certain that it was to terminate at the side, and not in the center of the street, and if this is so the course to the place of beginning was along the side of the avenue.'

In *White's Bank of Buffalo v. Nichols* (64 N.Y. 65) the court had before it for construction a description which read in part as follows: 'Beginning on the northwesterly line of Carolina street at its intersection with the northeasterly line of Garden street; thence northwesterly along the northeasterly line of said Garden street * * *; thence southeasterly along the line of said block to the State line; thence southerly along the State line to Carolina street, and thence southwesterly bounding on Carolina street to the place of beginning.' It was held that this description did not include title to the bed of Garden street, and while it is true that such description was somewhat stronger in rebutting the presumption of a conveyance to the center of the street because the line was described as running 'along the northeasterly line of Garden street,' still the decision of the court was based on broader principles than mere consideration of said line, and which are applicable to the present case. For instance, it was said: 'The grant under which the defendant claims title, describes the granted premises as commencing at the intersection of the exterior lines of two streets, of which Garden street is one, and so as necessarily to exclude the soil of the street. The point thus established is as controlling as any monument would have been, and must control the other parts of the description; all the lines of the granted premises must conform to the starting point thus designated, so that wile but for this designation of the commencement of the survey or boundary, the lines along Garden street and Carolina street might, within the general principles before referred to, be carried to the center of those streets respectively, they are necessarily confined to the exterior lines of the streets, so as to connect at this starting point. The precise point was decided by this court. (*English v. Brennan*, 60 N.Y. [Mem.] 609.)'

In *Kings Co. Fire Ins. Co. v. Stevens* (87 N.,Y. 287) the court considered a deed which conveyed land described as, 'Beginning at a point on the southerly side of the Wallabout bridge road,' and which description after running certain courses and distances ran a certain distance 'to the Wallabout bridge road,' and from thence along said road a certain distance to the place of beginning. Again with full recognition of the general principles invoked by the respondent it was held that this description commencing on the side of the road rebutted the presumption of an intent to convey to the center, the court saying: 'In the case before us, the starting pint of the description is on the southerly side of the Wallabout bridge road, and the exact point of beginning, is fixed by the reference to the lands of Skillman. The other lines are described by courses and distances, and the third course gives the length of that line in feet, to the road, which we think fairly imports, that the measurement is to the side of the road, and the fourth course as along the road, etc., to the place of beginning.' (p.292.)

In *Deering v. Reilly* (167 N.Y. 184) it was held that a conveyance of lands described as 'beginning on the northeasterly corner of Blackburry alley, on the southeasterly side of the Bloomingdale road and thence along that road to the place of beginning, did not include any part of said road or alley in the conveyance. (See, also, *Holloway v. Southmayd*, 139 N.Y. 390.)

It seems to me clear that the descriptions involved in the authorities cited are so similar in the important features to the one before us that they lead us with compelling force to the conclusion that the present description does not run to the center of the street, unless their force has been weakened by later controlling decisions. I find none such. Counsel for the respondent thinks that he has found such in the cases of *Woolf v. Woolf* (131 App.Div. 751); *Van Winkle v. Van Winkle* (184 N.Y. 193) and *Trowbridge v. Ehrich* (191 N.Y. 361), and therefore, I shall proceed to consider them briefly.

The Woolf case seems to sustain his contention, and if controlling would be an authority for his contention that his client's deed took him to the center of the street. That decision, however, was based on a construction of the Trowbridge case, which in my opinion was erroneous, and to which I shall next refer.

The description in the Trowbridge case ran as follows: 'Beginning at a point where the northerly line of One Hundred and Sixty-third street intersects the easterly line of Stebbins Avenue; running thence easterly along the northerly line of One Hundred and Sixty-third street 30 feet; thence northerly and parallel with Stebbins Avenue 128.71 feet; thence westerly and parallel with One Hundred and Sixty-third Street 30 feet, and thence southerly and along the easterly line of Stebbins Avenue 128.71 feet to the point or place of beginning.'(p.364.)

It was held that this description indicated an intention not to include the fee of the street because the grant dealt with the boundary lines rather than the center lines of the streets, and to this extent the decision confirms the on which we are now making. In the course of his opinion Judge HAIGHT said, 'Had she (the grantor) commenced at the intersection of the two streets and thence ran along the street it would have been apparent that she intended to convey to the center of the street,' (p. 365) and it was this expression which was made the basis of the decision in the Woolf case. It is to be noted in the first place that what was said was purely a dictum, but in addition to this the terms of description suggested by way of illustration were entirely different than those which were and are being construed in the Woolf case and the present case. In both of these cases the starting point is concededly at the intersection of the exterior lines of two streets. In the illustration which was used in the Trowbridge case there was assumed a starting point at 'the intersection of the two streets,' and such point would be formed by the intersection of the two center lines, and hence the controlling point of commencement would be in the center of the street instead of at the side thereof.

In the Van Winkle case the conveyance was of a certain lot, and the commencing point of the description which was employed was stated in one conveyance as being 'at a stake by the fence on the cross road leading from Harlem fifty links from the southeastern corner of the fence,' and in another conveyance as being 'at a stake by the fence on the public cross road, the corner of Mr. Jauncey's land, 'and the determination whether the conveyance ran to the center of the highway largely turned on the effect of this statement of the point of commencement of the description. The case was regarded as a close one, and two principles which do not exist in this case were invoked for the purpose of carrying title to the center of the road. The first one was that the starting point was of uncertain location, and that, therefore, the other lines in the description were to be considered in determining the intent of the parties, and that doing this such other lines carried the description 'to the public road and thence south along the road to the place of beginning,' and, therefore, under general principles, included title to the middle of the street. The second point was that a stake or monument placed upon the boundary of a highway is not necessarily to be regarded as fixing the starting point in the description at the side instead of the center of the street because of the impossibility of locating such monument in the bed of the highway, and that, therefore, such a point of commencement under such circumstances would not be controlling upon the other lines of description.

Our conclusion, therefore, is that the conveyances in question carried title to the exterior and not to the center line of Walnut street, and that the orders appealed from were erroneous and should be reversed, and the award be directed to be paid to the appellants, with costs in all courts.

CULLEN, Ch. J., GRAY, CHASE, HOGAN and MILLER, JJ., concur; WILLARD BARTLETT, J., absent.

Ordered accordingly.

CASE #7 RIGHT OF WAY CREATED BY ESTOPPEL

CASELLA v. SNEIERSON
325 MASS. 85 (1949)

The plaintiff brings this bill in equity to restrain the defendants from erecting a garage on land over which the plaintiff claims to have a right of way. The evidence is reported and the judge made a report of the material facts found by him [see Figure 14.13].

The pertinent facts may be summarized as follows: The plaintiff is the owner of a lot of land in Waltham on which there is a dwelling. The defendants own land which adjoins the plaintiff's lot to the west and to the north and is designated as lot "B" on the plan, material features of which are shown on page 87, post. The defendants' lot is bounded on the north by a narrow strip of land owned by the Commonwealth, which runs to the Charles River. The southern boundaries of both lots form a continuous line. Wall Street lies to the south of the lots and adjoins them so that its easterly boundary if extended coincides with the westerly boundary of the plaintiff's lot, forming a line perpendicular to the southern boundary of the lots. Wall Street is a private way thirty-three feet wide and is paved up to the point where it adjoins the land of the plaintiff and the defendants. Beyond the point where the pavement ends

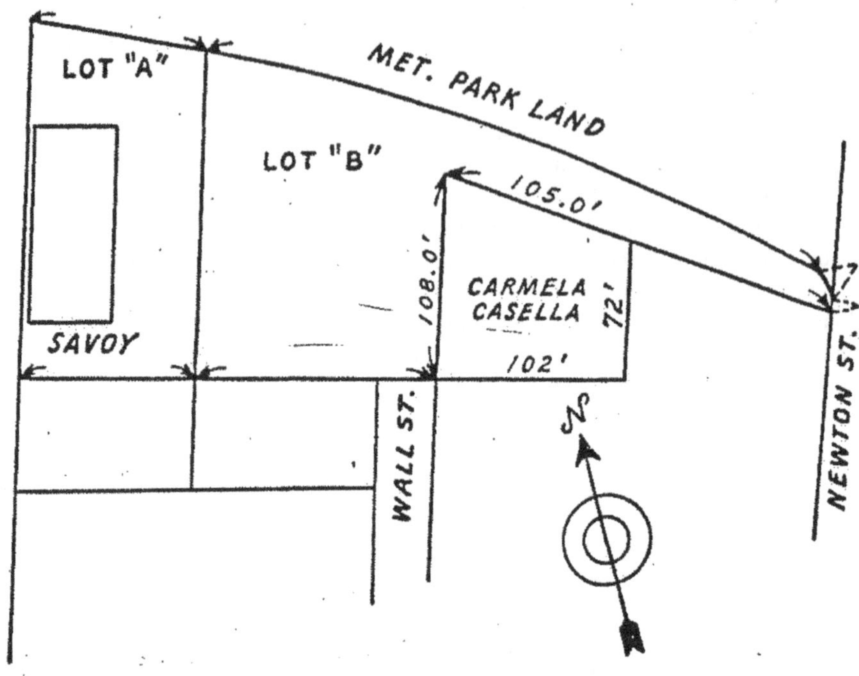

Figure 14.13

"no way had been laid out, the land being rough and uncultivated and having thereon grass and stones."

Both parcels of land involved in this litigation were at one time owned by one Wood. In 1922 what is now the plaintiff's lot was conveyed to one Durkiwiez by a deed which referred to the property as "land in Waltham situated on the easterly side of Wall Street, a private way." That part of the description contained in the deed to Durkiwiez here pertinent reads: "Beginning at the southwesterly corner of the granted premises at a point in the easterly line of said Wall Street at land of Hughes; thence running northerly on the easterly line of said Wall Street one hundred eight (108) feet more or less to a point……." In 1924 Durkiwiez conveyed the premises to the plaintiff by a deed describing the property in the same words as those just quoted. In August, 1948, Wood conveyed the land comprising lots "A" and "B" on the above mentioned plan[1] to Albany R. and Joseph G. Savoy, the defendants' predecessors in title. In October, 1948, the defendants acquired lot B from the Savoys.

In November, 1948, the defendants commenced the erection of a garage. The site of the proposed garage is twelve feet from the west line of the plaintiff's lot and fifty feet back from Wall Street. The plaintiff contends that she has a right of way over the strip of property to the west of her land corresponding to a northerly extension of

[1] The plan that bears the date of September 30, 1948, was not in existence when the conveyances to Durkiwiez and the Savoys were made and was not referred to in either of the deeds.

Wall Street and that this way is not only coextensive with her property but extends on northerly down to the river. If only the first part of the plaintiff's contention is correct, then, obviously, the proposed garage would interfere with the plaintiff's right of passage over the way, for it would be located on a continuation of Wall Street at a point opposite the plaintiff's land. The judge concluded that the oral and documentary evidence did not warrant a finding or ruling that any way was created by grant, and that if there was an attempt to create one it was "too indefinite and uncertain to establish any rights thereunder." He found, however, that the plaintiff had a right "both by prescription and by necessity" to pass over land of the defendants along her western boundary from Wall Street to a point opposite the northerly line of her residence. This way, which the judge found to be twenty feet wide and forty feet in length, did not embrace the area where the defendants' garage was to be built. A decree was entered accordingly, from which the plaintiff appealed.

Whether the plaintiff acquired a right of way by grant depends on the effect of the deed from Wood to Durkiwiez, her predecessor in title.[2] The significant words in that deed are the following: "Beginning at the southwesterly corner of the granted premises at a point in the easterly line of said Wall Street at land of Hughes; thence running northerly *on the easterly line of said Wall Street* one hundred eight (108) feet more or less to a point: (emphasis supplied). The plaintiff rightly does not contend that she acquired the fee in any part of the way. It is to be noted that the description here does not bound the property "on" or "by" Wall Street. Ordinarily a deed which bounds the premises "on" or "by" a way with no restricting or controlling words conveys title to the middle of the way if the way belongs to the grantor. *Gray* v. *Kelley*, 194 Mass. 533, 536-537. *Pinkerton* v. *Randolph*, 200 Mass. 24, 26-27. *Salem* v. *Salem Gas Light Co.* 241 Mass. 438, 441. *Erickson* v. *Ames*, 264 Mass. 436, 442-445. But the rule is otherwise where the deed describes the boundary as being "on" or "by" the side line of a way. Such a description ordinarily indicates that the grantor did not intend to part with title to any portion of the way. *Smith* v. *Slocomb*, 9 Gray, 36. *Holmes* v. *Turner's Falls Co.* 142 Mass. 590, 592. *McKenzie* v. *Gleason*, 184 Mass. 452, 458. *Hamlin* v. *Attorney General,* 195 Mass. 309, 312. *Wood* v. *Culhane*, 265 Mass. 555, 557. The description in the deed under consideration belongs to this class. Thus Wood's grantee, Durkiwicz, did not acquire a fee to the center of the way and, of course, the plaintiff, whose title stems from Durkiwicz, acquired no better right.

The question remains whether the deed to Durkiwicz, although it conveyed no fee in any part of the way, created an easement of way. The plaintiff invokes the familiar rule that, when a grantor conveys land bounded on a street or way, he and those claiming under him are estopped to deny the existence of such street or way, and the right thus acquired by the grantee (an easement of way) is not only conextensive with the land conveyed, but embraces the entire length of the way, as it is then laid

[2] The deed from the Savoys to the defendants contains the recital "subject to a right of way in Wall Street, a private way of record." But this gave no rights to the plaintiff, who was a stranger to that deed. *Haverhill Savings Bank v. Griffin*, 184 Mass. 419, 421. *Hodgkins v. Bianchini*, 323 Mass. 169, 172. (Both of these footnotes are part of the quoted judgment, and both are numbered "1" in the judgment.

out or clearly indicated and prescribed. *Tobey* v. *Taunton*, 119 Mass. 404, 410. *Ralph* v. *Clifford*, 224 Mass. 58, 60. *Oldfield* v. *Smith*, 304 Mass. 590, 595-596. *Frawley* v. *Forrest*, 310 Mass. 446, 451. *Daviau* v. *Betourney*, *ante*, 1, 3.

Although there is some authority to the contrary (see *McKenzie* v. *Gleason*, 184 Mass. 452, 458-459; *Wood* v. *Culhane*, 265 Mass. 555, 558-559), we think it must be regarded as settled in this Commonwealth that a description which bounds property by the side line of a way is no less effective to give the grantee an easement in the way, under the principle just stated, than a description which bounds the property *by* or *on* a way. *Gaw* v. *Hughes*, 111 Mass. 296. *Cole* v. *Hadley*, 162 Mass. 579. *Driscoll* v. *Smith*, 184 Mass. 221. *Hill* v. *Taylor*, 296 Mass. 107, 116. Thus had Wall Street been in existence along the westerly boundary of the plaintiff's lot at the time of the conveyance to Durkiwicz there could be no doubt, under the decisions just cited, that he and those claiming under him would have acquired an easement of way over it. That Wall Street was not then in existence would not necessarily preclude the creation of such an easement. *Lemay* v. *Clifford*, 224 Mass. 58, 60. *Tufts* v. *Charlestown*, 2 Gray, 271, 273. In the case last cited it was said, "When a grantor conveys land, bounding it on a way or street, he and his heirs are estopped to deny that there is such a street or way. This is not descriptive merely, but an implied covenant of the existence of the way" (page 272). Applying these principles here we think that the judge erred in holding that the deed under which the plaintiff claims gave her no easement over Wall Street bas continued beyond the point where it joins her lot and that of the defendants. While the judge based his conclusion on the oral as well as the documentary evidence, we find nothing in the oral evidence which overcomes the effect of the deed on which the plaintiff's rights are based. To be sure, Wall Street along the plaintiff's westerly boundary had not been laid out at the time of the conveyance to Durkiwicz; nor was it shown on any plan referred to in his deed. But it was sufficiently designated by the reference in the deed so that the grantor and those claiming under him would be estopped to deny its existence, at least as far north as the plaintiff's westerly boundary extends. To that point the way could be ascertained with reasonable certainty. Moreover, a way to some extent was necessary to a reasonable enjoyment of the property conveyed, for the judge found that the plaintiff had a way not only by prescription but by necessity. A way created by estoppel, of course, "is not a way by necessity, and the right exists even if there be other ways either public or private leading to the land." *New England Structural Co.* v. *Everett Distilling Co.* 189 Mass. 145, 152. *Hill* v. *Taylor*, 296 Mass. 107, 116. We have considered the factor of necessity only as it may bear on the intent of the parties at the time of the conveyance to Durkiwicz. It is reasonable to infer, therefore, that the reference in the deed to the side line of Wall Street was intended to have its ordinary effect and would estop the grantor and those claiming under him from denying the existence of Wall Street along the plaintiff's westerly boundary.

The extent of the plaintiff's rights beyond the limits of her land "will depend upon, and may be shown by, extrinsic facts, as they existed at the time of the conveyance." *Frawley* v. *Forrest*, 310 Mass. 446, 451. *Fox* v. *Union Sugar Refinery*, 109 Mass. 292, 295-296. The question is one of fact. *Driscoll* v. *Smith*, 184 Mass. 221, 223. Since all of the evidence is before us, we are in a position to decide it. We are of opinion that the plaintiff's easement ought not to extend beyond the limits of her land. In the closely analogous situation involving the rights of grantees of lots bounded on a way shown on a plan referred to in a deed, the court in determining the extent of the easement has taken into consideration what was necessary for the enjoyment of the premises granted (although the necessity need not be an absolute or physical one) and whether the way referred to at its distant end connected directly or indirectly with a public highway. *Wellwood* v. *Havrah Mishna Anshi Sphard Cemetery Corp.* 254 Mass. 350, 355. See *Prentiss* v. *Gloucester*, 236 Mass. 36, 53. Here, if the extension of Wall Street were to be continued beyond the limits of the plaintiff's land, it would lead not directly or indirectly to any public way, but only to the narrow strip along the river owned by the Commonwealth. It would not be necessary to any reasonable enjoyment of the granted premises. It is hardly conceivable that an extension of the way to that point was within the contemplation of the parties when the conveyance to Durkiwicz was made. Beyond the limits of the plaintiff's land it cannot be said that the way was "clearly indicated and prescribed." *Frawley* v. *Forrest*, 310 Mass. 446, 451.

It follows that the decree of the court below is reversed and a new decree is to be entered based on a right of way appurtenant to the plaintiff's land by the prolongation of Wall Street, at its original width, along the entire length of the plaintiff's western boundary. The plaintiff is to have costs of this appeal. *So ordered.*

CASE #8 THE MARGINAL ROAD, A SPECIAL CASE

TORREY v. PEARCE
92 ARIZ. 12, 373 P.2D 9 (1962)

The appellants, who were the plaintiffs below, filed a quiet title action in the Superior Court. Both sides agreed that the matter was one of law and both sides moved for summary judgment. The trial court ruled in favor of the defendants and this appeal followed.

The various land areas will be referred to as noted on the attached partial tracing of the subdivision of the land involved. This tracing was made for the court from a large copy of the official map of Why Worry Farms Amended. A copy of the plat was included in the record on appeal. Portions of the material appearing on the official plat have been eliminated and other matters have been added.[1]

[1] The designations 'North Alley' and 'East Alley' have been inserted by the court. The 9-foot offset between the extended common boundary line between Parcels 15 and 16 on the one hand and the extended common boundary line between Lots 1 and 2 on the other has been inserted. The designations 'Goodnight—Parcel 16' and 'Pearce—Parcel 15' have been inserted by the court. The distance underneath the name 'Goodnight,' that is to say, the distance of 354 feet, has been inserted, as has the distance from the west line of Lot 2 to the east line of the land included within the plat of the subdivision. The northerly boundary of Why Worry Farms Amended extends easterly from the northwest corner of the southeast quarter of the northeast quarter of Section 31, Township 3 North, Range 3 East, and all matters northerly therefrom have been inserted by the court as being matters which are not contained in the subdivision tract as recorded. The north-south line at the easterly extremity of Parcel 15 is not placed with accuracy as that boundary line was not disclosed in the record.

The court will use the last names of the person and a singular name includes husband and wife. We will omit recitation as to mesne conveyances where the same do not affect the legal problems involved.

The following is the important chronology:

Prior to November 9, 1948, Goodnight owned Parcels 15 and 16 as well as the land adjacent thereto and immediately to the south thereof, which later became Why Worry Farms Amended, the whole thereof being one continuous piece of land.

On November 9, 1948, the map or plat of Why Worry Farms Amended was filed in the office of the County Recorder of Maricopa County. The subdivision included nine lots. The entire northerly 5 feet of the subdivision, which is adjacent to the south boundary of Parcels 15 and 16, was designated as Tract A. There was an unnamed 20-foot strip along the north side of the subdivision adjacent to and southerly from Tract A, which the court herein arbitrarily calls the North Alley. This 20-foot strip continued down the east side of the subdivision, which the court arbitrarily calls the East Alley, and at the southerly extremity of the subdivision the 20-foot strip turns westerly to 11th Avenue. In the interior of the subdivision there is a 45-foot U-shaped area bearing the name of Why Worry Lane which touches upon all of the lots and gives access to all of the lots. The subdivision plat contains the following 'dedication.'

'And hereby declare that the accompanying plat sets forth the location and gives the dimensions of the lots, streets and alleys * * * and that the streets and alleys shown thereon are hereby dedicated to the public for its use and benefit. The easements shown thereon are for the use of irrigation ditch or pipe.'

By a warranty deed recorded on March 23, 1949, Goodnight conveyed Lots 1 and 2 to Converse. These were described by the reference to the numbered lots in the recorded plat. The deed recited that the conveyance was 'subject to' certain matters of record expressly recited therein.

By a warranty deed recorded on March 20, 1952, Goodnight conveyed Parcel 15 to Converse, and in addition thereto the east 378 feet of Tract A. Tract 'A' was subject to easements as recited in the dedication which is a part of the subdivision plat. 378 feet is the total lineal distance of the combined north lines of Lots 1 and 2, plus the 20-foot width of the alley east therefrom between the east line of Lot 1 and the east line of the subdivision.

On the 25th day of August, 1952, the Board of Supervisors vacated the entire alley, that is, the 20-foot strip on the north, on the east and on the south of and included within the Why Worry Farms Amended subdivision. As of the moment of this action by the Board of Supervisors Goodnight owned Parcel 16 and Converse owned Parcel 15, together with the east 378 feet of Tract A, together with Lots 1 and 2. All parties agree that after vacation and abandonment of the north alley by the Board of Supervisors, Converse owned the unencumbered fee in the east 378 feet thereof. This is the area presently in dispute.

By deed recorded on February 18, 1955, Converse conveyed Lot 1 (and a portion of Lot 2 which is not material to this litigation), making reference to the lot by the book and page of the recording of the map or plat of the subdivision. This deed carried the same 'subject to' provisions which were included in the

Goodnight deed to Converse in relation to Lots 1 and 2 in March of 1949. This deed contained no reference to the vacating of the alley by the Board of Supervisors. Title eventually became vested in Torrey by a series of deeds not varying in essential terms.

By deed recorded on August 31, 1955, Converse conveyed the balance of Lot 2, making reference to the book and page of the recording of the map. In this deed, however, we find not only the recitations contained in the Torrey deed but in addition thereto we find the following: 'Subject to * * * the effect of the abandonment of the 20-foot easement for roadway bordering the north side of Lot 2, as shown by Resolution, * * * . Title eventually vested in the plaintiff Hunter.

By deed recorded on September 17, 1957, Converse conveyed to Pearce, this deed containing three separate property descriptions, being, first, the land called Parcel 15; second, the east 378 feet of Tract A; and, third, the east 378 feet of the north alley (the disputed area).

Hunter and Torrey claim ownership of that portion of the north alley adjacent to and northerly from their north lot lines. Their theory is that the deeds of Converse conveying 'Lot 1, Why Worry Farms' and 'Lot 2, Why Worry Farms' conveyed the abandoned alley as a matter of law, and that the deed to Pearce, although it described the disputed area, did not convey it because Converse then had nothing left to convey. Hunter and Torrey filed their suit to quiet title against Pearce and Pearce claims the same land.

It is an almost universal rule that if land abutting on a public way is conveyed by a description covering only the lot itself, nevertheless, the grantee takes title to the center line of the public way if the grantor owned the underlying fee, unless the contrary intention sufficiently appears from the granting instrument itself, or the circumstances surrounding the conveyance.[2] The rule is one of construction, and applies in the absence at the time of the grant of any indication as to the grantor's intent in dealing with his interest in the public way, *Croker v. Cotting*, 166 Mass. 183, 44 N.E. 214, 33 L.R.A. 245 (1896); *Hobson v. City of Philadelphia*, 150 Pa. 595, 24 A. 1048 (1892); *Huff v. Hastings Express Co.*, 195 Ill. 257, 63 N.E. 105 (1902). Cf. *Overland March. Co. v. Alpenfels*, 30 Colo. 163, 69 P. 574 (1902). In the circumstances where the grantor conveys a parcel abutting a marginal roadway, if the grantor owns the underlying fee in the full width of the road, but owns no interest in the lands abutting the road on the side opposite the land conveyed, the rule gives to the grantee the grantor's interest in the full width of the roadway. *Lamprey v. American Hoist & Derrick Co.*, 197 Minn. 112, 266 N.W. 434 (1936); *Haines v. McLean*, 154 Tex. 272, 276 S.W.2d 777 (1955); *Rowe v. James*, 71 Wash. 267, 128 P. 539 (1912).

In its rationale, the rule recognizes that grantors tend to think of their holdings in terms of the areas within their lot lines, and to overlook any fee interest they might have in abutting roads or alleys. The sound presumption is that if the grantor had

[2] See, e.g. In re Parkway in the City of New York, 209 N.Y. 344, 103 N.E. 508, 2 A.L.R. 1 (1913) [case # 10] and the cases collected in annotations at 2 A.L.R. 6 and 49 A.L.R. 982.

been cognizant of his interest in the abutting public ways, he would have included it in his conveyance to the grantee, rather than retain an interest in a strip or gore of land of no utility to himself, *Mott v. Mott*, 68 N.Y. 246 (1877); *Henderson v. Hatterman*, 146 Ill. 555, 34 N.E. 1041 (1893).

The question now before this court is whether this same presumption or rule of construction should apply when a grantor conveys land described by lot number where that land abuts on a vacated alleyway. The question is not novel,[3] although it is one of first impression before this court.

Representative of the early opinions considering this issue is *White v. Jefferson*, 110 Minn. 276, 124 N.W. 373, 125 N.W. 262, 32 L.R.A., N.S., 778, 784 (1910), where the facts were essentially similar to the case now before us. There the court said:

'On vacation of a street like the one at bar, [the owner of adjacent lots] owns lots 23 and 24 and the space between in fee simple. He can transfer the whole tract, or any part of it, or transfer lot 23 to any person, and lot 24 to another person, and the space between the two to a third person. * * * The land which had been a street assumed exactly the same legal status as any other land which had not been impressed with a public easement. There is neither mystery nor magic in the word 'street.' The easement of use is the significant fact. When the easement ceases, there is no occasion nor justification for any imputation of intention. * * *' 110 Minn. at 284, 124 N.W. at 375.

The same line of reasoning was adopted in *Sanchez v. Grace M.E. Church*, 114 Cal. 295, 46 P. 2 (1896); *Brown v. Taber*, 103 Iowa 1, 72 N.W. 416 (1897); and a series of Washington cases including *Hagen v. Bolcom Mills, Inc.*, 74 Wash. 462, 133 P. 1000, 134 P. 1051 (1913) and *Turner v. Davisson*, 47 Wash.2d 375, 287 P.2d 726 (1955).

CASE #9 ROAD CONSTRUCTED OUTSIDE OF LAYOUT

SPROUL v. FOYE
55 ME. 162 (1867)

When a call in a deed bounds one side of the land therein conveyed "by the new county road leading from" a place named "to" another place named, and the road as located by the commissioners, and that as actually wrought and traveled, are not identical, the latter alone will answer the call.

On Exceptions

Writ of Entry The land demanded in the writ, was described as "beginning on the southerly line of the *existing road* leading from Wiscasset to Dresden, at," &c., thence in a certain direction named, to a "birch tree marked;" thence in a certain other direction named, to a monument named; thence in a certain other direction, "to the

[3] Cases which discuss this issue are collected in annotations at 32 L.R.A., N.S., 778, 2 A.L.R. 33 and 49 A.L.R.2d 1002.

road aforesaid; thence south-easterly, by the road *as it existed and has been traveled since March* 14, A.D. 1853, to the first mentioned bound," & c. [Figure 14.14]

The defendant seasonably filed a disclaimer to "all of said described premises, except that part between the road as laid out by the county commissioners and the road as the same was built," and, as to the strip this excepted, he pleaded *nul disseizin*. The plaintiff accepted the disclaimer and joined issue.

The deeds from the defendant to one Decker, dated March 14, 1853, and from the latter to the plaintiff, dated March 19, 1853, described the land as "beginning on the southerly side of the new county road leading from Wiscasset to Dresden," at, &c.; thence in a certain direction named, "to a birch tree marked;" thence in a certain other direction named, to a monument named; thence in a certain other direction named, to the aforesaid county road; "thence by said county road, to the bound began at," &c.

It was admitted that the defendant owned the premises in controversy, prior to his deed to Decker.

Several witnesses testified upon each side in regard to numerous facts connected with the land in controversy. The defendants contended and requested the presiding Judge to rule that there was no latent ambiguity in the deeds; that it was the duty of the Court to construe them; that the evidence did not warrant the Court in submitting the question of construction to the jury; and that, upon the legal construction of them, the line of the new county road as it was actually located and marked upon the face of the earth, was the one, and the only one, that answered the call in the deeds. But the presiding Judge declined to so rule and submitted the question to the jury.

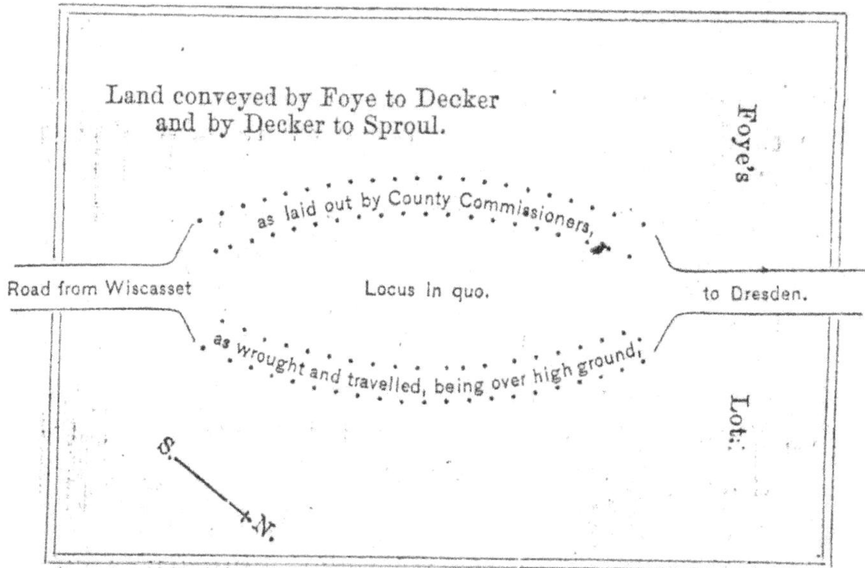

Figure 14.14 Diagram from court decision.

The verdict was for the demandant, and the defendant alleged exceptions, and also filed a motion to set aside the verdict as being against law and the manifest weight of evidence.

Walton, J.—This is a writ of entry. The question is one of boundary. The defendant conveyed a parcel of land, bounding it on one side by "the new county road leading from Wiscasset to Dresden." Some forty or fifty rods of this road was built outside of the original location of the county commissioners, and the question is whether the line of location or the road as actually built is to be regarded as the true boundary of the land conveyed.

The deed bounds the land "by the new county road leading from Wiscasset to Dresden." At this time the road had been open and used for public travel about three years. Did the parties refer to the road as located, or to the road as built? To a mere line of location not wrought, not in use for public travel, or to the road that was wrought and in actual use as a public highway?

When a road is referred to in a deed as one of the boundaries of the land conveyed, we should ordinarily suppose that something more than a mere location was meant. A road is way actually used in passing from one place to another. A mere survey or location of a route for a road is not a road. A mere location for a road falls short of a road as much as a house lot falls short of a house. "Bounded by the new county road leading from Wiscasset to Dresden." Can the proposition be maintained that an invisible and unwrought location answers such a call better than a visible wrought road over which the public travel is passing daily? We think not. We think such a call in a deed requires something more than a mere location to answer it.

Our conclusion is that the Judge who presided at the trial committed no error in declining to instruct the jury that the original location would alone answer the call I the deed; that, if he erred at all, it was in not ruling as matter of law that the road, as actually built, was the only boundary that would answer the call. This error, however, if it was one, operated favorably for the defendant, and against the plaintiff, and was one therefore of which the defendant cannot justly complain. It gave him a chance of winning from the jury if he could a point which should have been ruled against him as matter of law.

The verdict of the jury establishes the road as built and not the road as located, as the boundary of the land conveyed. We think the verdict is right.

Motion and Exceptions overruled.
Judgment on the verdict.

CASE #10 REVERSION OF A CEMETERY LOT

FROST v. COLUMBIA CLAY CO. ET AL.
130 S.C. 72, 124 S.E. 767 (1924)

Appeal from Common Pleas Circuit Court, of Richland County; J.W. De Vore, Judge.

Action by W.H. Frost against the Columbia Clay Company and others. Judgment for defendants, and plaintiff appeals. Reversed, and new trial ordered.

The defendants claimed that plaintiff had abandoned the burial ground and left it uncared for for a period of more than 20 years preceding the commencement of such action, and during such time had asserted no claim in the property and that defendants did not know that the land was a burial ground and discontinued excavating on discovery thereof. Witnesses for defendants testified that there was nothing to denote where burial ground was and that land had no characteristics of a cemetery. Defendants' employees testified that they did not know that land was a graveyard, that they had received no notice thereof, and that on discovery of bones they had offered to assemble bones, level up yard, put fence around it and maintain it, and would not have made excavations if they had known that land was burial ground.

Cothran, J., dissenting.

D.W. Robinson and D.W. Robinson, Jr., both of Columbia, for appellant.

Thomas & Lumpkin, of Columbia, for respondents.

FRASER, J.

The respondents make the following statement:

"As noted in the case for appeal, this was an action commenced in November, 1922, by the plaintiff against the defendants, for damages for the trespassing on and excavating in a graveyard which had formerly been the family burying ground of the plaintiff's family.

The trial court directed a verdict for the defendant on the ground that the alleged burial place had been abandoned.

It is admitted that excavations were made in what had formerly been the burying ground of the Frost family, and that some bones were unearthed.

In the motion for directed verdict for the defendant, a number of grounds were submitted, but the presiding judge stated that there were only two grounds which he would consider; that is to say, the question of notice to the defendant company of the existence of a graveyard and the question of abandonment, and the other grounds were overruled, but in order to properly controvert the exceptions raised by the appellant it is necessary to bring to the attention of this court the several grounds raised on the trial of the case by the respondents.

The court directed a verdict in favor of the defendant, on the primary ground that the alleged burial place had been abandoned. Using the words of the court in commencing on the testimony of the plaintiff, Mr. Frost, the judge, says: "The undisputed evidence of the plaintiff in this case, out of his own lips, is that that graveyard had not been used in over 20 years; and it was over 20 years ago since the last two people were buried in it. It has grown up in trees or shrubbery, and has been discontinued as a burial place, even for the family of the plaintiff.'"

The plaintiff moved for the direction of a verdict, leaving only the amount to the jury. The presiding judge refused the motion of the plaintiff and directed a verdict for the defendant, as above stated. It is not necessary or proper to discuss the evidence, as this case must go back.

1. The first question is: Is there indisputable evidence that this family graveyard had been abandoned? There was not. While there is no case in this state directly on point, yet Ex parte McCall, 68 S.C. 492, 47 S.E. 974, throws some light on it.

"It is also true, as a general proposition, that where ground has been dedicated to the public for use as a cemetery, the owner cannot afterward resume possession or remove the bodies interred therein, although he has received no consideration for its use, and the interments were made merely by his consent. [Cases cited.] This doctrine is somewhat anomalous, and is not to be extended beyond the principle upon which it is founded. That principle is that the most refined and sacred sentiments of humanity cluster around the graves of departed loved ones, and that when these sentiments have become associated and connected with a particular spot of ground, by the invitation or consent of the owner, he shall not, for any secular purpose, disturb them."

The laws do, or should, set forth the sentiment of the people who are subject to them. This is particularly true under a government like ours. From the time of Abraham, the places where the dead were buried have been considered sacred and inviolate. All nations respect the graves of the dead. The graves of ancestors are a subject of idolatrous worship by the heathen. Our literature is full of references to the last or final resting place. In some of our church literature we find the statement that the bodies of our dead "do rest in their graves until the Resurrection." These burial places are sometimes called "God's Acre." The idea of perpetuity runs through nearly all references to the grave. Of course, as in nearly every other matter, there are exceptions. As was said in Ex parte McCall, sometimes the tenderest love requires the removal to more suitable surroundings.

The abandonment of a burying place is accomplished by the removal of the remains to a more suitable place. The change of the place for the burial of those who die, or have died after a given time, does not constitute an abandonment of a graveyard, and his honor was in error in so holding.

2. There was no error in refusing to direct a verdict for the plaintiff. While it is true that a graveyard may not be abandoned except by the removal of the remains of the dead, that does not mean that a man may not abandon his right to damages, for the negligent or willful invasion of what should be a hallowed spot. The defendant pleaded estoppel and there was evidence upon which a verdict for estoppel might have been based. The injunction is not in issue. To this the respondent consents.

Besides this, the defendants claim that they did not know that there was a graveyard there, without fault of their own, they dug up the bones, and at once stopped when they found out their mistake. These were matters for the jury.

The judgment is reversed, and a new trial ordered.

WATTS and MARION, JJ., concur.

COTHRAN, J. (dissenting).

I think that Judge De Vore was entirely right in directing a verdict for the defendants, and I therefore dissent from the conclusion announced to the contrary, in the opinion of Mr. Justice FRASER.

This is an action for $50,000 damages, and for a permanent injunction, against the defendants, on account of an alleged trespass committed by them, in excavating through a graveyard in which relatives of the plaintiff had been buried, by which the bones of the dead were disturbed and scattered.

The graveyard in question is located a few miles north of Columbia, on a part of the old Faust or Frost lands, at one time owned by Sarah Faust, who was the grandmother of the plaintiff, having married John D. Frost, his grandfather. Later it was owned by J. Frost Walker, and by him it was conveyed to F.H. Hyatt, who conveyed a part of lit t the defendant Columbia Clay Company. The area of the graveyard, by survey is 100X150 feet, elliptical in form, and contains slightly less than one-third of an acre. The line between the defendant's land and the other part of the Frost lands divides the graveyard plot about equally. It had been established as a family graveyard many years ago, perhaps 100. Both of the plaintiff's grandparents were buried there, one of them in 1880, and the other many years before; a brother and sister were buried there about 45 years ago, also an infant daughter in May, 1901, and an uncle in July that year. The uncle was the last person to be buried there. Since that time, 1901, it appears to have been recognized by the older people simply as the place where the persons above named had been burier. During the time the land was owned by Hyatt, all of the trees in and around the graveyard were cut down, and the spot has long since grown up in weeds, bramble, and blackberry bushes. Not a single member of the family is shown to have visited it in more than 20 years; not a rake, hoe, or axe has broken the "solemn stillness"; no fence has inclosed it; not a monument, or even a rude headstone, marks a single grave; the mounds even have long since disappeared, leveled with the ground, as if emphasizing the consignment of "dust to dust"; it has presented a picture of abject neglect; the silent sleepers have become, indeed, "to dumb forgetfulness a prey." It may appropriately have been called "God's Acre," for He alone had visited it and hidden with undergrowth the human shame of neglect.

"Above the graves the blackberry hunt,
In bloom and green its wreath,
And harebells swung, a if they sung
The chimes of peace beneath."

The plaintiff himself could not locate a single grave and exhibited no interest until the alleged desecration occurred, when he called upon the negro gravediggers to assist him in locating the graves, which they could not do. He then gathered the scattered bones and skull of his relative together, with the remark, "These will be good evidence," and carried them home in preparation for the damage suit which quickly followed. His witness Elliott says:

"It looked like a waste piece of ground. I was there numbers of times shooting around it and playing around, and I didn't know it was a graveyard. * * * In fact, I have been there numbers of times before I knew it was a burying ground."

How it can be said that there is no indisputable evidence that the family graveyard has been abandoned is inconceivable to me.

In the summer and fall of 1922, the defendant, in order to lay a side track or spur track, made two excavations along the dividing line, and at both times exposed human bones, some of which the plaintiff, as stated, took care of. The plaintiff himself testified that, as soon as Smith discovered that they had gotten into the cemetery with the excavation, he immediately ceased operations; tried in every way possible to take care of the plot of ground, by fencing it in and taking care of the bones that had been uncovered. It is for this excavation, under these circumstances, that the plaintiff asks $50,000 damages!

The plaintiff makes no claim of title to the land where the graveyard was located; that is unquestionably in the defendant company. His right to maintain an action for the desecration of the graves must therefore be derived from some other source; either from a deed by which the graveyard or his right to burial was established, or from a dedication of the plot by the owner, for the purposes of a graveyard. He makes not claim under a deed, but does claim that a dedication was effected by the acts and conduct of John D. Frost, the former owner.

I do not think that there can be any doubt as to the proposition that a burial ground may be so dedicated, either by express instrument, by adverse use, or by permission or license to so use the property. Nor do I entertain the slightest doubt that the former owner established here a family graveyard and dedicated the plot to that use.

A beneficiary, either under a deed or a dedication, establishing a certain plot of ground as a graveyard acquires two distinct rights: 1. The right to bury his dead there; and, 2, the right to protect the graves of the buried dead from desecration. The rights, of necessity acquired from one or the other source, are coterminous with the source; they expire with its annihilation; if they come by deed, they may be destroyed by deed; if they come by either, they may destroyed by abandonment.

The plaintiff does not insist upon the right to bury his dead in the old burying ground, and therefore this right need not be considered in the case; he is only interested in the alleged breach of his right to demand that his buried dead by not disturbed.

The question in the case is whether or not the utter abandonment of the burial ground, by those for whose benefit it was dedicated, is an abandonment of the right to demand that the graves shall not be interfered with; in other words, whether or not this right is dependent upon the maintenance of the burial ground as such. It is declared, in the opinion of Mr. Justice FRASER:

> "The abandonment of a burying place is accomplished by the removal of the remains to a more suitable place. * * * While it is true that a graveyard may not be abandoned except by the removal of the remains of the dead, etc."

With this proposition I do not agree. On the contrary, I am of opinion that all rights connected with it may be abandoned, as all other rights may be; and that, where the beneficiaries of the dedication of land for a graveyard have conducted themselves towards it as the plaintiff and his relatives have done in this case, the abandonment is complete, regardless of the matter of removal of the bodies buried there.

The abandonment of a graveyard is not shown by the fact that no more burials may be accommodated within the space. The tender and commendable care for the resting place of the dead may be as clearly directed towards a cemetery that is fully occupied as towards one that is not so. So long as the space is maintained as a burial place, either with regard to those already buried or to those who may thereafter be buried there, it is holy ground and human instinct responds to this sentiment. Nor is the fact that interments which might be made there are no longer made conclusive of abandonment. Although the use of the graveyard for burials may have ceased, if it should be maintained and cared for out of respect for those already buried, it is none the less a cemetery and entitled to be treated as such.

> "So long as a cemetery is kept and preserved as a resting place for the dead, with anything to indicate the existence of graves, or so long as it is known or recognized by the public as a cemetery, it is not abandoned. But where a cemetery has been so neglected as entirely to lose its identity as such, and is no longer known, recognized, and respected by the public as a cemetery, it may be said to be abandoned." 11 C.J. 58.

In 5 R.C.L. 242, it is said:

> "On the other hand, it may contain the remains of the dead and yet be abandoned. If no interments have for a long time been made, and cannot be made, there, and in addition thereto the public and those interested in its use have failed to keep and preserve it as a resting place for the dead, and have permitted it to be thrown out to the commons, the graves to be worn away, gravestones and monuments to be destroyed (or, as in the case at bar, never to have been placed), and the graves to lose their identity, and, if it has been so treated and used, or neglected by the public as entirely to lose its identity as a graveyard, and is no longer known, recognized, and respected by the public and those interested in its use as a graveyard, then it has been abandoned."

In *Hunter v. Sandy Hill*, 6 Hill (N.Y.) 407, it is said:

> "When these graves shall have worn away, when they who now weep over them shall have found kindred resting places for themselves, when nothing shall remain to distinguish this spot from the common earth around, and it shall be wholly unknown as a graveyard, it may be that some one who can establish a good 'paper title,' will have a right to its possession; for it will then have lost its identity as a burial ground, and, with that, all right founded on the dedication must necessarily become extinct."

In 5 A. & E. Enc. L. 797, it is said:

> "Where land has been dedicated to cemetery purposes, and there has been a lawful and effectual abandonment of the cemetery as such, the land will revert to the original owner."

In *Tracy v. Bittle*, 213 Mo. 302, 112 S.W. 45, 15 Ann. Cas. 167, it is said:

> "In the latter class of cases [dedication], if there comes a time when the bodies are all removed, or when by other conditions a clear abandonment of the graveyard is made apparent, then the right of the public, which is somewhat in the nature of an easement, ceases, and the land reverts to the original owner or his grantees. By mere ceasing to make further interments does not abandon the graveyard, as we have seen, so long as it is kept in condition to be known, and is known, as a burying ground."

In *Campbell v. City of Kansas*, 102 Mo. 326, 13 S.W. 897, 10 L.R.A. 593, it is said:

> "The use of a graveyard is twofold—for the purpose of continuous burials, and for the purpose of preserving the remains and memory of those who have been buried. The original uses can be continued only by the public continuing to bury, or by continuing to protect the remains already buried, and to preserve the identity and memory of the persons who have left them. * * * The public may cease to bury in the dedicated ground whenever it pleases. It may also refuse or neglect to either erect or preserve any monuments to indicate the identity of those already buried, or to give and continue to the place the character and name of a graveyard. When this happens, the original use terminates and the fee vests in the original donors or their legal representatives, free from it."

In *Trefry v. Younger*, 226 Mass. 5, 114 N.E. 1033, it is held that the interest of the owner of a burial lot is in the nature of an easement; it is not an absolute right in the property, but the right of burial, so long as the place continues to be used as a burial ground. It may have been added that it was also the right, so long as the place continues to be used as a burial ground, to prevent a disturbance of the bodies. Both rights spring from the same easement, and if the easement should be lost by abandonment both rights would necessarily perish with the destruction of the easement. In *Grinnan v. Lodge*, 118 Va. 588, 88 S.E. 79, Ann. Cas. 1918D, 729, it is said:

> "The courts are much divided as to the character of the estate one may have in a burial lot in a cemetery. It is certain that it not a fee. The weight of authority is, and we think the better view, that it is a mere privilege or license to make interments in the lot exclusively of others as long as the burying ground or cemetery remains as such."

Coupled with this right, and referable to the same source, conveyance, or dedication, is the right to prevent the desecration of the graves of relatives. This right, as the other, can continue only as long as the plot remains as a cemetery and necessarily is lost by the loss of the dedication.

It is stated, in *Stewart v. Garrett*, 119 Ga. 386, 46 S.E. 427, 64 L.R.A. 99, 100 Am. St. Rep. 179, and in *Anderson v. Acheson*, 132 Iowa 744, 110 N.W. 335,

9 L.R.A. (N.S.) 217, that the purchaser of a lot in a public cemetery, though under a deed absolute in form, does not take any title to the soil, but that he acquires only a privilege or license to make interments in the lot purchased, exclusive of others, so long as the ground remains a cemetery. The point was not at issue in these cases as to the right to prevent a desecration of the graves of those who had been buried under such privilege or license, but unquestionably it flowed from the privilege or license equally with the right of burial, and, upon logical grounds, would disappear along with a destruction of the privilege or license. In my opinion, the case of *Hines v. Tennessee*, 127 Tenn. 1, 149 S.W. 1058, 42 L.R.A. (N.S.) 1138, expresses the law clearly and correctly. The facts thus stated are in striking contrast with those of the case at bar:

> "W. Crawford, more than 60 years ago, set apart about one acre of his farm, in one of the cultivated fields, as a family burial ground or cemetery, and it was so used by him during his life, several of his family being then and there buried, and when he died he was there buried. His descendants, since his death, have used it as a family burying ground, and many of them are there buried. Monuments and gravestones have been erected and maintained over several of the graves, and from time to time have been repaired, and the cemetery put in order and otherwise cared for."

A subsequent purchaser of the farm was indicted by the descendants of Crawford for desecrating the graves, under a statute. The defense was that he was not bound by the dedication and that the rights of the Crawfords were barred by the statute. The court held that the purchaser took with notice of the dedication and was bound to respect it. As to the bar of the statute, the court held:

> "Nor is the right barred by the statute of limitations, so long as the lot is kept inclosed, or, if unenclosed, so long as the monuments and gravestones marking the graves are to be found there, or other attention is given to the graves, so as to show and perpetuate the sacred object and purpose to which the land has been devoted. No possession of the living is required in such cases, and there can be no actual ouster or adverse possession, to put in operation the statute of limitations, so long as the dead are there buried, their graves are marked, and any acts are done tending to preserve their memory and mark their last resting place."
>
> "When a cemetery association or church sells particular lots in a cemetery, the purchaser becomes the owner of the soil, and manifestly his right to its possession protects interments made by him from disturbance." Ex parte McCall, 68 S.C. 489, 47 S.E. 974; *Vance v. Ferguson*, 101 S.C. 125, 85 S.E. 241.

If the right of a purchaser of a cemetery lot to protect interments made by him from disturbance, is derived from his ownership of the soil, certainly the same right in one who claims as the donee of a privilege or license must be derived from such privilege or license secured by the dedication. If the dedication should fail, the right goes with its failure. In *Kelly v. Tiner*, 91 S.E. 41, 74 S.E. 30, the court, quoting from a case stated (*Davidson v. Reed*, 111 Ill. 167, 53 Am. Rep. 613), says:

> "If one has been permitted to bury his dead in a cemetery by the express or implied consent of those in proper control of it, he acquires such possession in the spot of ground in which the bodies are buried, as will entitle him to maintain an action of trespass 'quare clausum fregit' against the owners of the fee or strangers who, without his consent, negligently or wantonly disturb it."
>
> The "burial of the dead body in a cemetery lot is the only possession, when claimed and known, necessary to ultimately create complete ownership of the easement so as to render it inheritable; and as long as gravestones stand, marking the place as burial ground, the possession is actual, adverse, and notorious." *Hook v. Joyce*, 94 Ky. 450, 22 S.W. 651, 21 L.R.A. 96.
>
> "When one is permitted to bury his dead in a public cemetery, by the express or implied consent of those in proper control of it, he acquires such a possession in the spot of ground in which the bodies are buried as will entitle him to action against the owners of the fee or strangers, who, without his consent, negligently or wantonly disturb it. This right of possession will continue as long as the cemetery continues to be used." *Bessemer Co. v. Jenkins*, 111 Ala. 135, 18 South. 565, 56 Am. St. Rep. 26.
>
> "When one buries his dead, therefore, in soil to which he has the freehold right, or to the possession of which he is entitled, it would seem there is no difficulty in his protecting their graves from insult or injury, by an action of trespass against a wrongdoer. But bodies are most commonly interred in public cemeteries, where the parties whose duty it is to give them burial are not the owners of the soil by deed properly executed, and have no higher right than a mere easement or license. Of such it is held that they do so under a mere license and their exclusive right to make such interments in a particular lot would be limited to the time during which the ground continued to be used for burial purposes; and, upon it ceasing to be so used, all they could claim would be that they should have due notice and an opportunity to remove the bodies to some other place of their own selection, if they so desire, or, on failure to do so, that the remains should be decently removed by others." *Bessemer Co. v. Jenkins*, 111 Ala. 135, 18 South. 565, 56 Am. St. Rep. 26.
>
> "Such right of burial is not an absolute right of property, but a privilege or license, to be enjoyed so long as the place continues to be used as a burial ground." *Page v. Symonds*, 63 N.H. 17, 56 Am. Rep. 481.
>
> "Where one is permitted to bury his dead in a public cemetery, by the express or implied consent of those in proper control of it, he acquires such a possession in the spot of ground in which the bodies are buried as will entitle him to action against the owners of the fee or strangers, who, without his consent, negligently or wantonly disturb it. His right of possession will continue as longs as the cemetery continues to be used." *Bessemer Co. v. Jenkins*, 111 Ala. 135, 18 South. 565, 56 Am. St. Rep. 26.

It is held, in the case of Ex parte McCall, 68 S.C. 489, 47 S.E. 974, that in the case of a dedication, even when the cemetery has not been abandoned, circumstances may arise which will justify the owner of the land in requiring the removal of bodies; in other words, a revocation of the license. That which may be revoked certainly may be abandoned.

> "Trespass quare clausum fregit cannot be sustained by the owner of property, not in possession, nor entitled to the possession thereof, at the time of the alleged trespass." Note, 56 Am. St. Rep. 37.

If, therefore, at the time of the alleged trespass, the graveyard had been abandoned, the plaintiff's right of possession under the privilege or license, either to burial in that space or to protect the bodies already there, had come to an end ad his right to maintain trespass no longer existed.

In *Badeaux v. Ryerson*, 213 Mich. 642, 182 N.W. 22, it is held (quoting syllabus):

> "A common-law dedication does not pass the fee, but only an easement, and where land is conveyed subject to easement in public to use land as a cemetery, the possession and beneficial use of the land, on public's abandonment of cemetery, reverts to grantee."
>
> "Right of burial in a cemetery is not absolute right of property but merely a privilege * * * to be enjoyed so long as the place continues to be used as a burial ground." *Brown v. Hill*, 284 Ill. 286, 119 N.E. 977.
>
> "Where land is dedicated for a burying ground, whether by a common-law dedication, under which the fee remains in the owner, or pursuant to Acts Ohio * * * abandonment of the land as a burying ground restores the former owner to his right of possession." *Mahoning v. Young*, 59 Fed. 96, 8 C.C.A. 27.

For these reasons, it appears perfectly clear to me that, as the graveyard has for years been abandoned by those interested in its preservation as such, the dedication is extinguished, and the only rights which the plaintiff ever enjoyed, the right to bury his dead and the right to recover damages for the desecration of the graves, were lost to him by the abandonment of the dedication, the source of his rights while they existed.

CASE #11 DETERMINING TITLE TO LAND PARCEL WHEN A ROAD IS RELOCATED

CARPENTER v. LUKE
225 W.VA. 35, 689 S.E.2D 247 (W.VA., 2009)

PER CURIAM: The defendant below and appellant herein, Shirley Blaniar Luke (hereinafter referred to as "Ms. Luke"), appeals from an order entered December 18, 2007, by the Circuit Court of Harrison County. By that order, the circuit court denied Ms. Luke's motion to alter or amend judgment and motion for a new trial. In the underlying case, the circuit court entered judgment as a matter of law in favor of the plaintiff below and appellee herein, Betty Lou Zirkle Carpenter (hereinafter referred to as " Ms. Carpenter"), on the issue of the disputed ownership of certain real estate. The remaining issues of Ms. Luke's claims of adverse possession, a prescriptive easement, and unjust enrichment were submitted to the jury and resulted in a jury verdict in favor of Ms. Carpenter. Based upon the parties' arguments, the record designated for our consideration, and the pertinent authorities, we affirm the decisions of the circuit court.

I. Factual and Procedural History This case involves a dispute over the ownership of real property. Ms. Carpenter filed a declaratory judgment action in 2006, seeking a judicial determination that she is the owner of certain real estate. Both Ms. Carpenter and Ms. Luke claim ownership of the tract of land.

As evidenced by the record submitted in this case, the disputed property is located along State Route 3 in Harrison County, West Virginia. In 2005, Ms. Carpenter inherited property on the eastern side of the route when her father died. Ms. Carpenter's father had owned his property since 1960. The 1960 deed described the land as

[b]eginning at the middle of the public road and running thence N. 77.5 E. 5.48 poles to a stake; S. 15 E. 10.72 poles to the middle of said road; thence with the meanders of said road S. 75 W. 4.88 poles; thence N. 17.5 W. 10.32 poles to the beginning, containing 55 square poles, and being the same lot that was conveyed . . . by deed dated April 10, 1947[.]

Ms. Luke owned and occupied property situated opposite from Ms. Carpenter's property. The property inhabited by Ms. Luke was on the western side of State Route 3. Ms. Luke's father had purchased the property in 1972 and had conveyed it to her in 1988. The land description in the deed conveyed " a certain tract or parcel of land situate in the Village of Peora, Eagle District, Harrison County, West Virginia, containing two acres and twenty-six square poles, and being the same real estate which was conveyed . . . by deed dated September 21, 1949[.]" The testimony introduced at trial averred that no deed in Ms. Luke's chain of title after 1922 contained a metes and bounds description of her property. The only descriptions were in terms of area and by reference to the prior deed in the chain.

Complicating this case is the fact that the public road's location was moved in 1922. There is no dispute that Ms. Carpenter owns the property on the eastern side of current State Route 3 and that Ms. Luke owns a portion of property on the western side of current State Route 3. However, the disagreement arises over a portion of real estate that is also located on the western side of State Route 3, and is adjacent to Ms. Luke's undisputed portion of property. Ms. Luke claims that the disputed property is hers because she received it from her father who was a bona fide purchaser of the same. Alternatively, Ms. Luke claims that she acquired the disputed property through adverse possession and/or a prescriptive easement. Conversely, Ms. Carpenter claims title to the disputed property through the description in the deeds in her chain of title. While the land description in her chain of title references a property boundary line in the middle of the public road, Ms. Carpenter argues that the public roadway was moved in 1922 and that the last deed in the chain of title that referenced the property by a metes and bounds description was in 1916. Ms. Carpenter contends that because the current land descriptions are the same as those contained in the 1916 deed, prior to the 1922 relocation of the road, the reference to the " public road" in the current deeds references the public road as it existed prior to 1922.

During the underlying case, both parties moved for summary judgment following discovery. The lower court denied both motions for summary judgment, and the case proceeded to a jury trial. At the conclusion of the testimony and evidence, the lower court entered judgement as a matter of law in favor of Ms. Carpenter, finding that she was the title owner of the real estate. The issues remaining for jury consideration were Ms. Luke's alternative claims of adverse possession, prescriptive easement, and unjust enrichment. There was conflicting testimony at trial regarding which party actually used the disputed tract. The jury returned a verdict in favor of Ms. Carpenter on these claims, finding that Ms. Luke had not acquired ownership of the property through either adverse possession or a prescriptive easement. Further, the jury found no unjust enrichment. Ms. Luke filed a motion to alter or amend judgment and a motion for a new trial, both of which were denied. It is from these rulings that Ms. Luke now appeals to this Court.

II. Standard of Review This case is before this Court on appeal from the circuit court's order denying Ms. Luke's motion to alter or amend judgment and a motion for a new trial pursuant to West Virginia Rule of Civil Procedure 59(e). As this Court has previously explained,

> " ' [t]he standard of review applicable to an appeal from a motion to alter or amend a judgment, made pursuant to W. Va. R. Civ. P. 59(e), is the same standard that would apply to the underlying judgment upon which the motion is based and from which the appeal to this Court is filed.' Syllabus point 1, *Wickland v. American Travellers Life Insurance Co.,* 204 W.Va. 430, 513 S.E.2d 657 (1998)." Syllabus point 2, *Bowers v. Wurzburg,* 205 W.Va. 450, 519 S.E.2d 148 (1999).

Syl. pt. 1, *Alden v. Harpers Ferry Police Civil Serv. Comm'n,* 209 W.Va. 83, 543 S.E.2d 364 (2001).

One of the underlying rulings involved granting a motion for judgment as a matter of law under Rule 50 of the West Virginia Rules of Civil Procedure. We have held:

> "The appellate standard of review for the granting of a motion for a [judgment as a matter of law] pursuant to Rule 50 of the West Virginia Rules of Civil Procedure is de novo. On appeal, this court, after considering the evidence in the light most favorable to the nonmovant party, will sustain the granting of a [judgment as a matter of law] when only one reasonable conclusion as to the verdict can be reached. But if reasonable minds could differ as to the importance and sufficiency of the evidence, a circuit court's ruling granting a [judgment as a matter of law] will be reversed.' Syllabus Point 3, *Brannon v. Riffle,* 197 W.Va. 97, 475 S.E.2d 97 (1996)." Syl. pt. 5, *Smith v. First Community Bancshares, Inc.,* 212 W.Va. 809, 575 S.E.2d 419 (2002).

Syl. pt. 1, *Estep v. Mike Ferrell Ford Lincoln-Mercury, Inc.*, 223 W.Va. 209, 672 S.E.2d 345 (2008). In addition to the trial court's grant of a judgment as a matter of law, we are asked to review the lower court's denial of the motion for a new trial. In that respect, this Court applies an abuse of discretion standard:

Although the ruling of a trial court in granting or denying a motion for a new trial is entitled to great respect and weight, the trial court's ruling will be reversed on appeal when it is clear that the trial court has acted under some misapprehension of the law or the evidence.

Syl. pt. 4, *Sanders v. Georgia-Pacific Corp.*, 159 W.Va. 621, 225 S.E.2d 218 (1976). Mindful of these applicable standards, we proceed to consider the arguments set forth by the parties.

III. Discussion On appeal before this Court, Ms. Luke argues that the circuit court erred in denying her motion to alter or amend judgment and her motion for a new trial. Ms. Luke asserts that Ms. Carpenter's claim to the property through an unrecorded 1922 document is improper and, further, Ms. Luke contends that her father was a bona fide purchaser from whom she received her claim to the land in question.[11] In response, Ms. Carpenter avers that the circuit court was correct in its decision that Ms. Carpenter is the record owner of the disputed portion of land.

Dispositive of this issue is the interpretation of the two deeds through which each party claims ownership of the subject property. In this case, after all of the evidence was submitted and the testimony was heard, the trial judge determined that Ms. Carpenter was the title owner to the disputed property. Upon review of the record, this Court agrees with the decision of the lower court.

At the outset, we note that the general rule is that " ' [a] valid written instrument which expresses the intent of the parties in plain and unambiguous language is not subject to judicial construction or interpretation but will be applied and enforced according to such intent.' Syl. pt. 1, *Cotiga Development Company v. United Fuel Gas Company*, 147 W.Va. 484, 128 S.E.2d 626 (1963)." Syl. pt. 1, *Sally-Mike Props. v. Yokum*, 175 W.Va. 296, 332 S.E.2d 597 (1985). " In the construction of a deed or other legal instrument, the function of a court is to ascertain the intent of the parties as expressed in the language used by them." *Davis v. Hardman*, 148 W.Va. 82, 89, 133 S.E.2d 77, 81 (1963) (internal citations omitted). Further guidance is provided in the principle that,

> " [i]n construing a deed, will or other written instrument, it is the duty of the court to construe it as a whole, taking and considering all the parts together, and giving effect to the intention of the parties wherever that is reasonably clear and free from doubt, unless to do so will violate some principle of law inconsistent therewith." Pt. 1, syllabus, *Maddy v. Maddy*, 87 W.Va. 581[, 105 S.E. 803 (1921)].

Syl. pt. 5, *Hall v. Hartley*, 146 W.Va. 328, 119 S.E.2d 759 (1961). *Cf.* Syl. pt. 9, *Paxton v. Benedum-Trees Oil Co.*, 80 W.Va. 187, 94 S.E. 472 (1917) (" Extrinsic evidence will not be admitted to explain or alter the terms of a written contract which is clear and unambiguous.").

Ms. Carpenter's father had owned his property since 1960, and it was inherited by Ms. Carpenter in 2005 when her father died. The 1960 deed, which is the source of Ms. Carpenter's ownership claim of the subject land, states as follows:

Beginning at the middle of the public road and running thence N. 77.5 E. 5.48 poles to a stake; S. 15 E. 10.72 poles to the middle of said road; thence with the meanders of said road S. 75 W. 4.88 poles; thence N. 17.5 W. 10.32 poles to the beginning, containing 55 square poles, and being the same lot that was conveyed . . . by deed dated April 10, 1947[.]

An examination of the deeds in the chain of title to this land illustrates that the description of the land is identical in all of the deeds submitted into the record. The oldest deed submitted is dated 1919, contains the identical land description, and references that it is the same property conveyed in a 1916 deed. Likewise, Ms. Luke's father had purchased his property in 1972, and conveyed it to her in 1988. The 1972 deed that is the origin of Ms. Luke's claim to ownership purports to convey " a certain tract or parcel of land situate in the Village of Peora, Eagle District, Harrison County, West Virginia, containing two acres and twenty-six square poles, and being the same real estate which was conveyed . . . by deed dated September 21, 1949[.]"

It is clear that Ms. Carpenter owns land on the eastern side of the public road and that Ms. Luke owns property on the western side of the public road. This case concerns a small piece of land, also on the western side of the public road, whose ownership is claimed by both parties. Based on the descriptions in the deeds, Ms. Luke argues that the shared boundary line is the " middle of the public road" and that no document states that any property lines cross the public road as would have to be the case if this Court accepted Ms. Carpenter's description of the property lines. Ms. Carpenter responds that the public road was moved in 1922 and, because the land descriptions in the deeds do not change after 1922, the " middle of the public road" described in the deeds is the public road as it existed prior to 1922. In its simplest terms, this case revolves around the context associated with the phrase " middle of the public road" set forth in the deed description.

To illustrate the change in the road location, Ms. Carpenter submitted a December 1922 plat titled " Section of Robinson Run-Peora Road through Property of Chas L. Ashcraft and Others," which was completed by the Office of the County Road Engineer, Clarksburg, West Virginia. Ms. Luke seizes on this document and argues that it was an unrecorded document and, as such, it cannot defeat her claim that her chain of title resulted from her father's status as a bona fide purchaser who had no way of knowing of the existence of the 1922 plat. However, her reliance is misplaced. W.Va.Code § 40-1-9 (1963) (Repl. Vol. 2004) states that

> [e]very such contract, every deed conveying any such estate or term, and every deed of gift, or trust deed or mortgage, conveying real estate shall be void, as to creditors, and subsequent purchasers for valuable consideration without notice, until and except from the time that it is duly admitted to record in the county wherein the property embraced in such contract, deed, trust deed or mortgage may be.

The document showing the relocation of the road was not a contract or deed conveying real estate. It transferred no property. It was merely a document showing a public road in its old location and new location. Thus, there was no requirement that it be recorded. The movement of the road did not change or affect the boundary lines of the property owned by each party either through devise or conveyance. Thus, Ms. Luke's argument that this unrecorded document cannot be used to thwart her claim that her father was a bona fide purchaser without notice is completely without merit.

Moreover, we note that Ms. Luke's father did not qualify as a bona fide purchaser of the disputed tract of land. A bona fide purchaser of land is " one who purchases for a valuable consideration, paid or parted with, without notice of any suspicious circumstances to put him upon inquiry." *Stickley v. Thorn,* 87 W.Va. 673, 678, 106 S.E. 240, 242 (1921) (internal citations omitted). *See also Simpson v. Edmiston,* 23 W.Va. 675, 680 (1884) (" [A] *bona fide* purchaser is one who buys an apparently good title without notice of anything calculated to impair or affect it [.]"); *Black's Law Dictionary* 1355 (9th ed. 2004) (defining " bona fide purchaser" as " [o]ne who buys something for value without notice of another's claim to the property and without actual or constructive notice of any defects in or infirmities, claims, or equities against the seller's title; one who has in good faith paid valuable consideration for property without notice of prior adverse claims."). As previously held by this Court, and more recently reiterated, " ' [a] *bona fide* purchaser is one who actually purchases in good faith.' Syl. pt. 1, *Kyger v. Depue,* 6 W.Va. 288 (1873)." *Subcarrier Communications, Inc. v. Nield,* 218 W.Va. 292, 300, 624 S.E.2d 729, 737 (2005). As this opinion explains, Ms. Luke's father purchased only the land described in his deed of conveyance. The disputed portion of land was not part of his deed description but, instead, was part of the land described in Ms. Carpenter's chain of title. Thus, Ms. Luke's father could not have been a bona fide purchaser of land that was not included in his property description but was, instead, included in the property description of another tract of land.[2]

In support of her position before the lower court, Ms. Luke submitted a survey showing her boundary line as being the middle of the presently existing public road. However, Ms. Luke did not present any expert testimony, nor did she call the surveyor to testify. Before the lower court, Ms. Carpenter also submitted a property line survey, showing that the current road crosses through her land, leaving a 2,234 square foot portion of her property on the western side of the current road and adjacent to Ms. Luke's undisputed land. Additionally, Ms. Carpenter presented her expert, David L. Jackson, who had completed the survey on her behalf.

During Mr. Jackson's testimony, he explained that " when we trace the deeds back the deed for the Carpenters went back prior to 1922 with basically the same description in it." He stated that it is quite common for a surveyor to find that a road used as a marker or call in a deed has either been moved or no longer exists. Relying on the calls that they were able to find still in existence from the description of the land, Mr. Jackson testified that " the deed calls of the Carpenter[s] [were] consistent with the location of the old roadbed and no[t] with the relocated position for the road." He further elaborated his point as follows:

> Q. When you have, when you encounter something like this say for instance that you hadn't had that [1922] plat and you go and someone calls for the location of a road or a monument or something like that as one of the boundary lines, do you just take that as a way to start?
> A. Basically you are getting into an area where what is called for is the record call. Record calls which is basically when a deed calls for a road location, is calling for the road location as the date of that deed and basically you are trying to look for consistencies for where you believe that road was as of the date of that deed and for the deed calls.
> Q. Okay, so if you notice that the deed called for the same location throughout time would you check to see whether or not anything happened to that road or monument, I mean I guess it is not just a road, it could be a maple tree or a rock or anything like that, is that fair?
> A. Yeah, you are trying to find the location of the road and in this particular case it was without the plat showing that the old roadbed was located to the southwest of the current Route 3. The only thing that you would have had would have had to be some distances in the Carpenter description placing it over there and it may have been a difficult task.

Later in his testimony, Mr. Jackson was asked whether, " in cases like this where the road has in fact moved[,] would that change the boundary lines?" He responded as follows:

> No, in my opinion the boundary would be where the road was at the time of the deed so in the case you are talking about a deed that was dated 1960 or 16 or earlier and the road was relocated in 1922 so the boundary would be where it was called for in 1916 and in my opinion again the only thing that would change would be if there was a conveyance.
>
> Q. And as you just discussed if the descriptions, you know, as far as right here at the public road and the area are the same throughout the years and the boundary line didn't change, is that what you basically indicated? That when the road moved it didn't change the boundary line?
>
> A. That is correct. You try to reestablish the road in a position that is called for as of the date of the deed.
>
> Q. Okay and that 1916 deed conveying the same property as the 1960 deed, is that correct?
>
> A. That is correct.

Further explanatory testimony was elicited on cross-examination by opposing counsel with the following exchange:

> Q. Mr. Jackson, when you testified earlier that if you did not have that [1922] plat that had been provided to you by Ms. [Carpenter] then the common practice is to start the survey or define the boundary where it calls for a roadway, to use the roadway that was placed as of the day of the deed was written, that is correct, isn't it?

A. In this case the-you have something that is in conflict with the current location of the roads which are basically the record distances for the Carpenter description would not fit with where the current road was. It is pretty clear that it takes it across the road. What helps clarify that is the [1922] plat because then you have two pieces of information that are consistent. You have the bearings and distances on the Carpenter description as well as where the plat showing the location of the older road. Lacking that plat would have-it would have been [sic] the chore more difficult but when you have the two pieces of information that is what a surveyor does. He looks at everything he can get his hands on and you take it as a whole. You evaluate it as a whole.

Q. Is it common practice, sir, to begin a survey when the boundary calls for-to begin at the middle of the public road is it common practice to begin with or do the survey based upon the public road that is in place on the deed-on the date the deed is written?

A. On the date the deed is written. That is correct. That is what you are trying to reestablish is the road position at the date of the deed.

Q. Thank you sir. When was this property deeded to Mr. Zirkle?

A. I don't know for sure. It may have been trying to get off memory off of that last deed it may have been in the ' 60's.

Q. Okay, does 1960 sound right?

A. It sounds right.

Mr. Jackson further explained that, "you have to look at the entire chain of title because when I am referring to a date of the deed is when that description originally came out of a larger piece of ground which again is prior to 1922."

Based on the record evidence submitted and the testimony of the only expert, the lower court was correct to enter judgment as a matter of law, finding that the deeds of title show that Ms. Carpenter is the owner of the disputed piece of property. The fact that the road was moved in 1922 did not change the property boundary lines and does not subtract from the land that is Ms. Carpenter's. Because the recent deeds refer only to the start of the property line and then continue to convey that which had been previously conveyed, the deeds do not reference the public road as it currently exists. Rather, their reference to the public road is the public road as it existed prior to the move in 1922. Thus, the circuit court was correct in finding Ms. Carpenter to be the title owner of the land.

There simply is no way to find Ms. Luke owns the land under any deed construction, and there was no reason to submit the issue to the jury. *See* Syl. pt. 5, *Davis Colliery Co. v. Westfall,* 78 W.Va. 735, 90 S.E. 328 (1916) ("Where the right of a party to recover, in ejectment, depends solely upon the construction of his deed, in the light of the undisputed facts, the question is one of law for the court, and not one of fact for the jury."), *see also* Syl. pt. 2, *Prickett v. Frum,* 101 W.Va. 217, 132 S.E. 501 (1926) (" Where the description of a parcel of land in a deed is certain, and buttressed by documentary evidence, parol testimony cannot be used to enlarge the scope of the descriptive words so as to include another and distinct parcel owned by the grantor. In such case the construction of the deed is for the court, and not for the jury."). The circuit court correctly denied the motion to alter or amend the judgment and the motion for a new trial; therefore, the lower court's rulings are affirmed.

IV. Conclusion Based on the foregoing, the decisions of the circuit court are affirmed.

Affirmed.

NOTES:

[1] In the underlying case before the circuit court, Ms. Luke also argued, in the alternative, that if it was determined that she was not the title owner of the land, that she had acquired the land through adverse possession or a prescriptive easement. She also asserted a claim for unjust enrichment due to the improvements she alleged she made to the property. After hearing all of the evidence, the jury found adverse to Ms. Luke and in favor of Ms. Carpenter on all of the claims submitted to it. However, these issues were not assigned as error before this Court, were not briefed, and were only mentioned during oral arguments before this Court as a result of direct questions from the bench. Significantly, when asked at oral argument before this Court if this was a case about adverse possession, Ms. Luke's counsel responded that it was not. Rather, counsel stated that it was a case of a bona fide purchaser.

While some of the relevant facts regarding the alternative claims are intertwined within her legal proposition surrounding title ownership pursuant to a deed, the alternative legal issues are not set forth or adequately briefed so as to bring them before this Court for consideration. *See State v. LaRock,* 196 W.Va. 294, 302, 470 S.E.2d 613, 621 (1996) (" Although we liberally construe briefs in determining issues presented for review, issues which are not raised, and those mentioned only in passing but are not supported with pertinent authority, are not considered on appeal. *State v. Lilly,* 194 W.Va. 595, 605 n. 16, 461 S.E.2d 101, 111 n. 16 (1995) (' casual mention of an issue in a brief is cursory treatment insufficient to preserve the issue on appeal'). We deem these errors abandoned because these errors were not fully briefed."). *Accord In re Edward B.,* 210 W.Va. 621, 625 n. 2, 558 S.E.2d 620, 624 n. 2 (2001) (" The defendants' petition for appeal cited as error the circuit court's application of the five year statute of limitations to this case. However, the defendants did not address that issue in their brief and therefore have abandoned that assignment of error") (internal quotations and citation omitted); Syl. pt. 6, *Addair v. Bryant,* 168 W.Va. 306, 284 S.E.2d 374 (1981) (" Assignments of error that are not argued in the briefs on appeal may be deemed by this Court to be waived."). Thus, the issues of adverse possession, prescriptive easement, and unjust enrichment are not before this Court and will not be addressed by this opinion.

[2] As evidenced by the record in this case, Ms. Luke's father's deed merely described the land as " containing two acres and twenty-six square poles, and being the same real estate which was conveyed[.]" This same description was followed in the chain of title prior to the relocation of the road in 1922.

CASE #12 EASEMENT BY AGREEMENT RESULTING IN CESSATION OF NECESSITY

ROBNETT v. TENISON
NO. M2007-02490-COA-R3-CV

COURT OF APPEALS OF TENNESSEE, NASHVILLE

SEPTEMBER 23, 2008

OPINION

FRANK G. CLEMENT, JR., JUDGE. Vickie Robnett and Edward H. Tenison, Jr., are neighbors in a rural area of Lewis County. Each owns and resides on property adjacent to the other. Mr. Tenison's property fronts Tutor Lane. Ms. Robnett's property, however, is landlocked, meaning her property does not have access to a public road without the benefit of a way of easement to and from a public road.

To remedy this problem, in 1995 the Chancery Court for Lewis County decreed a "permanent implied easement of necessity" across Mr. Tenison's property thereby giving Ms. Robnett ingress and egress to Tutor Lane. The easement of necessity was established for the sole purpose of giving Ms. Robnett ingress and egress to her personal residence. Predating the easement of necessity and appearing in Ms. Robnett's chain of title, is a recorded express easement that affords Ms. Robnett ingress and egress to Highway 412 via another neighbor's property.[1] Ms. Robnett does not cross Mr. Tenison's property when she and her guests use the express easement for ingress and egress from her property to Highway 412. Until recently, however, the express easement was not amenable to vehicular traffic due to many factors. Accordingly, Ms. Robnett and her guests have always used the easement by necessity from Tutor Lane for ingress and egress to her residence.

In an effort to alter the manner and minimize the frequency with which Ms. Robnett, her husband, their guests, and trucks of their garbage collection company utilized the easement of necessity, Mr. Tenison began to interfere with and otherwise obstruct their use of the easement. As a consequence of Mr. Tenison's interference with her right to use the easement, Ms. Robnett filed a Petition for Contempt in October of 2006 alleging, *inter alia*, that Mr. Tenison had unreasonably interfered with her use of the easement by digging a ditch across the easement, by parking vehicles on the roadway to block her use of the easement, by harassing her visitors using the easement, and by not allowing her to pave the easement.

Mr. Tenison filed an Answer to the Petition and a Counter-Complaint against Ms. Robnett. In his Answer, he denied the allegations in the Petition and asserted several affirmative defenses. In the Counter-Complaint, Mr. Tenison sought to terminate the easement of necessity on alternative grounds. He contended the easement should be terminated because it was no longer necessary due to substantial improvements to the pre-existing express easement. In the alternative, he contended that Ms. Robnett was violating the easement by necessity by operating a garbage hauling business at her residence, that Ms. Robnett and her visitors were misusing the easement by driving excessively fast and in a reckless manner, that Ms. Robnett was using her residence to distribute marijuana, and that Ms. Robnett allowed her dogs to run loose and cause damage to his property.

A bench trial was held on August 31, 2007. At the close of all the proof, the trial judge went to the parties' properties to personally view the easements at issue. Ms. Robnett then withdrew her petition for contempt, leaving only Mr. Tenison's Counter-Complaint at issue. Thereafter, the trial court issued its ruling. The court denied Mr. Tenison's request to terminate the easement by necessity based on a finding that termination "would place an undue burden" on Ms. Robnett; however, it permanently enjoined Ms. Robnett from operating a commercial business from her property. The court also ruled that Ms. Robnett has the right to improve the easement by necessity, and it ordered Mr. Tenison to repair the ditch he had dug across the easement to reduce speeding. This appeal followed.

Standard of Review The standard of review of a trial court's findings of fact is *de novo* and we presume that the findings of fact are correct unless the preponderance of the evidence is otherwise. Tenn. R. App. P. 13(d); *Rawlings v. John Hancock Mut. Life Ins. Co.*, 78 S.W.3d 291, 296 (Tenn. Ct. App. 2001). For the evidence to preponderate against a trial court's finding of fact, it must support another finding of fact with greater convincing effect. *Walker v. Sidney Gilreath & Assocs.*, 40 S.W.3d 66, 71 (Tenn. Ct. App. 2000); *The Realty Shop, Inc. v. R.R. Westminster Holding, Inc.*, 7 S.W.3d 581, 596 (Tenn. Ct. App. 1999). Where the trial court does not make findings of fact, there is no presumption of correctness and we "must conduct our own independent review of the record to determine where the preponderance of the evidence lies." *Brooks v. Brooks*, 992 S.W.2d 403, 405 (Tenn. 1999). We also give great weight to a trial court's determinations of credibility of witnesses. *Estate of Walton v. Young*, 950 S.W.2d 956, 959 (Tenn. 1997); *B & G Constr., Inc. v. Polk*, 37 S.W.3d 462, 465 (Tenn. Ct. App. 2000). Issues of law are reviewed *de novo* with no presumption of correctness. *Nelson v. Wal-Mart Stores, Inc.*, 8 S.W.3d 625, 628 (Tenn. 1999).

Analysis Although both parties have presented issues for our consideration, we have determined the dispositive issue on appeal is whether the trial court erred in denying Mr. Tenison's petition to terminate the easement by necessity.

The trial court denied Mr. Tenison's petition to terminate the easement by necessity based upon a finding that termination of the easement by necessity would place an *undue burden* on Ms. Robnett. The fact Ms. Robnett may be burdened if she loses the benefit of the easement by necessity is a justifiable concern to her; however, it is not the correct legal standard for the court's consideration.

The existence of an easement by necessity is dependent on the necessity that created it. 28A C.J.S. *Easements* § 161 (2008). Therefore, "a way of necessity continues as long, but *only as long, as a necessity for its use continues*." *Id.* (emphasis added). "If an easement for a particular purpose is granted, when that purpose no longer exists, there is an end of the easement." *McGiffin v. City of Gatlinburg*, 260 S.W.2d 152, 154 (Tenn. 1953) (quoting Washburn, *Treatise on Easement*, 654 (3d ed.)). "The fact that a former way of necessity continues to be the most convenient way will not prevent its extinguishment when it ceases to be absolutely necessary." 28A C.J.S. *Easements* § 161 (2008).

The trial court's decision to deny Mr. Tenison's petition to terminate the easement was based upon an incorrect legal standard. Therefore, it is incumbent on this court to apply the correct standard to the facts in the record and to determine the parties' rights accordingly.

It is undisputed that Ms. Robnett has two easements, each of which affords her ingress and egress to her residence from a public road. One is the easement by necessity, which goes through, and thus encumbers, Mr. Tenison's property. That easement affords her ingress and egress from Tutor Lane. The other is an express easement that affords her ingress and egress to her residence from Highway 412 via another neighbor's property. The express easement was obtained by agreement of the respective property owners. The easement by necessity was court ordered in 1995, over Mr. Tenison's objections, upon a finding that the easement was necessary to afford ingress and egress to a public road.

At trial, Ms. Robnett acknowledged that she had an express easement of record that afforded her ingress and egress to her residence from Highway 412. She, however, contended that the express easement continues to be impassable and that Mr. Tenison had failed to show that circumstances had changed since 1995 sufficient to prove that the express easement, which was found to be essentially impassable in 1995, afforded her reasonable ingress and egress. We respectfully disagree.

Mr. Tenison testified that the express easement had been improved "a lot" since the court-ordered easement by necessity was mandated in 1995. He testified that culverts had been installed where the road would previously "wash out" and the road and land around it had been graded, which improved the condition of the road and minimized problems from drainage. Mr. Tenison also testified that he had seen vehicles using the express easement for ingress and egress to Ms. Robnett's property. For her part, Ms. Robnett admitted that culverts had been put in on both sides of the express easement to prevent flooding of the express easement. She also admitted that immediately prior to trial the portion of the express easement "going up the hill" had been graded with a bobcat and a gate that was wide enough for a vehicle had been installed. Photographs of the express easement show a recently graded roadway that resembles a driveway.

Significantly, Ms. Robnett did not testify that the express easement was not suitable for ingress and egress, she merely stated that the express easement was not "as reasonable" as the easement by necessity. In an effort to explain why the express easement was not as reasonable as the easement by necessity, Ms. Robnett testified that to use the express easement she would have to come down the paved driveway used by another neighbor and then split off onto her unpaved portion of the easement. When asked why she was seeking the court's permission to spend money to pave the easement by necessity rather than use her express easement, Ms. Robnett could only state the easement by necessity "has always been our access."

We acknowledge that the trial judge visited the properties at issue at the close of all the proof and that the trial judge noted that he was "glad" he went out to the properties because "[i]t really put everything into perspective." The trial court did not, however, state on the record how the on-site visit put things in perspective. The court merely announced that "[h]aving gone out to the property, I think it would be an undue burden."

A trial judge has "the inherent discretion to take a view of the site of a property dispute . . . where such a view will enable the judge to assess the credibility of witnesses, to resolve conflicting evidence, or to obtain a clearer understanding of the issues." *Tarpley v. Hornyak*, 174 S.W.3d 736, 748-49 (Tenn. Ct. App. 2004). However, the trial judge's viewing is not a substitute for the evidence presented in court, *Id.* at 749, and without the benefit of knowing how the viewing put everything into perspective, our analysis is limited to the evidence in the record.

As we stated earlier, an easement by necessity may continue "*only as long, as a necessity for its use continues*." 28A C.J.S. *Easements* § 161 (2008) (emphasis added). The fact the way of necessity continues to be the most convenient is not sufficient to prevent its extinguishment if it ceases to be absolutely necessary. *Id.* The easement was created for the purpose of affording Ms. Robnett ingress and egress from a public road. Due to recent improvements to the easement that leads to Highway 412, the easement through Mr. Tenison's property is no longer absolutely necessary. Because the purpose for which the easement through Mr. Tenison's property was created no longer exists, there is an end to the easement by necessity through Mr. Tenison's property. *See McGiffin*, 260 S.W.2d at 154.

Based upon the evidence in the record, we find the preponderance of the evidence shows that it is no longer necessary for Ms. Robnett to use Mr. Tenison's property for ingress and egress to her residence.

In Conclusion The judgment of the trial court is reversed, and this matter is remanded with instructions for the trial court to enter an order terminating the easement by necessity that encumbers Mr. Tenison's property and for such other proceedings as the trial court may deem necessary.

NOTE:

[1] Tutor Lane is on the east boundary line of Mr. Tenison's property. Highway 412 does not touch either of the parties' properties; it is located just south of their properties. Tutor Lane connects with Highway 412 via Indian Creek Road.

CASE #13 ROAD SHOWN ON SUBDIVISION PLAT NOT A PUBLIC WAY

LATTIN, ET AL. v. ADAMS COUNTY
149 IDAHO 497, 236 P.3D 1257 (IDAHO, 2010)

W. JONES, JUSTICE.

I. Nature of the Case Adams County appeals from the district court's ruling that an access road running through Respondents' property is not a public road. Adams County claims that it has acquired the road both by dedication and by prescription.

II. Factual and Procedural Background Kathleen and Dale Lattin, Kathy and Tyler Chase, and Taffy and Kenneth Stone, Respondents, are three married couples who own properties in the Reico Subdivision of Adams County, Idaho. Sometime in the 1920s, when their parcels were part of a single tract, their predecessor in interest permitted a local logger to construct a temporary access road on the property. The road became known as Old Sawmill Road, but is now commonly known as Burch Lane. Today, Burch Lane connects a public highway with a forest-service road in the Payette National Forest. On the way, it crosses a number of privately owned parcels in the Reico Subdivision, three of which are owned by Respondents, who rely on the road to access their homes. Although the road has always crossed private property, for decades some locals have used the road to reach the forest for recreational purposes. Adams County (the "County"), however, has never expended any resources in maintaining the road or improving it.

Reico Subdivision was first subdivided into thirty-four individual lots in 1974, some of which were sold off, and most of which do not touch Burch Lane. In 1983, the County involuntarily compelled the subdivision developers to record a plat.[1] At the time, Burch Lane was overgrown with trees and difficult to use. The plat depicts Burch Lane but does not identify it as a public easement or right-of-way.

After the subdivision was recorded, some of the new residents improved the road to enable themselves to access their properties. Christy Ward, the Lattins' predecessor in interest, as well as Idaho Power, improved Burch Lane beginning in 1984. At some point, parts of the road were also relocated, although some portions of the original road are still visible. Kathy Chase, one of the landowners in this suit, improved portions of the road in 1998. In 2002, Idaho Power obtained an easement from Respondents to use the road to maintain a power substation located beyond their properties.

At some point, Respondents installed signs on their properties declaring the road to be private. The County responded by threatening Respondents with criminal sanctions if they did not remove the signs. It also informed the landowners that it would classify Burch Lane as a county road and begin maintaining it. The landowners responded by filing a lawsuit against the County requesting declaratory and injunctive relief and seeking to quiet title. The County responded that it had acquired the road by dedication in the Reico Subdivision plat and that, even if it had not, the road had passed into public control by prescription. The district court granted summary judgment to the landowners, finding that the County had failed to follow the statutory protocols necessary to establish a public road. It did not expressly address the County's prescription claim. On appeal, the County again raises its dedication and prescription theories.

III. Issues on Appeal

1. Whether the County acquired the road by dedication in the subdivision plat.
2. Whether the County acquired the road by prescription under I.C. § 40-202(3).
3. Whether Respondents are entitled to attorney fees on appeal.

IV. Standard of Review When reviewing a grant of summary judgment, this Court applies the same standard of review the district court did when ruling on the motion. *Doe v. City of Elk River,* 144 Idaho 337, 338, 160 P.3d 1272, 1273 (2007). Summary judgment is proper if " the pleadings, depositions, and admissions on file, together with the affidavits, if any, show that there is no genuine issue as to any material fact and that the moving party is entitled to a judgment as a matter of law." I.R.C.P. 56(c). " All disputed facts are to be construed liberally in favor of the non-moving party, and all reasonable inferences that can be drawn from the record are to be drawn in favor of the non-moving party." *Estate of Becker v. Callahan,* 140 Idaho 522, 525, 96 P.3d 623, 626 (2004).

> "If there is no genuine issue of material fact, only a question of law remains, over which this Court exercises free review." *Indian Springs LLC v. Indian Springs Land Inv.,* 147 Idaho 737, 746, 215 P.3d 457, 466 (2009) (quoting *Cristo Viene Pentecostal Church v. Paz,* 144 Idaho 304, 307, 160 P.3d 743, 746 (2007)). This Court likewise freely reviews the construction of a statute. *Gibson v. Ada County,* 142 Idaho 746, 751, 133 P.3d 1211, 1216 (2006).

V. Analysis

A. The Subdivision Plat Does Not Dedicate the Road to Public Use The County asserted that Burch Lane had been dedicated to public use, but it has been unclear throughout the course of litigation whether the County is relying on a statutory or common-law theory. Under either theory, however, the claim fails because the Reico Subdivision plat does not unequivocally dedicate Burch Lane to public use.

A common-law dedication requires the county to show "that 1) there has been a valid offer to dedicate real property to a public use, and 2) lots were subsequently sold or otherwise conveyed by instruments which specifically refer to such plat." *Worley Highway Dist. v. Yacht Club of Coeur D'Alene, Ltd.,* 116 Idaho 219, 225, 775 P.2d 111, 117 (1989). Filing and recording a plat or map can establish the owner's intent to make a dedication to the public. *Id.* at 224, 775 P.2d at 116. The dedication in the plat, however, must be clear and explicit. *State v. Fox,* 100 Idaho 140, 147, 594 P.2d 1093, 1100 (1979).

Similarly, a statutory dedication under I.C. § 50-1309 requires a plat that unequivocally dedicates a road to public use. The Reico Subdivision plat was recorded in 1983. Section 50-1309 at that time permitted landowners to dedicate a portion of their property by recording a plat depicting "all streets and alleys shown on said plat" that is accepted by the appropriate public entity. I.C. §§ 50-1309, -1313 (1982).[2] Although this statute has since been amended several times, the plat has always had to demark public rights-of-way in order to dedicate them. *Harshbarger v. Jerome County,* 107 Idaho 805, 806-07, 693 P.2d 451, 452-53 (1984); *see* Act of Mar. 26, 1988, ch. 175, sec. 2, 1988 Idaho Sess. Laws 306, 307 (amending I.C. § 50-1309 to specifically require government approval, but still requiring a marked plat). In any event, there is no reason why a statutory dedication in a plat would have to be less clear and unequivocal than a dedication by common law. *Cf. Harshbarger,* 107 Idaho at 806, 693 P.2d at 452 (stating that a dedication must be "by an unequivocal act of the owner of the fee" (quoting 26 C.J.S. *Dedication* § 1 (1956))). Thus, this Court has previously held that where the plat depicts no public streets or rights-of-way, there can be no statutory dedication. *Armand v. Opportunity Mgmt. Co.,* 141 Idaho 709, 714, 117 P.3d 123, 128 (2005).

The parties do not dispute the fact that a subdivision plat for the Reico Subdivision was properly recorded. The dispute instead is whether there was an unequivocal dedication of Burch Lane in the plat. Because the plat does not clearly dedicate Burch Lane to public use, it is unnecessary to determine whether, under a statutory theory, the County correctly accepted the dedication.[3]

To determine whether there was a dedication, this Court will interpret the plat like a deed, giving effect to the intent of the parties. *Sun Valley Land & Minerals, Inc. v. Hawkes,* 138 Idaho 543, 547, 66 P.3d 798, 802 (2003); *Daugharty v. Post Falls Highway Dist.,* 134 Idaho 731, 735, 9 P.3d 534, 538 (2000). The inquiry begins by determining whether a deed is ambiguous, which is a question of law subject to free review. *Union P. R. Co. v. Ethington Family Trust,* 137 Idaho 435, 438, 50 P.3d 450, 453 (2002). A document is ambiguous if it is reasonably subject to conflicting interpretations. *Neider v. Shaw,* 138 Idaho 503, 508, 65 P.3d 525, 530 (2003). Where a deed is unambiguous, "the parties' intent must be ascertained from the language of the deed as a matter of law." *C & G, Inc. v. Rule,* 135 Idaho 763, 766, 25 P.3d 76, 79 (2001).

The Reico Subdivision plat, as a matter of law, simply does not dedicate any roads to public use.[4] The map legend in this case depicts a style of dotted line for easements and public rights-of-way that is different from the style indicating a centerline of a private road. The line portraying Burch Lane on the map is a road centerline, not a public street or right-of-way. The map also specifically labels access or utility easements applicable to other private roads, but does not include any such label for Burch Lane. Accordingly, there has been no public dedication.

The plat map does state, however, that "any right, title and interest that [the developers] may have in the road rights-of-way as shown on this plat . . . is hereby dedicated to the use of the public." The plat nonetheless does not contain a dotted line anywhere representing a public easement or right-of-way. The County contends that the parties to the subdivision plat must have intended for there to be a public easement or right-of-way somewhere on the map because of the plat's language dedicating rights-of-way to public use and the fact that the legend contains a separate style of dotted line for such roads.

The County's interpretation is unreasonable. The County is correct that deeds must be interpreted to give effect to every part of the document. *Daugharty,* 134 Idaho at 735, 9 P.3d at 538 (citing *Twin Lakes Village Prop. Ass'n v. Crowley,* 124 Idaho 132, 137, 857 P.2d 611, 616 (1993)). Nonetheless, "[t]he intent of the owner to dedicate his land to public use must be clearly and unequivocally shown and must never be presumed." *Fox,* 100 Idaho at 147, 594 P.2d at 1100. Other private-road centerlines appearing on the map are identical to the one depicting Burch Lane. Even if this Court were to determine that there must be a public easement or right-of-way somewhere on the plat map, the County cannot explain why Burch Lane, and not some other road marked as private, would have to be a public road. Moreover, the legend appearing on the plat map almost certainly is boilerplate included on many similar maps. It contains distinct symbols for fences, canals, and stone monuments, none of which appear on this map either. The County's interpretation would require this Court to arbitrarily assume these landmarks also exist somewhere in the subdivision. The subdivision plat, as a matter of law, does not dedicate Burch Lane as a public right-of-way.

The County also argues that a woman named Mrs. Stover orally dedicated the road sometime between 1942 and 1974 when she owned the tract that later became the Reico Subdivision.[5] The County relies on an affidavit from an area resident claiming that Mrs. Stover told the affiant she intended for the road to remain open to the public. The County, however, admitted that there is no evidence in the record that it had accepted any dedication by Mrs. Stover. Even if there had been an acceptance, this paraphrased statement attributed to an absent witness does not unequivocally create a public dedication. There is no evidence that Mrs. Stover stated her present intent to dedicate the road to the County, but rather that she simply intended to allow the public to continue using the road for recreation. Her statement, therefore, would not generate a genuine fact issue as to dedication.

B. There Is No Material Issue of Fact Supporting the County's Claim that Burch Lane Was Acquired by Prescription A county may also gain control of a roadway by prescription under I.C. § 40-202(3) if the road is publicly used and maintained for a sufficient length of time.[6] "[A] public road may be acquired: (1) if the public uses the road for a period of five years, and (2) the road is worked and kept up at the expense of the public." *Ada County Highway Dist. v. Total Success Inv., L.L.C.,* 145 Idaho 360, 365, 179 P.3d 323, 328 (2008) (citing I.C. § 40-202(3)). The County must prove these elements by a preponderance of the evidence. *Floyd v. Bd. of Comm'rs,* 137 Idaho 718, 724, 52 P.3d 863, 869 (2002).

First, there is insufficient evidence as to whether there was sufficient public use of the road. This Court has repeatedly found that casual or sporadic use is not enough-the use must be regular and continuous. *See Total Success Inv.,* 145 Idaho at 367, 179 P.3d at 330 (finding public use where a number of people routinely used an alley to access their businesses); *Floyd,* 137 Idaho at 724, 52 P.3d at 869 (stating that a road acquired by prescription was " regularly and continuously used by the public" for recreation); *Kirk v. Schultz,* 63 Idaho 278, 282-84, 119 P.2d 266, 268-69 (1941) (holding that "casual and desultory" use by "miners, hunters, fishermen, and persons on horseback" even of a well-marked road was not public use). The County provided affidavits from three Adams County residents attesting to the fact that, for at least twenty years, local residents used Burch Lane to access the Payette National Forest for recreational or personal purposes such as hunting, berry picking, and wood gathering. There is also evidence of individual incidents when construction equipment traversed the road, such as an occasion when Idaho Power used it to build power lines in the area. Even so, nothing in the record indicates whether this activity occurred frequently or with any consistency, especially over a five-year span.

Furthermore, the record does not suggest that any public access was hostile to Respondents' ownership. "[W]here the owner of real property constructs a way over it for his use and convenience, the mere use thereof by others which in no way interferes with his use will be presumed to be by way of license or permission." *Chen v. Conway,* 121 Idaho 1000, 1005, 829 P.2d 1349, 1354 (1992) (quoting *Simmons v. Perkins,* 63 Idaho 136, 144, 118 P.2d 740, 744 (1941)). Respondents apparently never did anything to keep the public off the road prior to putting up the signs that triggered this lawsuit, nor does the County suggest that occasional traffic into the national forest interfered with Respondents' ownership of the road. Respondents very well could have permitted such access to the forest. In short, there is simply not enough evidence to create a fact issue as to public use.

There is also no issue of material fact to support the County's claim that it has maintained the road for any five-year span of time. There is absolutely no evidence anywhere that the County has ever performed any maintenance on or improved Burch Lane before Respondents filed this lawsuit. Even if such evidence existed, nothing in the record suggests that the county maintained the road at the same time the public was using it. The County nonetheless contends that, since the public has apparently been able to use the road for so long, no public maintenance was ever necessary. It asserts that it can still acquire a road by prescription without maintaining it if no repairs were needed to keep the road safe for travel.

The County is correct that there is "no mandated level of maintenance other than ' as necessary.' "*Floyd,* 137 Idaho at 725, 52 P.3d at 870; *accord State v. Berg,* 28 Idaho 724, 726, 155 P. 968, 969 (1916). This Court has specifically held that "[v]ery few roads require work throughout their entire length,"and that "[o]ur statute does not require work to be done upon a part of a highway not needing work, in order to acquire a right of way by prescription."*Gross v. McNutt,* 4 Idaho 286, 289, 38 P. 935, 936 (1894); *see also Cox v. Cox,* 84 Idaho 513, 521, 373 P.2d 929, 933 (1962) (reaffirming *Gross*).

Even so, the County can point to nothing in the record suggesting that the road did not need maintenance over a period of any length prior to this lawsuit. A nonmoving party cannot resist a motion for summary judgment by resting on "mere allegations or denials of his pleadings." I.C.R.P. 56(e). The moving party is entitled to summary judgment if the nonmoving party cannot raise a question of fact for an element that it would have to establish at trial. *Baxter v. Craney,* 135 Idaho 166, 170, 16 P.3d 263, 267 (2000). The County provides no evidence that road maintenance was unnecessary over the twenty-five years preceding this lawsuit. It never sent any crews to inspect the road and determine if it needed any work and, consequently, it cannot even say what condition the road was in over the years prior to the events in this lawsuit. Instead, the record contains affidavits from current and former landowners attesting that they had actually maintained the road. Prior to the mid-1980s, the road was overgrown and hardly accessible by vehicle. At that time, the Lattins' predecessor in interest, along with Idaho Power, first cleared out the road. Respondent Kathy Chase also attested that she also improved the road in 1998. Respondents alone have maintained and improved the road since the Reico Subdivision was created.

Since the County has the burden at trial to prove that it would have maintained the road if such work was needed, there is no issue of material fact in support of this element. *Total Success Inv.,* 145 Idaho at 365, 179 P.3d at 328. Neither the County, nor the "public" at large, could gain a prescriptive easement without the public maintenance required by statute. *State v. Fox,* 100 Idaho 140, 146, 594 P.2d 1093, 1099 (1979). The district court therefore correctly granted summary judgment to Respondents.[7]

C. Respondents Are Entitled to Attorney Fees on Appeal Respondents do not cite any statutes in support of their request for attorney fees. *PHH Mortgage Servs. Corp. v. Perreira,* 146 Idaho 631, 641, 200 P.3d 1180, 1190 (2009) (stating that statutory authority is required for an award of attorney fees). Instead, they request attorney fees under I.A.R. 11.2, which in relevant part provides: [8]

The signature of an attorney or party constitutes a certificate that the attorney or party has read the notice of appeal, petition, motion, brief or other document; that to the best of the signer's knowledge, information, and belief after reasonable inquiry it is well grounded in fact and is warranted by existing law or a good faith argument for the extension, modification, or reversal of existing law, and that it is not interposed for any improper purpose, such as to harass or to cause unnecessary delay or needless increase in the cost of litigation. If the notice of appeal, petition, motion, brief, or other document is signed in violation of this rule, the court, upon motion or upon its own initiative, shall impose upon the person who signed it, a represented party, or both, an appropriate sanction, which may include an order to pay to the other party or parties the amount of the reasonable expenses incurred because of the filing of the notice of appeal, petition, motion, brief or other document including a reasonable attorney's fee.

This Court has held that " [a]lthough an attorney's purpose in filing an appeal may not always appear clear from the record, this Court can infer intent and purpose from the attorney's actions and the surrounding circumstances." *Fritts v. Liddle & Moeller Constr., Inc.,* 144 Idaho 171, 176, 158 P.3d 947, 952 (2007) (quoting *Neihart v. Universal Joint Auto Parts, Inc.*, 141 Idaho 801, 803, 118 P.3d 133, 135 (2005)). A signed legal document violates Rule 11.2 if (1) it is not well grounded in fact; (2) it is not warranted by existing law or a good-faith extension, modification, or reversal of existing law; and (3) it was interposed for an improper purpose. *Read v. Harvey,* 147 Idaho 364, 371, 209 P.3d 661, 668 (2009).

First, as discussed above, the County utterly lacked any factual basis for bringing this appeal. It had no prescription claim because it never maintained the road. For its dedication claim, it relied on a plat map that unequivocally did not portray Burch Lane as a public road. Next, the County admitted that it had never even checked to see if road maintenance was necessary, undermining any reasonable legal argument in favor of its prescription theory or of extending existing law to apply to this case. Last, although Respondents initiated this lawsuit, they did so in response to the County's threats to forcibly remove their signs and to pursue criminal sanctions against them. In such a stark absence of any reasonable factual or legal justification for its actions, this Court must conclude that the County has acted with an improper purpose under I.A.R. 11.2 by harassing Respondents and needlessly prolonging this litigation. Attorney fees are therefore awarded against the County to Respondents.

VI. Conclusion The district court correctly granted Respondents' motion for summary judgment because there is no genuine issue of material fact to support the County's claims that Burch Lane is a public highway. Respondents are entitled to attorney fees on appeal because the County has pursued this case for an improper purpose. The judgment of the district court is affirmed. Costs to Respondents.

Chief Justice EISMANN, Justices BURDICK, HORTON, and Justice pro tem TROUT concur.

NOTES:

[1] The record is unclear as to which or how many parcels were sold off prior to 1983.

[2] In 1983, I.C. § 50-1309 provided:

The owner or owners of the land included in said plat shall make a certificate containing the correct description of the land, with the statement as to their intentions to include the same in the plat, and make a dedication of all streets and alleys shown on said plat, which certificate shall be acknowledged before an officer duly authorized to take acknowledgments and shall be endorsed on the plat. The surveyor or engineer making the survey shall certify the correctness of the plat. Act of Apr. 12, 1967, ch. 429, sec. 227, 1967 Idaho Sess. Laws 1249, 1322.

[3] Respondents also claim that the subdivision developers did not have the legal authority to dedicate Burch Lane because some of the parcels had already been sold off to third parties before the subdivision plat was recorded. The landowners are correct that the only party able to dedicate a road is someone who actually owns

the land. *Allen v. Blaine County,* 131 Idaho 138, 141, 953 P.2d 578, 581 (1998). Here, however, the record is silent as to which parcels had actually been sold at the time the subdivision plat was recorded, making it impossible to determine if the developers still held legal title to all the land underlying Burch Lane. Even if the developers still had legal title to all the lots over which Burch Lane travels, the plat did not dedicate that road to public use.

[4] Although the map does depict Fruitvale Road, which was a preexisting public highway, it uses the line labeled " property boundary." It presumably uses the "property boundary" line because Fruitvale Road forms the subdivision's southerly property boundary.

[5] Mrs. Stover's first name does not appear in the record, nor did it surface during oral argument.

[6] The Idaho Code provides:

Highways laid out, recorded and opened . . . by order of a board of commissioners, and all highways used for a period of five (5) years, provided they shall have been worked and kept up at the expense of the public, or located and recorded by order of a board of commissioners, are highways.

I.C. § 40-202(3); *see also id.* § 40-109(5) (providing an identical definition for a public "highway").

[7] The County also contends that summary judgment was improper because the district court made no specific findings of fact regarding its prescription theory. Regardless of what the district court decided about the County's prescription claim, its decision " should be affirmed if the grant of the summary judgment order was proper on any ground." *Richard B. Smith Real Estate, Inc. v. Knudson,* 107 Idaho 597, 599, 691 P.2d 1212, 1214 (1984).

[8] In their brief on appeal, Respondents incorrectly cited I.A.R. 11.1 instead of 11.2. They apparently made this error merely because I.A.R. 11.2 had previously been numbered I.A.R. 11.1, but was renumbered several months prior to Respondents filing their brief. Respondents also cited *Read v. Harvey,* a case dealing specifically with sanctions under this rule just before the renumbering became effective. 147 Idaho 364, 370-71, 209 P.3d 661, 667-68 (2009). Thus, the authority for Respondents' request is clear despite the citation error.

CASE #14 RAILROAD AS ABUTTER NOT RECEIVING ONE-HALF OF VACATED HIGHWAY

TAYLOR, ET AL., v. HAFFNER, D/B/A TASCO GRAIN, ET AL.
114 P.3D 191 (KAN.APP., 2005)

Editorial Note: This case does not have precedential value under Kansas supreme court rule 7.04 (f) and may only be cited as persuasive authority on a material issue not addressed by a published Kansas appellate court decision.

Appeal from Sheridan District Court; Glenn D. Schiffner, judge. Opinion filed July 1, 2005. Affirmed.

Memorandum Opinion

Per Curiam. The plaintiffs-appellees, Marvin Taylor; his wife, Kathleen Taylor; and Harold Taylor ("Taylors"), filed a quiet title action claiming ownership of all of the property abandoned by the Union Pacific Railroad Company (Union Pacific) in the South Half of Section Fourteen, Township Eight South, Range Twenty-seven West of the Sixth Principal Meridian (S/2 of 14-8S-27W of 6th P.M.), Sheridan County, Kansas (hereafter called "half-section"). The defendants-appellants, Pat Haffner and Patricia Adams, also claimed ownership of a portion of the abandoned railroad property by virtue of their respective ownership of certain tracts which had been previously carved out of the half-section. The parties both filed motions for summary judgment. Haffner and Adams appeal summary judgment in favor of the Taylors. We affirm.

In 1888, through three condemnation proceedings, the predecessor to Union Pacific acquired railroad rights-of-way and easements through the half-section, the total width of which was 400 feet. The original railroad right-of-way was the standard 100 feet in width, with 50 feet on each side of the railroad track centerline. The track ran diagonally across the half-section, from southwest to northeast. The railroad's additional easements were situated on both sides of the original right-of-way. An additional 50 feet was acquired on the southeast side of the original right-of-way, making a total of 100 feet lying on the southeast side of and parallel to the track centerline. An additional 250 feet was acquired on the northwest side of the original right-of-way, making a total of 300 feet lying on the northwest side of and parallel to the track centerline.

The record reflects that the Receiver of U.S. Land Office issued a receipt on at least a portion of the half-section in 1889, subsequent to the railroad condemnations. Thus, although not abundantly clear, the parties' respective chains-of-title appear to commence after the establishment of the railroad's interests.

The Taylors directly acquired their interest in the half-section from Orval Taylor by warranty deed, dated December 21, 1973. The abstract page in the record describes the Sheridan County, Kansas, real estate being conveyed as: "South Half (S1/2) of Section Fourteen (14), Township Eight (8), Range Twentyseven (27) subject to existing public roads, railway right of way and townsite of Tasco, Kansas." The abstract indicates the previous owner, Orval Taylor, had acquired ownership by a deed which excepted from the half-section certain tracts that had been previously conveyed by deeds which were identified by the book and page in which the recorded deeds appeared. Ten book and page recitations were excepted, suggesting there had been 10 tracts previously carved out of the half-section. The excepted tracts apparently comprised the townsite of Tasco. The deed to Orval Taylor also recited that the conveyance was "subject to rights of way and condemnation reports shown at entries 1, 2 and 3 of part 1, entries 15 and 16 of part 2, entries 15 and 16 of part 3, and entries 2, 6 and 7 of the supplemental abstract; and other land."

Haffner owns one tract and Adams owns five tracts on the northwest side of the railroad property which are contiguous and which apparently form at least a part of the Tasco townsite. Attached as an exhibit to Haffner's answer and counterclaim is a boundary survey, reciting the legal description of Haffner's tract and showing the location of the tract in relation to the railroad property. Specifically, the survey shows that Haffner's property lies 40 feet to the northwest of the northwest boundary of the railroad property with the 40-foot strip being labeled as a vacated county road. To his answer and counterclaim, Adams attached the legal descriptions of his five tracts, together with an aerial map. Those exhibits do not pictorially depict the exact location of the tracts in relation to the railroad property.

Sometime after 1993, Union Pacific abandoned the railroad property. In 2003, the Taylors filed a petition to quiet title, claiming ownership of the entire railroad right-of-way and naming as defendants Pat Haffner, d/b/a Tasco Grain; Tasco Grain Elevator, Inc .; Union Pacific Railroad Company; and Floyd Adams. The railroad disclaimed any remaining interest in the property. The grain elevator apparently is no longer involved in the property, leaving Haffner and Adams as the only participating defendants. Patricia Adams was substituted for her husband, Floyd. They each filed answers and counterclaims, seeking, *inter alia,* to quiet title to the portion of the railroad right-of-way directly related to the frontage of their respective tracts from the northwest right-of-way boundary line to the center of the original track centerline, *i.e.,* the northwest 300 feet of the abandoned right-of-way.

The oral arguments on the summary judgment motions might be characterized as obtuse, ill-defined, and fluid. We perceive the Taylors' principal argument to be that they could trace their title to the original owner of the land from which the railroad right-of-way was taken, and, therefore, the entire tract reverted to them upon the railroad's abandoning its use of the property. Haffner and Adams argued that their tracts abutted the right-of-way, entitling them to an accretion of the abandoned property to the centerline of the original track. Both parties supported their arguments by citation to *Stone v. U.S.D. No. 222,* 278 Kan. 166, 91 P.3d 1194 (2004). In a supplemental argument, the Taylors contended that Haffner and Adams were not adjoining landowners.

Complicating the analysis was a reference to a grant to the Board of County Commissioners of Sheridan County, Kansas, of a 66-foot right-of-way for a public road, which apparently ran between the Tasco townsite and the railroad right-of-way. Inexplicably, the record does not contain that conveyance or any other information from which we can know the pertinent details. Haffner and Adams seize upon this omission to argue that there is no record that a road was ever officially created between their tracts and the railroad property, but if it was so created, it was subsequently vacated. Curiously, however, they acknowledge "that there is in fact a roadway used by the public running northeasterly to southwesterly along the north side of the railroad property and along the southeast side of the Appellants' property," *i.e.,* between the Haffner and Adams tracts and the abandoned right-of-way.

The Taylors also pointed out to the district court that the abandoned right-of-way had been added to their property for purposes of assessing ad valorem taxes. They claimed to have paid the real estate taxes on the entire abandoned right-of-way for 5 years.

The district court took the matter under advisement and permitted the parties to submit supplemental written arguments. By memorandum opinion, the district court ruled in the Taylors' favor. The principal rationale was that the defendants' tracts were separated from the railroad right-of-way by a county road, depriving defendants of a claim to be abutting landowners. The court relied heavily upon *Danielson v. Woestemeyer,* 131 Kan. 796, 293 Pac. 507 (1930). The court also found that, because the metes and bounds descriptions of the defendants' tracts did not include any portion of the railroad right-of-way property, the defendants never acquired an ownership interest in the real estate which was subject to the railroad right-of-way and the defendants were not successors in title to the original landowner from whom the easement was taken.

On appeal, Haffner and Adams first argue that the district court erroneously found that the county road that separates their tracts from the right-of-way was owned in fee simple so as to prevent the tracts from adjoining or abutting the right-of-way. Next, they contend that the vacating of the public road allowed their tracts to touch the right-of-way boundary, which in turn gave the tracts the right of accretion to the northwest 300 feet of the abandoned right-of-way. Finally, the appellants challenge the district court's finding that the omission of any railroad property in the tracts' metes and bounds descriptions precluded their claim to the abandoned property. At oral argument, counsel for Haffner and Adams asserted that the tracts' original legal descriptions placed them immediately adjacent to and abutting the right-of-way.

In response, the appellees continue their argument that their predecessor in title owned all of the servient estate in all of the railroad property, entitling them to a reversionary interest in all of the right-of-way upon the railroad's abandonment. Additionally, the Taylors assert that neither Haffner nor Adams is an abutting landowner.

Standard of Review Although our standard of review on summary judgment is familiar, we will repeat it, emphasizing an important aspect:

"Summary judgment is appropriate when the pleadings, depositions, answers to interrogatories, and admissions on file, together with the affidavits, show that there is no genuine issue as to any material fact and that the moving party is entitled to judgment as a matter of law. The trial court is required to resolve all facts and inferences which may reasonably be drawn from the evidence in favor of the party against whom the ruling is sought. *When opposing a motion for summary judgment, an adverse party must come forward with evidence to establish a dispute as to a material fact.* In order to preclude summary judgment, the facts subject to the dispute must be material to the conclusive issues in the case. On appeal, we apply the same rules and where we find reasonable minds could differ as to the conclusions drawn from the evidence, summary judgment must be denied. [Citation omitted.]" (Emphasis added.) *Bracken v. Dixon Industries, Inc.*, 272 Kan. 1272, 1274-75, 38 P.3d 679 (2002).

Determinative Issues First, we will eliminate the surplusage in the parties' briefs and endeavor to refine what must be decided to resolve this dispute.

The Taylors did not seek to quiet their title in and to the abandoned right-of-way as against all potential claimants; their petition sought to resolve the respective claims of Union Pacific, Haffner, and Adams, who were the only named defendants. Haffner and Adams concede that the Taylors' property abuts the right-of-way on the southeast and acknowledge that the Taylors are entitled to ownership of a portion of the right-of-way by virtue of that abutting ownership. Thus, if none of the named defendants have a factual or legal basis to exert a claim to any portion of the right-of-way, we can affirm the district court's ruling in favor of the Taylors, even though the record may not definitively establish that the Taylors are the owners of all the abandoned property, *i.e* ., that they own the land abutting the right-of-way on the northwest boundary. In other words, the Taylors' title can be quieted vis-a-vis the named defendants, even though an unnamed person or entity might have a claim to the northwest portion of the right-of-way that is superior to the Taylors' claim.

The district court found that the parties had agreed that the railroad only acquired an easement in the disputed property; that the railroad did not hold a fee simple interest in any of the property; and that the railroad had properly abandoned its use of the property. On appeal, the parties do not dispute this finding, and, therefore, Union Pacific is eliminated as a claimant. See *Abercrombie v. Simmons,* 71 Kan. 538, 546, 81 Pac. 208 (1905) (an interest taken for use as a right-of-way is limited to that use and must revert when the use is abandoned).

The railroad condemnation occurred before the receipt and patent were issued. Therefore, the Taylors' argument that the railroad easement was taken entirely from their exclusive predecessor in title is unpersuasive, given that the easement existed prior to their root title. However, Haffner and Adams appear to concede that their conveyances did not purport to convey any portion of the servient estate in the land being used for railroad purposes. See, *e.g., Gauger v. State,* 249 Kan. 86, 90-91, 815 P.2d 501 (1991) (when a dominant estate abandons a right-of-way, the property reverts to the servient estate). Therefore, to be entitled to ownership of any of the abandoned railroad right-of-way, Haffner and Adams must be owners of land adjacent to and abutting the right-of-way. See *Danielson v. Woestemeyer,* 131 Kan. 796, Syl. ¶ 3 (owner of land separated from railroad right-of-way by intervening strip of land is deprived of rights of an abutting owner).

First, we must determine whether Haffner's and Adams' claim at oral argument that their original conveyances placed their tracts adjacent to and abutting the right-of-way has any basis in fact. If not, then we must decide whether, by operation of law, their boundary lines were enlarged to the point of adjoining the right-of-way.

Original Location of Tracts One of the Taylors' exhibits is a hand-drawn map of the half-section, apparently copied from an abstract of title, showing the railroad property and the defendants' six tracts. The map shows the northwest boundary of the right-of-way being 300 feet from the center of the original railroad track and a road separating the right-of-way from the townsite tracts. Attached as an exhibit to Haffner's answer and counterclaim is a boundary survey, prepared by a licensed surveyor, which also reflects the northwest boundary as 300 feet perpendicular from the railroad track. The survey describes Haffner's tract as follows:

"Commencing at a point that is 325 feet West of the East front of the Union Pacific Depot when measured on a line that is parallel with the Center line of the main track of the Colby Branch of the Union Pacific Railroad and 340 feet Northerly from said center line when measured at right angles thereto; thence Westerly and parallel with said center line 490 feet to the place of beginning; thence Easterly and parallel with said center line a distance of 390 feet; thence Northerly and at right angles with said center line 340 feet; thence Westerly and parallel with said center line 300 feet; thence Southerly from said point to the place of beginning."

The southerly (southeast) boundary of Haffner's property is described as running parallel to the track centerline and being a perpendicularly measured distance of 340 feet from the centerline; the right-of-way boundary line is 300 feet away and parallel to the railroad track centerline; and, therefore, Haffner's property is 40 feet outside of and parallel to the railroad right-of-way boundary line. The survey's scale drawing confirms that Haffner's property is separated from the right-of-way by an intervening strip of land that is 40 feet in width.

As noted in our standard of review, Haffner was required to come forward with evidence to dispute the material fact that Haffner's tract did not originally abut the right-of-way. To the contrary, Haffner presented evidence, in the form of a boundary survey, which corroborated that fact. Therefore, the district court was correct in proceeding with summary judgment on the basis that Haffner's property did not originally abut the right-of-way.

Adams did not submit a boundary survey. However, the legal descriptions Adams proffered on her five tracts mirror Haffner's boundary survey in placing the starting points at right angles to the track centerline and describing the southerly (southeast) boundary lines as running parallel to the track centerline. Three tracts are 340 feet from the railroad track centerline and two tracts are 360 feet away. Therefore, Adams' own evidence also corroborates that her tracts were originally separated from the railroad right-of-way by a strip of land.

Therefore, neither Haffner nor Adams can claim the rights of an abutting landowner based upon the property they claim to have originally acquired. They must find some way in which to extend their respective southeast boundaries across the intervening strips of land.

Vacated Road The appellants repeatedly point out that there is nothing in the record to establish that a road was ever created between the right-of-way and their tracts. One might find that argument to be curious, because if there is no road, then there would be no means by which they could jump their boundary lines over the separating tract and come in contact with the right-of-way boundary.

However, Haffner and Adams do point out that there is currently a road in use in the area between their boundaries and the right-of-way boundaries. Further, the Haffner survey designates the 40-foot separating strip of land as a vacated county road. Resolving all facts and inferences in appellants' favor, we will consider the legal effect of the intervening strip of land being a vacated road.

The appellants argue that a public road, however acquired, is merely a right-of-way and upon being vacated, the land returns to the owner from whom the right-of-way was taken, or, if the dominant estate was taken from a common predecessor in title, then the current adjacent landowners each take to the center of the road. We agree. However, this argument only takes Haffner and Adams halfway there, literally. If the 40-foot wide county road was vacated, the landowners on the northwest boundary of the roadway would only take 20 feet of the land, leaving them 20 feet short of abutting the right-of-way.

Appellants attempt to overcome this result by arguing as follows:

"If by some chance there was a legally established roadway, it has since been vacated [citation omitted], and in that event, one-half (1/2) of the road right-of-way reverted to Appellants as adjoining land owners on the northwest and *the other one-half (1/2) would revert to the railroad and adjoin its easement that it acquired in 1888.*" (Emphasis added.)

The appellants' assertion that the railroad acquired 1/2 of the road is belied by authority cited in their own brief. A railroad right-of-way easement only grants the right to exclusively use the land surface for railroad purposes. *Harvest Queen Mill & Elevator Co. v. Sanders,* 189 Kan. 536, 542, 370 P.2d 419 (1962). The railroad does not acquire a fee simple absolute title or a fee simple determinable title. 189 Kan. at 544. All incidents of ownership not inconsistent with the railroad's use remain in the owner of the servient estate. "[A]ny right not inconsistent with the easement remains in the abutting owner." 189 Kan. at 542. Therefore, the railroad could not utilize its restricted use easement to acquire fee simple title to the 1/2 of the roadway lying adjacent to the right-of-way. That interest would pass to the owner of the servient estate.

The appellants simply cannot stretch their tract boundary lines to abut the right-of-way, and, therefore, they have no legal basis to claim an interest in the abandoned railroad property. The district court was correct in granting judgment against Haffner and Adams.

Affirmed.

CASE #15 OVERBURDENING AN EASEMENT CAUSING ITS TERMINATION

SHOOTING POINT, L.L.C., ET AL. v. WESCOAT
WESCOAT v. SHOOTING POINT, L.L. C., ET AL.
265 VA. 256; 576 S.E.2D 497 (VA. 2003)

Opinion

Barbara Milano Keenan, Justice* In this appeal, we primarily consider whether the chancellor erred in determining the location of an easement and in ruling that the proposed use of the dominant estate as a residential subdivision would not overburden the servient estate.

John W. Wescoat owns a tract of land in Northampton County (the Wescoat parcel) that is subject to a recorded easement in favor of a 176-acre tract owned by Shooting Point, L.L.C. (the Shooting Point parcel). The easement, which is 15 feet wide and 0.3 mile in length, is the only means of ingress and egress between the Shooting Point parcel and a nearby state highway. In response to a plan by Shooting Point, L.L.C. (Shooting Point) to develop its parcel into a residential subdivision, Wescoat filed a bill of complaint alleging, among other things, that Shooting Point's proposed use of its parcel would "impose an additional and unreasonable burden on the easement" over Wescoat's land.

After hearing the evidence ore tenus, the chancellor ruled that use of the Shooting Point parcel as a residential subdivision would not overburden the servient estate. The chancellor also determined that the actual location of the easement was as shown on certain survey plats. Both Wescoat and Shooting Point appeal.

The evidence before the chancellor showed that the Shooting Point parcel is separated from State Highway Route 622 (Route 622) by the Wescoat parcel. The easement, which follows a dirt road over the Wescoat parcel, is located between a field on one side and woods on the other side. The dirt roadway has three 90-degree turns, including two turns that are "blind" where the wooded areas obscure approaching traffic.

In 1974, Wescoat's predecessors in title executed and recorded a written grant of easement establishing the right-of-way. The grant described the location of the easement in the following terms:

[S]aid right-of-way easement to follow the present road leading from Virginia State Highway Route 622 to lands . . . known as Shooting Point Farm, said present road running generally in a northerly direction from a point in a turn of said Virginia State Highway Route 622 to a point at or near a corner of a certain woods, thence turning in a generally easterly direction and running along the northern edge of said woods to a point at or near the edge of said woods, thence turning in a generally northerly direction and following along the edge of said woods to a point at or near a corner of said woods, thence turning in a generally easterly direction and running along the edge of said woods until the boundary line separating Shooting Point Farm from the [Wescoat parcel] is reached, at which boundary line the said right-of-way easement terminates.

The grant further described the right-of-way as "the only easement to provide a means of ingress and egress" from Route 622 to the Shooting Point parcel. The grant did not contain a clause limiting use of the easement.

At the time the easement was established, both the servient estate and the dominant estate were used primarily for agricultural and recreational purposes. In June 1979, Shooting Point's predecessors in title conveyed 13.2 acres at the southern border of the Shooting Point parcel to Richard E. Meekins, Sr. The deed conveyed to Meekins the right to use the easement as shown on a plat prepared in May 1979 by Bonifant Land Surveys (the Bonifant plat).

In December 1999, Shooting Point purchased the dominant estate and began planning the development of a residential subdivision. The proposed subdivision has 18 residential lots, each averaging over five acres, which border a 50-acre lot to be preserved as "open space."

Shooting Point recorded a plat in the circuit court clerk's office, prepared by Baldwin & Gregg Surveyors (the Gregg plat), that showed the proposed subdivision and the 15-foot-wide easement connecting the Shooting Point parcel to Route 622. The Gregg plat incorporated the Bonifant plat and, in depicting the easement, adopted the Bonifant plat's courses, distances, measuring points, and centerline.

Shooting Point also recorded a declaration of protective covenants that incorporated the Gregg plat, and later used that plat to describe the easement in a deed of trust conveying a subdivision lot to a trustee. Shooting Point conveyed certain other subdivision lots in five separate deeds, each conveying the right to use the easement and referencing the Gregg plat's depiction of the right-of-way.

In January 2000, Wescoat sent a letter to some of the subdivision lot purchasers advising them that the easement was restricted to a width of 15 feet. Wescoat further informed the purchasers that the right-of-way would be "clearly marked" to make them aware of the easement's width. Wescoat's son placed two stakes 15 feet apart at the easement's entrance near Route 622 that straddled the existing usage of the easement. A large sign was placed near the stakes that read, "Begin 15 Foot Right of Way."

In February 2000, Wescoat filed a bill of complaint against Shooting Point alleging that Shooting Point's proposed use of its parcel as a residential subdivision was not reasonable and would create "an additional and unreasonable burden" on the easement. Wescoat asked the chancellor, among other things, to enjoin Shooting Point from selling and conveying the remaining lots in the proposed subdivision.

In January 2001, Wescoat employed George E. Walters, a certified land surveyor, to survey the easement and to place markers delineating its course. After Walters situated the markers on the property, Wescoat's son placed wooden posts outside those markers along the roadway to designate the easement's course. In general, the pathway created by the posts followed the line of the woods more closely than the existing roadway and resulted in "sharper" 90-degree turns.

In February 2001, Wescoat filed a bill of complaint for declaratory judgment against Shooting Point, L.L.C., Shooting Point Property Owners Association, Inc. (collectively, Shooting Point), and others, seeking various rulings concerning Shooting Point's use of its property. The chancellor consolidated Wescoat's two suits for trial.

Before trial, Shooting Point requested leave to file a cross-bill in Wescoat's declaratory judgment suit. In its proposed cross-bill, Shooting Point sought a determination of the easement's location and removal of the posts that Wescoat's son had placed along the course of the easement. The chancellor denied Shooting Point's motion.

Shooting Point also filed a motion in limine to exclude from evidence Walters' testimony and the two revised plats he prepared depicting the easement (the Walters plats) on the ground that this evidence was not timely disclosed. Shooting Point did not receive copies of Walters' revised plats until the day before trial.

In response to the motion in limine, Wescoat noted that no order had been entered regulating discovery in the case, and that Shooting Point also was not timely in its disclosures, having designated an expert witness only the day before trial. The chancellor denied Shooting Point's motion in limine.

On the first day of trial, Wescoat moved the chancellor to continue the case on the ground that the issue of the easement's location was not properly before the court. Shooting Point opposed the motion, arguing that the issue was "directly" before the court. The chancellor denied the continuance motion.

At trial, the chancellor received evidence from expert witnesses indicating that the proposed residential subdivision would generate daily about ten vehicle trips per lot. Thus, the proposed subdivision would result in an additional 180 trips daily over the easement.

Wescoat's son, John W. Wescoat, Jr., testified that vehicles traveling in opposite directions on the easement could not pass at the same location. John stated that the worn roadway remained the same from 1977 to 1999, and that, after Shooting Point purchased its parcel, the traffic on the easement increased and the roadway became wider as motorists drove around "mudholes" in the easement and "cut" corners at the turns in the roadway.

Curtis Jones, Jr., Wescoat's cousin, leased both the Wescoat and Shooting Point parcels for farming purposes. Jones testified that he and his employees make heavy use of the easement when they plant, maintain, irrigate, and harvest the crops.

Wescoat presented the testimony of Walters, who qualified as an expert witness on the subject of land surveying. He testified that the easement was first surveyed in 1979 by P. Bonifant, and that Walters created his plats in an attempt to "resurvey" the easement shown on the Bonifant plat.

Walters stated that he first chose a buried survey pin, or "rebar," that he discovered in the middle of the existing roadway near Route 622 to mark the centerline of the easement, and that he originally used that centerline in his plat to delineate the easement's course. However, after a consultation with Bonifant, Walters later concluded that a " 'bent rebar' marker," located approximately nine feet east of the other "rebar," was the marker indicating the correct location of the easement's centerline. Walters testified that he revised his plats to reflect the "bent rebar" as the centerline of the easement, which resulted in a nine-foot eastward shift of the easement's entrance onto Route 622.

Walters stated that the course designated on his revised plats reproduced the easement as shown on both the Gregg and the Bonifant plats. Walters opined that the present usage of the easement had moved westward since the time of Bonifant's survey and explained that the paths of farm roads "tend to wander" as motorists drive vehicles around potholes and tree limbs that protrude into roadways.

Shooting Point presented the expert testimony of James B. Latimer, II, a licensed land surveyor, who testified that the easement's centerline in the Bonifant plat "is closer to the east, closer to the woods than the physical road that's there." He also stated that the roadway has always been in its present location.

Shooting Point also submitted expert testimony from Millison E. Duff, Jr., a licensed surveyor and president of Baldwin & Gregg Surveyors. Duff stated that the Gregg plat adopted the Bonifant plat's depiction of the easement because that depiction "appeared to follow generally along the road that we had evidence of being in existence at that time."

Duff further testified that the "bent rebar" located about nine feet east of the center of the existing roadway was the survey pin that Bonifant used to mark the easement's centerline. Duff stated that if Bonifant's centerline were followed, the eastern border of the easement would "go right through an 18-inch pine tree," and motorists traveling on the easement would "scrape" the right side of their vehicles against the tree. Duff said that he did not believe that anyone presently could determine the precise location of the roadway in 1974 "short of doing a soils analysis." However, he concluded that the Bonifant plat was the "best evidence" available concerning the easement's location when the Gregg plat was prepared in 1999.

At the conclusion of the evidence, the chancellor held that the Bonifant plat, the Gregg plat, and the Walters plats were the "best evidence" of the easement's location, and that the easement's location was accurately depicted on those plats. The chancellor stated that "[a]ny attempt to establish an alteration [of the designated easement] would simply amount to no more than guesswork or speculation on the part of the Court." The chancellor also held that use of the Shooting Point parcel as a residential subdivision would not overburden the servient estate.

Shooting Point argues that the chancellor erred in denying its motion for leave to file a cross-bill and its motion in limine to exclude Walters' testimony and survey plats. In support of these contentions, Shooting Point advances the same arguments it made before the chancellor. We disagree with Shooting Point's arguments.

The chancellor's rulings on both pretrial motions were proper exercises of his discretion. First, Shooting Point did not need to file a cross-bill to raise the issue of the easement's location, which already was before the court as Shooting Point observed in its opposition to Wescoat's continuance motion. Moreover, the location of the easement was the subject of extensive evidence presented by both parties during trial and is before us in this appeal. Second, the chancellor's ruling denying the motion in limine is supported by the materiality of Walters' testimony and his plats to the issues being tried, and the absence of any order requiring earlier disclosure of discoverable information.

Shooting Point next argues that the chancellor erred in concluding that the Bonifant plat, the Gregg plat, and the Walters plats accurately depict the easement's location. Shooting Point contends that the chancellor improperly ignored evidence of existing usage and established a new easement location. Shooting Point asserts that a literal application of the Bonifant and Walters plats results in an easement that is unreasonably close to the line of woods, includes a pine tree over 18 inches in diameter, and contains sharp turns that impede the passage of larger vehicles. We disagree with Shooting Point's arguments.

An established standard of review governs our consideration of both this issue and the issue of the burden placed on the servient estate. The chancellor, as trier of fact, evaluated the witnesses' testimony and their credibility. *Tauber v. Commonwealth,* 263 Va. 520, 526, 562 S.E.2d 118, 120 (2002); *Johnson v. Cauley,* 262 Va. 40, 44, 546 S.E.2d 681, 684 (2001). Because he heard the evidence ore tenus, the chancellor's decree is entitled to the same weight as a jury verdict. *Chesterfield Meadows Shopping Ctr. Assocs., L.P. v. Smith,* 264 Va. 350, 355, 568 S.E.2d 676, 679 (2002); *Johnson,* 262 Va. at 44, 546 S.E.2d at 684; *Hoffman Family, L.L.C. v. Mill Two Assocs.*

P'ship, 259 Va. 685, 696, 529 S.E.2d 318, 325 (2000). Thus, on appeal, we will not set aside the chancellor's findings unless they are plainly wrong or without evidence to support them. *Tauber,* 263 Va. at 526, 562 S.E.2d at 120; *Hudson v. Pillow,* 261 Va. 296, 302, 541 S.E.2d 556, 560 (2001).

Here, the chancellor received substantial evidence supporting his determination of the easement's location. The Bonifant Plat, the first plat depicting the easement, was prepared only five years after the easement was established. The Gregg plat and the revised Walters plats placed the easement at the same location detailed in the Bonifant plat.

The chancellor's determination also is supported by Shooting Point's own extensive use of the Bonifant plat's location of the easement. Shooting Point implicitly agreed to the accuracy of this location by referring to the Gregg plat in five deeds conveying lots to subdivision purchasers, in one deed of trust, and in Shooting Point's declaration of protective covenants. In addition, Shooting Point's expert, Duff, testified that the Bonifant plat was the "best evidence" available of the easement's location when the Gregg plat was prepared.

We disagree with Shooting Point's assertion that a literal application of these plats incorrectly would place the easement too close to the woods. Although Duff initially testified that the eastern border of the easement, as shown in the plats, would "go right through an 18-inch pine tree," he effectively modified this statement when he later testified that a vehicle traveling on the easement would merely "scrape" its side against the tree. Also, the chancellor received testimony indicating that the path of the worn roadway had "migrated" away from the woods since the time of Bonifant's survey.

We also find no merit in Shooting Point's contention that evidence of existing usage showed that Wescoat consented to a change in the easement's location. We initially observe that, generally, when a fixed location of a granted easement is established, that location may be changed only with the express or implied consent of the persons interested. *Buxton v. Murch,* 249 Va. 502, 508, 457 S.E.2d 81, 84 (1995); *Fairfax County Park Auth. v. Atkisson,* 248 Va. 142, 148, 445 S.E.2d 101, 104 (1994); *Wagoner v. Jack's Creek Coal Corp.,* 199 Va. 741, 746, 101 S.E.2d 627, 630 (1958). Thus, in the present case, evidence of existing usage of the easement was competent evidence for the chancellor's consideration.

Here, however, the evidence of usage did not establish consent by Wescoat to a new easement location. Although Curtis Jones, Wescoat's cousin and tenant, sometimes drove his vehicles outside the defined course of the easement as a matter of convenience, his actions did not indicate that Wescoat consented to a different course of the roadway. Similarly, Wescoat's consent cannot be inferred from evidence that after Shooting Point purchased its parcel, the worn pathways in the road widened as motorists drove their vehicles around mud holes and "cut" corners to ease the sharp turns along the roadway.

The placement of stakes at the easement entrance, and posts along the course of the easement, also did not establish Wescoat's consent to a different fixed location for the right-of-way. Although Wescoat's letter to the subdivision lot owners advised that the easement would be "clearly marked," only two stakes were placed at the easement's entrance near Route 622. The following month, Wescoat initiated the present suit against Shooting Point.

While some posts later were placed along the course of the worn roadway, the evidence showed that the path marked by the posts generally followed the line of woods more closely than the existing roadway and resulted in "sharper" 90-degree turns. In addition, both Shooting Point and Wescoat disputed that the pathway created by the posts was the true easement location. At trial, Shooting Point asserted that the posts improperly restricted its use of the easement. Wescoat argued that the revised Walters plats, which shifted the easement about nine feet to the east of the posts at the entrance onto Route 622, depicted the correct location of the right-of-way. Thus, we conclude that the record did not establish an express or implied agreement by Wescoat to effect a change in location of the easement, and we hold that the chancellor did not err in his determination of the easement's location.

Wescoat assigns error to the chancellor's ruling that Shooting Point's use of its parcel as a residential subdivision would not overburden the servient estate. Wescoat argues that this use would create an additional burden on his property that would adversely impact his ability to use the easement. He alternatively contends that even if Shooting Point's use would only result in an increase in degree of the existing burden, that increase would have the practical effect of imposing an additional burden on the servient estate. We disagree with Wescoat's arguments.

A party alleging that a particular use of an easement is unreasonably burdensome has the burden of proving his allegation. *Shenandoah Acres, Inc. v. D.M. Conner, Inc.*, 256 Va. 337, 342, 505 S.E.2d 369, 371 (1998); *Hayes v. Aquia Marina, Inc.*, 243 Va. 255, 259, 414 S.E.2d 820, 822 (1992). Generally, when an easement is created by grant or reservation and the instrument creating the easement does not limit its use, the easement may be used for "any purpose to [576 S.E.2d 503] which the dominant estate may then, or in the future, reasonably be devoted." *Id.* at 258, 414 S.E.2d at 822 (quoting *Cushman Virginia Corp. v. Barnes,* 204 Va. 245, 253, 129 S.E.2d 633, 639 (1963)); *see also Collins v. Fuller,* 251 Va. 70, 72, 466 S.E.2d 98, 99 (1996). However, this general rule is subject to the qualification that no use may be made of the easement, different from that established when the easement was created, imposes an additional burden on the servient estate. *Id.; Hayes,* 243 Va. at 258-59, 414 S.E.2d at 822; *Cushman,* 204 Va. at 253, 129 S.E.2d at 639-40.

In the present case, the 1974 grant did not restrict use of the easement. Therefore, we consider whether the evidence supports a conclusion that Shooting Point's subdivision of the dominant estate is a reasonable use of the parcel that would not overburden the servient estate.

Our decisions in *Hayes* and *Cushman* illustrate the nature of this inquiry. In *Hayes,* an operator of a marina on the dominant estate, a 2.58-acre tract, proposed to expand its marina facility from 84 to 280 boat slips. The easement providing access to the marina was a private roadway about 1,120 feet long and 15 feet wide along its entire course. The agreement creating the easement did not restrict its use. 243 Va. at 256-59, 414 S.E.2d at 820-22.

We held that the record supported the chancellor's conclusion that the proposed expansion would not unreasonably burden the servient estate, although the "degree of burden" would be increased. We assumed, without deciding, that an expanded use of a dominant estate could be of such degree as to create an additional burden on a servient estate, but concluded that the proposed marina expansion was not shown to create such an additional burden. *Id.* at 260, 414 S.E.2d at 823.

Similarly, in *Cushman,* the instrument creating the easement did not contain any language limiting the easement's use. When the easement was established, the dominant estate, a 126.67-acre tract, had two dwelling houses and was used as a farm. The owner of the dominant estate proposed to subdivide his land for a residential and commercial development that would include 34 residential lots. 204 Va. at 252-53, 129 S.E.2d at 639-40.

We reversed the chancellor's decree limiting the easement to its original uses, stating:

The fact that the dominant estate is divided and a portion or portions conveyed away does not, in and of itself, mean that an additional burden is imposed upon the servient estate. The result may be that the *degree* of burden is increased, but that is not sufficient to deny use of the right of way to an owner of a portion so conveyed.

Id. at 253, 129 S.E.2d at 640. Emphasis added.

Applying these principles to the present case, we hold that the subdivision of the 176-acre Shooting Point parcel into 18 residential lots is, in the language of *Cushman,* a purpose to which the dominant estate may be reasonably devoted. *See id.,* 129 S.E.2d at 639. Moreover, the record supports the chancellor's conclusion that Shooting Point's proposed use of the easement would not impose an unreasonable burden on the servient estate. Although the number of vehicles using the easement would increase substantially as a result of the proposed use, this fact demonstrates only an increase in degree of burden, not an imposition of an additional burden, on the servient estate. Like the facts underlying our decision in *Hayes,* the facts here do not support consideration of a further question whether an increased degree of burden could be so great as to impose an additional burden on the servient estate.

For these reasons, we will affirm the chancellor's judgment.

Affirmed.

NOTE:

* Chief Justice Carrico presided and participated in the hearing and decision of this case prior to the effective date of his retirement on January 31, 2003.

CASE #16 MAJOR EXPANSION OF DEVELOPMENT NOT CAUSING AN OVERBURDEN

HAYES, ET AL. v. AQUIA MARINA, INC., ET AL.

243 VA. 255; 414 S.E.2D 820 (VA. 1992)

Present: All the Justices.

STEPHENSON, Justice.

The principal issue in this appeal is whether an easement across the servient estates will be overburdened by the proposed expanded use of the dominant estate.

Robert C. Hayes and others [1] (collectively, Hayes) brought a chancery suit against Aquia Marina, Inc., Warren E. Gnegy, and Cynthia Gnegy (collectively, Gnegy). [2] Hayes alleged, inter alia, that a proposed expansion of a marina located on Gnegy's land (the dominant estate or marina property) would overburden the easement across Hayes's lands (the servient estates). Hayes, therefore, sought to have the trial court enjoin the proposed expanded use of the dominant estate.

The cause was referred to a commissioner in chancery. Following an ore tenus hearing, and after taking a view of the subject properties, the commissioner filed a report containing the following findings: (1) a perpetual easement exists across the servient estates for ingress to and egress from the dominant estate; (2) the easement is not limited solely for domestic use, but may be used commercially by the marina and its customers and by boat owners and their guests; (3) the proposed expansion of the marina from 84 to 280 boat slips is a reasonable use of the dominant estate; (4) the resulting increase in traffic over the easement will not change the type, only the degree, of use and will not overburden the easement; and (5) paving the easement is reasonable and a proper means of maintenance.

By a final decree, entered March 5, 1991, the trial court overruled all of Hayes's exceptions to the commissioner's report and confirmed the report in all respects. Hayes appeals.

We must view the evidence in the light most favorable to Gnegy, the prevailing party at trial. The marina property is a 2.58-acre tract situate on Aquia Creek in Stafford County. The easement is the sole means of land access to the marina property.

The litigants' predecessors in title entered into a written agreement, executed February 3, 1951, for "the establishment of a certain roadway or right of way beginning at the Northern terminus of State Highway No. 666, and terminating at the property division line between [the servient estates], and where [the dominant estate] adjoins the same on the North side thereof" and for "the continuation of said right of way." The agreement recited that "the State Department of Highways will be requested . . . to take over into the State Highway System the present roadway beginning at the North terminus of said State Highway No. 666, and leading through [the servient estates]." The roadway that was intended to be taken into the state highway system was "approximately something less than one-half mile in length." The "newly established private roadway" was "approximately 1,120 feet in length" and "fifteen feet wide along its entire distance." The agreement provided that the parties thereto "shall have an easement of right of way over the entire length [thereof]."

The record indicates that the portion of the easement, beginning at the northern terminus of State Highway No. 666, became a part of the state highway system in 1962. The record also indicates that the "private roadway" is constructed of dirt and gravel.

By 1959, three residential buildings and a wooden pier were located on the dominant estate. The pier was approximately 30 feet long and contained about 10 boat slips. This small marina was operated commercially.

Between 1961 and 1962, the current marina was constructed. This marina has been operated commercially for the general public from 1964 until the present. The marina consists of 84 boat slips, a travel lift station,[3] a public boat launch, and a gas dock. Boats and boat parts are sold at the marina. Boats also are repaired on the marina property.

In September 1989, the Board of Supervisors of Stafford County granted Gnegy a special use permit to expand the marina by increasing the number of boat slips to 280. After the proposed expansion, the marina will continue to provide the same services it has provided since 1964.

There has never been a "traffic problem" with the easement. An expert witness on emergency services testified that there never had been a problem with access to the marina property and none was anticipated if the proposed expansion occurred. On weekends, a time of maximum use of the marina property, Gnegy anticipates that only 20 to 30 percent of the boat owners will make use of the marina.

As a general rule, when an easement is created by grant or reservation and the instrument creating the easement does not limit the use to be made of it, the easement may be used for "any purpose to which the dominant estate may then, or in the future, reasonably be devoted." *Cushman Corporation v. Barnes,* 204 Va. 245, 253, 129 S.E.2d 633, 639 (1963). Stated differently, an easement created by a general grant or reservation, without words limiting it to any particular use of the dominant estate, is not affected by any reasonable change in the use of the dominant estate. *Savings Bank v. Raphael,* 201 Va. 718, 723, 113 S.E.2d 683, 687 (1960) (citing Ribble, 1 Minor on Real Property § 107, at 146 n. 2 (2d ed. 1928)). However, no use may be made of the easement which is different from that established at the time of its creation and which imposes an additional burden upon the servient estate. Cushman Corporation, 204 Va. at 253, 129 S.E.2d at 639-40.

Hayes contends that, by using the phrase, "private roadway," in the easement agreement, the parties to the agreement intended to limit the use of the easement to domestic purposes, thereby prohibiting commercial uses.[4] Gnegy contends, on the other hand, that the agreement created an easement for access without limitation. The commissioner and the trial court adopted Gnegy's contention.

When the agreement is read as a whole, it is clear that the phrase, "private roadway," was used to distinguish that portion of the easement that would not become a part of the state highway system from that portion of the easement that could be taken into the system. Thus, the phrase is descriptive, not restrictive.

Consequently, we hold that the agreement creating the easement for access contains no terms of limitation upon the easement's use. Additionally, the record supports the conclusion that the operation of a marina is a use to which the dominant estate reasonably can be, and has been, devoted.[5]

Hayes further contends that the proposed expansion of the marina will impose an additional and unreasonable burden upon the easement. Having alleged that the proposed expansion will impose an additional burden upon the easement, Hayes has the burden of proving this allegation. *Holt v. Holt,* 174 Va. 120, 123, 5 S.E.2d 504, 505 (1939).

A contention similar to the one advanced by Hayes was presented in Cushman Corporation, supra. In Cushman Corporation, as in the present case, the instruments creating the easement contain no language limiting the easement's use. 204 Va. at 253, 129 S.E.2d at 640. When the easement was established, the dominant estate, a 126.67-acre tract, was used as a farm and contained two single-family dwellings with appurtenant servant and tenant houses. Id. at 252, 129 S.E.2d at 639. A controversy arose when the dominant owner proposed to subdivide the tract for residential and commercial uses. Id. at 249, 129 S.E.2d at 637. The trial court limited the easement to its original uses. Id. at 247, 129 S.E.2d at 635. We reversed the ruling, stating, inter alia:

The fact that the dominant estate is divided and a portion or portions conveyed away does not, in and of itself, mean that an additional burden is imposed upon the servient estate. The result may be that the degree of burden is increased, but that is not sufficient to deny use of the right of way to an owner of a portion so conveyed.

Id. at 253, 129 S.E.2d at 640. (Emphasis added.)

Here, after weighing the evidence, both the commissioner and the trial court concluded that the proposed expansion would not unreasonably burden the easement. On appeal, a decree confirming a commissioner's report is presumed to be correct and will be affirmed unless plainly wrong. *Bain v. Bain,* 234 Va. 260, 263, 360 S.E.2d 849, 851 (1987); *Seemann v. Seemann,* 233 Va. 290, 293, 355 S.E.2d 884, 886 (1987).

In the present case, we cannot say that the trial court's conclusion is plainly wrong. Indeed, we think that it is supported by the evidence and by well-established principles of law. Here, as in Cushman Corporation, the proposed expansion will not, "in and of itself," impose an "additional burden" upon the easement, even though the "degree of burden" may be increased. Therefore, assuming, without deciding, that an expanded use of the dominant estate could be of such degree as to impose an additional and unreasonable burden upon an easement, such is not the situation in the present case.

Finally, Hayes contends that Gnegy does not have the right to pave the easement. Hayes acknowledges, and we agree, that the owner of a dominant estate has a duty to maintain an easement. *Pettus v. Keeling,* 232 Va. 483, 490, 352 S.E.2d 321, 326 (1987); *Oney v. West Buena Vista Land Co.,* 104 Va. 580, 585, 52 S.E. 343, 344 (1905). However, Hayes reasons that, because the owner of a dominant estate has a duty to maintain an easement, it follows that the owner does not have a right to improve the easement. We agree that there is a distinction between maintenance and improvement. See *Montgomery v. Columbia Knoll Condo. Council,* 231 Va. 437, 344 S.E.2d 912 (1986). However, we do not agree that the owner of a dominant estate does not have the right to make reasonable improvements to an easement.

Although we previously have not addressed the "improvement" issue, courts in other jurisdictions have held that the owner of a dominant estate has the right to make reasonable improvements to an easement, so long as the improvement does not unreasonably increase the burden upon the servient estate. *See, e.g., Stagman v. Kyhos,* 19 Mass.App.Ct. 590, 476 N.E.2d 257 (1985); *Glenn v. Poole,* 12 Mass.App.Ct. 292, 423 N.E.2d 1030 (1981); *Schmutzer v. Smith,* 679 S.W.2d 453 (Tenn.App.1984). Such improvement may include paving a roadway. See, e.g., Stagman, supra; Schmutzer, supra. Ordinarily, the reasonableness of the improvement is a question of fact. *Guillet v. Livernois,* 297 Mass. 337, 340, 8 N.E.2d 921, 340, 8 N.E.2d 921, 922 (1937). We adopt these principles of law.

In the present case, the commissioner and the trial court found that the proposed paving of the roadway by Gnegy, under the existing facts and circumstances, is reasonable. We will affirm this finding; it is supported by the evidence and is not plainly wrong.

Accordingly, the trial court's judgment will be

Affirmed.

NOTES:

[1] The other complainants/appellants are Irmgard E. Hayes, Nathan L. Fendig, Charlotte Fendig, G.K. Massie, Jr., Mary Massie, and Jacqueline Davis.

[2] The Board of Supervisors of Stafford County initially was a party defendant but is not a party in this appeal.

[3] A "travel lift" is a device for moving boats out of the water for repairs.

[4] Hayes also contends that, as a result of this allegedly restrictive language, Gnegy, "at best, acquired a prescriptive easement" for the existing marina. In fact, Hayes endeavors to concede that such a prescriptive easement was established. By making this concession, Hayes, relying upon McNeil v. Kingrey, 237 Va. 400, 406, 377 S.E.2d 430, 433 (1989), seeks to place on Gnegy the burden of showing that the proposed change in use of the easement imposes no additional burden on the servient estates. We summarily reject Hayes's prescriptive easement theory. The record simply does not support it.

[5] Significantly, Hayes stated in oral argument that Gnegy had a "right to access the present-day marina."

CASE #17 PROPRIETOR'S WAY

CAMPBELL [1] *& OTHERS* [2] *v. NICKERSON & OTHERS* [3]
73 MASS.APP.CT. 20 (2008)

Heard May 19, 2008.

CIVIL ACTION commenced in the Land Court Department on August 25, 2003. The case was heard by *Charles W. Trombly, Jr.,* J.

CYPHER, J.

The plaintiffs appeal from a Land Court judgment declaring that their land in Orleans is subject to an easement in favor of the defendants for access to their otherwise landlocked parcels in neighboring Brewster. The easement at issue runs over Eli Rogers Road in Orleans to the properties of the defendants along the road's continuation as Clay Hole Road in Brewster. The Land Court judge determined that the easement was created in the early Eighteenth Century in ancient documents not found in the registry of deeds. This case requires us to consider whether the practice, some three hundred years ago, of recording the conveyance of land in a book of proprietors, rather than with the registry of deeds, should be recognized as valid. We conclude that it is valid and affirm the judgment of the Land Court.

1. Background The plaintiffs filed a complaint in the Land Court on August 25, 2003, seeking a comprehensive adjudication and declaration of the rights of the parties in Eli Rogers Road; and annulment of a special permit granted to the defendant Carrie L. Nickerson, by the planning board of Brewster, which would permit the construction of one residential structure on her parcel.

Following trial in January, 2006, the judge ruled that (1) the plaintiffs were without standing to challenge the special permit,[4] and (2) the defendants have an easement and a right to use Eli Rogers Road in Orleans to access their property on Clay Hole Road in Brewster.[5], [6]

The plaintiffs appealed, challenging the validity of the easement only on the ground that it was not preserved in conformity with the recording statute, G.L. c. 183, § 4. [7]

2. The Origin of the Easement The defendants' claim of easement is founded on a conveyance by deed on July 20, 1711, made by John and Tom Sipson, then residents of the town of Harwich. [8] That instrument conveyed shares in a large parcel of land to fourteen grantees, hereafter known as the "proprietors." On January 25, 1713/1714, [9] the proprietors met to divide the land among themselves into seventeen lots, and recorded that action and description of the lots in a proprietors' book. At the same meeting, the proprietors recorded the following action:

"The proprietors . . . considering it will be convenient to them and their heirs and assigns forever to have and receive . . . privilege of passing through one or another of said lots . . . therefore conclude and vote that notwithstanding said lots are laid out as they are or may be in each respective division that each and every of said proprietors of said lands and their heirs and assigns shall forever have and enjoy free liberty of carting, driving and passing over any of said lots both for egress and regress." [10]

3. The Land Court Decision The principal issue at trial was whether the defendants, having the burden of proof, could prove their claim of an easement by grant, and not by prescription or necessity. The plaintiffs argued that (1) their land was not within the parcel conveyed by the Sipson deed and therefore was free of the defendants' claim of easement, and (2) in any event, the proprietors could not create an easement over their own land.

Following a lengthy and detailed analysis of the evidence presented on whether the parcels of the plaintiffs and the defendants are located within the land described in the Sipson deed, the judge concluded that all the land involved in this case was included in the Sipson deed of 1711.[11]

The plaintiffs further asserted, however, that even if their land was within the Sipson deed, the proprietors held the land in common, and tenants in common already have a right to occupy and use the land and, therefore, cannot create an easement to cross over their own land.

The judge determined that not all the land was divided by the January 25, 1713/1714 vote, and that the rights of "egress and regress" were to apply to the lots not only as then laid out, but to lots "as they . . . may be [laid out] in each respective division." The judge correctly reasoned that the proprietors intended the rights of passing over the lots to apply to future divisions of the land, and that those rights were established in a manner comparable to the later "common scheme" doctrine. He concluded that the "votes [of the proprietors] amounted to a grant of easement to all those owning land located within the Sipson [deed], including land of both plaintiffs and defendants."

After reviewing the plaintiffs' considerable evidence, as well as their various arguments and memoranda, the judge concluded that the "plaintiffs' land is subject to the easement of Nickerson and Leckie to gain access to their land in Brewster over Eli Rogers Road in Orleans." The judge further determined that the "easement . . . has not been abandoned, in spite of the fact that the use thereof has been sporadic at best and that its precise location may have moved slightly." The plaintiffs on appeal do not challenge any of the judge's factual findings or the conclusions flowing therefrom.

4. The Plaintiffs' Appeal The plaintiffs assert that "[n]o easement created solely by an unrecorded proprietors' vote in the early 18th century is valid against a party who acquired the servient estate in the 20th century without actual knowledge of that vote." Further, relying on the provisions of G.L. c. 183, §§ 4 and 15, the plaintiffs assert that the encumbrance on their titles is not valid because the easement was never recorded. These assertions appear as argument made for the first time on appeal.[12] Because the parties have briefed the issue, and the ramifications of this argument have broader significance, we comment on it to a limited degree. See *Royal Indem. Co. v. Blakely,* 372 Mass. 86, 88 (1977). [See Figure 14.15.]

The plaintiffs' assertion that the easement is not valid because it was not recorded is entirely without relevant legal support. It ignores the long and unchallenged history of conveyances made by proprietors. It is significant that the recording statute, G.L. c. 183, § 4, and the provisions concerning the records of proprietors in G.L. c. 179, see note 10, *supra,* have had parallel existence for over three hundred years, each without express reference to, or exception from, the other. We have not found any instance, nor has any been cited to us, where the conveyancing actions of proprietors have been challenged because they were not carried out within the provisions of the recording statute. Moreover, we think it is significant that c. 179 first appears in 1712-1713, many years after the recording statute, which has an origin in Seventeenth Century colonial laws. Presumed to be aware of the recording statute, the Legislature nevertheless provided a different path for transactions of proprietors.

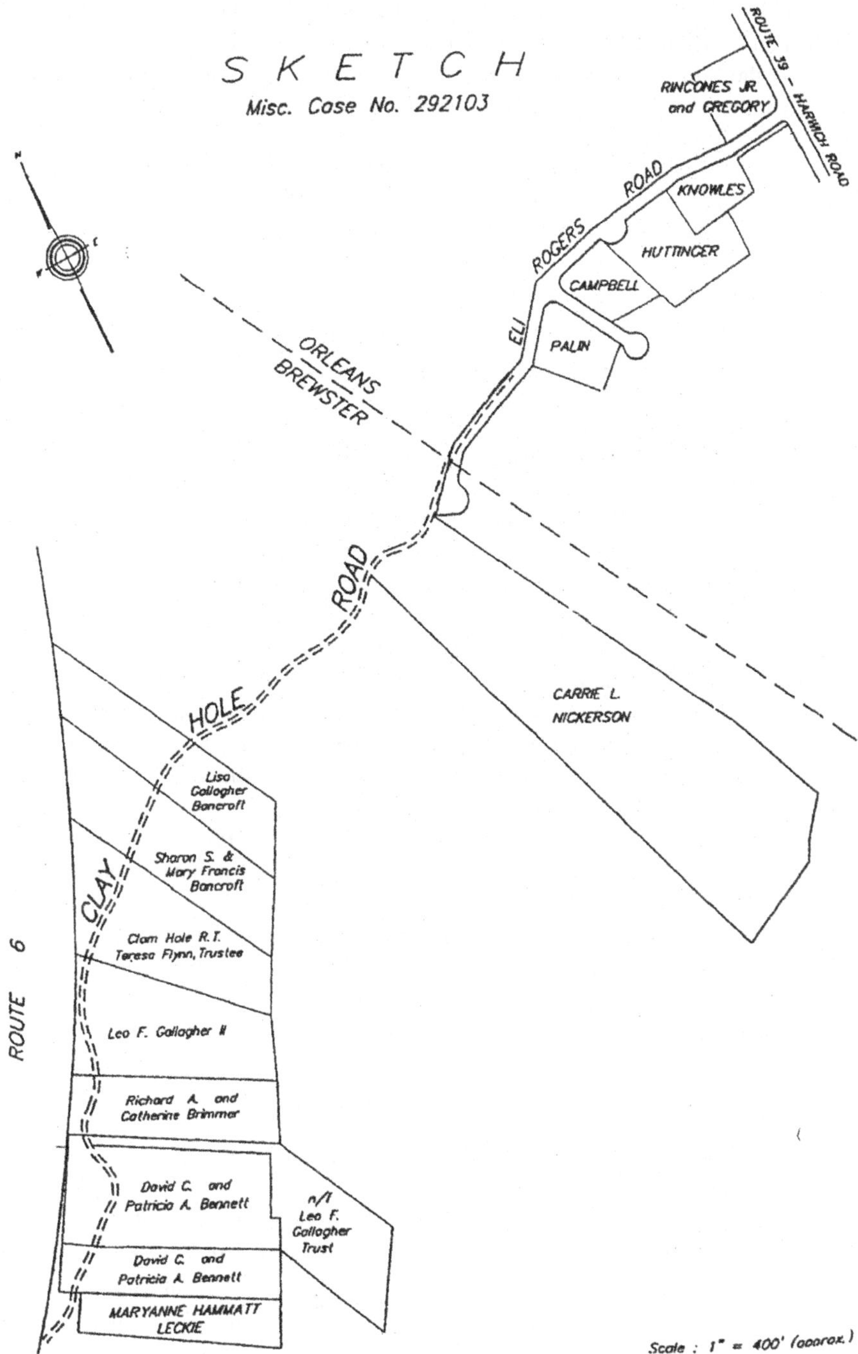
SKETCH
Misc. Case No. 292103
ROUTE 39 - HARWICH ROAD
RINCONES JR. and GREGORY
KNOWLES
ELI ROGERS ROAD
HUTTINGER
CAMPBELL
PALIN
ORLEANS
BREWSTER
CLAY HOLE ROAD
CARRIE L. NICKERSON
Lisa Gallagher Bancroft
Sharon S. & Mary Francis Bancroft
Clam Hole R.T. Teresa Flynn, Trustee
Leo F. Gallagher II
Richard A. and Catherine Brimmer
David C. and Patricia A. Bennett
n/f Leo F. Gallagher Trust
David C. and Patricia A. Bennett
MARYANNE HAMMATT LECKIE
ROUTE 6
Scale : 1" = 400' (approx.)

Figure 14.15

"It has long been the settled law of this commonwealth, that proprietors in common had authority in early times to alienate their lands by vote, which, if duly recorded on the books of the proprietary, passed the title and constituted competent evidence of the transfer." *Green v. Putnam,* 8 Cush. 21, 24-25 (1851), and cases cited. Our courts continue to respect the validity of those acts when they are ascertainable on a proper record. Cf. *Makepeace Bros. v. Barnstable,* 292 Mass. 518, 520-523 (1935); *Newburyport Redev. Authy. v. Commonwealth,* 9 Mass.App.Ct. 206, 213-214, 216, 229-230 (1980); *Sturdy v. Planning Bd. of Hingham,* 32 Mass.App.Ct. 72, 74-75 & n. 7 (1992).

There is no merit in the plaintiffs' assertion that the easement was not valid because they acquired their land without knowledge of the vote of the proprietors. The plaintiffs' knowledge of the proprietors' vote is irrelevant in view of the considerable evidence of their knowledge of the existence and use of modern day Eli Rogers Road. The plaintiffs do not dispute that the easement determined by the judge is applicable to the road in its present configuration. While the defendants cited evidence of the existence of Clay Hole Road in 1886 and 1887, we need not be concerned about how or when the path of the road evolved after the proprietors' vote of January 25, 1713/1714, circumstances no doubt lost in antiquity, because the parties have accepted its present path on the ground without objection. We think the present case falls within the following long-standing principle:

"Where a right of way, or other easement, is granted by deed without fixed and defined limits, the practical location and use of such way or easement by the grantee under his deed, acquiesced in by the grantor at the time of the grant and for a long time subsequent thereto, operate as an assignment of the right, and are deemed to be that which was intended to be conveyed by the deed, and are the same, in legal effect, as if it had been fully described by the terms of the grant."

Bannon v. Angier, 2 Allen 128, 129 (1861). See *Cotting v. Murray,* 209 Mass. 133, 139 (1911). The plaintiffs do not dispute the obvious conclusion that the road has had a very long and undisturbed existence in its present configuration.

Moreover, there is considerable evidence that the plaintiffs had notice that the road was used by the defendants and others over many years. The plaintiffs appear to have obtained their lands in the 1970's. At that time Eli Rogers Road was an unimproved dirt road which had been used by one or more of the original defendants, at least since the 1940's, to access their lands along Clay Hole Road. One of them testified that in the 1960's, in response to a request of the Brewster fire chief, he hired a bulldozer to widen the road and reshape the banks along the entire length of the road, and that he and members of his family have continued to perform maintenance to eliminate potholes.

The plaintiffs introduced in evidence three subdivision plans dated 1970, 1972, and 1973, which show vehicle tracks which represent the "original" path of the road. Superimposed on that path the plans show a forty-foot private way labeled Eli Rogers Road. The deed of plaintiff Campbell, then known as Pamela C. Osgood, is typical in referring to the 1970 plan, and states that the conveyance is made with a "right to pass and repass and to use a right of way as ways are commonly used in the Town of

Orleans both now and hereafter, on a certain way being a 40.00 foot private way" as shown on the plan cited.

We think the plaintiffs' failure at trial to challenge the use of these roads, and their long-time acquiescence in the existence of the roads, and circumstances of their use, completely undermine their assertion that the easement was not valid because they acquired their lands without knowledge of the vote of the proprietors.

Finally, the plaintiffs assert that modern day purchasers "can ill afford to bear the risk of ancient unrecorded easements by grant," and further assert that recognizing such easements would disrupt title examinations. [13] They add that the decision of the Land Court "puts into question between 50 to 100 other parcels of land, covering at least 1,000 acres of land on Cape Cod."

Nothing appears in this case to indicate that the titles of the lands in question have been clouded. There is no indication that the current owners have been subjected to any risks, nor have they indicated any insecurity in their titles. The plaintiffs offer no evidence that other land titles in Cape Cod are insecure. Accordingly, we reject the request of the plaintiffs for a reversal of the Land Court judgment. Such a request for changes in long-settled law is more properly addressed to the Legislature.[14]

Judgment affirmed.

NOTES:

[1] Also known as Pamela C. Osgood.

[2] Diane C. Gregory, Richard L. Huttinger, Claire E. Knowles, Harold E. Knowles, Jr., Dorothy L. Palin, and Ralph A. Rincones.

[3] Town of Brewster, planning board of Brewster (board), Mark D. Nickerson, Bernard A. Leckie, and Maryanne H. Leckie. The town and board did not participate in briefing this appeal. As entered in the Land Court, the plaintiffs' complaint originally named a number of other individual defendants. Only the Nickerson and Leckie defendants remain in the case on appeal. See note 6, *infra*.

[4] In challenging the special permit, the plaintiffs alleged that Nickerson had no access over Eli Rogers Road by claim of easement. The plaintiffs also claimed ownership of the fee in that road under G.L. c. 183, § 58. Noting that the board clearly avoided any consideration of the rights in Eli Rogers Road, the judge ruled that the plaintiffs failed to support their claims of aggrievement, that is, harm from increased traffic from construction of the proposed residence, as well as from use of the road by the defendants. He also found no authority permitting the plaintiffs, whose properties are in Orleans, to challenge the zoning by-law of Brewster, and therefore ruled the plaintiffs were without standing to pursue their action under G.L. c. 240, § 14A. The plaintiffs do not challenge these rulings in this appeal.

[5] A sketch of the locus made in the Land Court is reproduced in the Appendix of this case.

[6] The judge noted that prior to trial, settlement discussions between the plaintiffs and most of the individual defendants originally named in the plaintiffs' amended complaint resulted in agreements with all but defendants Carrie L. and Mark D. Nickerson, and Bernard A. and Maryanne H. Leckie, and the trial proceeded accordingly. The plaintiffs granted eight easements to the other original individual defendants for access over the plaintiffs' lands on Eli Rogers Road.

[7] We acknowledge the submission of a brief by amicus curiae William V. Hovey.

[8] The deed contains a statement by the register, dated July 23, 1711, that the deed was received to be recorded, and was "accordingly entered and compared." However, the deed could not be found in the Barnstable registry of deeds because a fire in 1827 destroyed virtually all of the documents which had been recorded up to that time. See Makepeace Bros. v. Barnstable, 292 Mass. 518, 520 (1935).
The record on appeal contains a copy of this handwritten deed as well as a copy of a typewritten transcription prepared by the defendants. The Land Court judge noted that the copy of the deed was obtained from the Suffolk Files Collection of the Supreme Judicial Court in the Massachusetts State Archives, and appears in volume 61, Superior Court of Judicature number 8331.

[9] The judge noted that the adoption of a new calendar, changing the start of a new year from March to January, often resulted in the notation of 1713/1714 on documents executed in the interim period. In this case, January 25 was only a few months after November, 1713.

[10] The proprietors' book, entitled "Records of the Proprietors of the [Seventeen] Share Purchase of the Lands of the Sipsons," long had been in the possession of the Harwich town clerk, who presented it at trial and testified. See G.L. c. 179, § 4, with an antecedent in St. 1712-1713, c. 9, § 1 (providing that the clerk of the proprietors "shall record all votes . . . in books which he shall keep for that purpose until they are delivered to the town clerk [in accordance with § 15, added by St. 1783, c. 39, § 9]"). See Thayer, Crocker's Notes on Common Forms c. 1, § 9 (9th ed.2006).

[11] While both parties presented expert testimony, the judge's decision follows the tracing of the boundaries described in the Sipson deed by the defendants' expert surveyor. This process was significantly aided by the boundary line agreement of 1705 between the towns of Harwich and Eastham (Eastham included Orleans until 1797). The judge also stated that he took judicial notice of Land Court records, cf. Miller v. Norton, 353 Mass. 395, 399 (1967), and cited A History of Harwich, by Josiah Paine (1937), which was admitted in evidence.
The judge appears to have had little difficulty in reaching a conclusion with the aid of these materials. Finally, the plaintiffs' counsel stipulated at trial that the land of all the parties is within the boundaries outlined as within the Sipson deed by the defendants' expert surveyor.

[12] There is no indication that this argument was set out in the plaintiffs' amended complaint, was articulated during the trial, or arose at trial as a live issue. It was presented in a posttrial memorandum submitted to the judge in lieu of closing argument, on April 18, 2006, nearly three months after the trial. There is no

indication that the judge based his decision on the argument as stated in the post-trial memorandum.

[13] Because of the loss of records of the Barnstable registry of deeds by fire, see note 8, supra, it could not have been determined in this case whether there were other transactions recorded following the Sipson deed which would have been relevant. In any event, the Land Court judge's determination that the defendants have an easement to use Eli Rogers Road hardly can be said to be "disruptive" in view of the considerable evidence of the plaintiffs' knowledge of the actual use of the road, and the documentary evidence available to them in the acquisition of their lands.

[14] The defendants' request for attorney's fees is denied.

CASE #18 EASEMENT BY CUSTOM

STATE OF OREGON, EX REL. THORNTON v. HAY
254 OR. 584, 462 P.2D 671 (OR. 1969)

William and Georgianna Hay, the owners of a tourist facility at Cannon Beach, appeal from a decree which enjoins them from constructing fences or other improvements in the dry-sand area between the sixteen-foot elevation contour line and the ordinary high-tide line of the Pacific Ocean.

The issue is whether the state has the power to prevent the defendant landowners from enclosing the dry-sand area contained within the legal description of their ocean-front property.

The state asserts two theories: (1) the landowners' record title to the disputed area is encumbered by a superior right in the public to go upon and enjoy the land for recreational purposes; and (2) if the disputed area is not encumbered by the asserted public easement, then the state has power to prevent construction under zoning regulations made pursuant to ORS 390.640.

The defendant landowners concede that the State Highway Commission has standing to represent the rights of the public in this litigation, ORS 390.620, and that all tideland lying seaward of the ordinary, or mean high-tide line is a state recreation area as defined in ORS 390.720. [1]

From the trial record, applicable statutes, and court decisions, certain terms and definitions have been extracted and will appear in this opinion. A short glossary follows:

ORS 390.720 refers to the 'ordinary' high-tide line, while other sources refer to the 'mean' high-tide line. For the purposes of this case the two lines will be considered to be the same. The mean high-tide line in Oregon is fixed by the 1947 Supplement to the 1929 United States Coast and Geodetic Survey data.

The land area in dispute will be called the dry-sand area. This will be assumed to be the land lying between the line of mean high tide and the visible line of vegetation.[2]

The vegetation line is the seaward edge of vegetation where the upland supports vegetation. It falls generally in the vicinity of the sixteen-foot-elevation contour line, but is not at all points necessarily identical with that line. Differences between the vegetation line and the sixteen-foot line are irrelevant for the purposes of this case.

The sixteen-foot line, which is an engineering line and not a line visible on the ground, is mentioned in ORS 390.640, and in the trial court's decree.

The extreme high-tide line and the highwater mark are mentioned in the record, but will be treated as identical with the vegetation line. While technical differences between extreme high tide and the highwater mark, and between both lines and the sixteen-foot line, might have legal significance in some other litigation, such differences, if any, have none in this case. We cite these variations in terminology only to point out that the cases and statutes relevant to the issues in this case, like the witnesses, have not a always used the same words to describe similar topographical features.

Below, or seaward of, the mean high-tide line, is the state-owned foreshore, or wet-sand area, in which the landowners in this case concede the public's paramount right, and concerning which there is no justiciable controversy.

The only issue in this case, as noted, is the power of the state to limit the record owner's use and enjoyment of the dry-sand area, by whatever boundaries the area may be described.

The trial court found that the public had acquired, over the years, an easement for recreational purposes to go upon and enjoy the dry-sand area, and that this easement was appurtenant to the wet-sand portion of the beach which is admittedly owned by the state and designated as a 'state recreation area.'

Because we hold that the trial court correctly found in favor of the state on the rights of the public in the dry-sand area, it follows that the state has an equitable right to protect the public in the enjoyment of those rights by causing the removal of fences and other obstacles.

It is not necessary, therefore, to consider whether ORS 390.640 would be constitutional if it were to be applied as a zoning regulation to lands upon which the public had not acquired an easement for recreational use.

In order to explain our reasons for affirming the trial court's decree, it is necessary to set out in some detail the historical facts which lead to our conclusion.

The dry-sand area in Oregon has been enjoyed by the general public as a recreational adjunct of the wet-sand or foreshore area since the beginning of the state's political history. The first European settlers on these shores found the aboriginal inhabitants using the foreshore for clam-digging and the dry-sand area for their cooking fires. The newcomers continued these customs after statehood. Thus, from the time of the earliest settlement to the present day, the general public has assumed that the dry-sand area was a part of the public beach, and the public has used the dry-sand area for picnics, gathering wood, building warming fires, and generally as a headquarters from which to supervise children or to range out over the foreshore as the tides advance and recede. In the Cannon Beach vicinity, state and local officers have policed the dry sand, and municipal sanitary crews have attempted to keep the area reasonably free from man-made litter.

Perhaps one explanation for the evolution of the custom of the public to use the dry-sand area for recreational purposes is that the area could not be used conveniently by its owners for any other purpose. The dry-sand area is unstable in its seaward boundaries, unsafe during winter storms, and for the most part unfit for the construction of permanent structures. While the vegetation line remains relatively fixed, the western edge of the dry-sand area is subject to dramatic moves eastward or westward in response to erosion and accretion. For example, evidence in the trial below indicated that between April 1966 and August 1967 the seaward edge of the dry-sand area involved in this litigation moved westward 180 feet. At other points along the shore, the evidence showed, the seaward edge of the dry-sand area could move an equal distance to the east in a similar period of time.

Until very recently, no question concerning the right of the public to enjoy the dry-sand area appears to have been brought before the courts of this state. The public's assumption that the dry sand as well as the foreshore was 'public property' had been reinforced by early judicial decisions. See *Shively v. Bowlby,* 152 U.S. 1, 14 S.Ct. 548, 38 L.Ed. 331 (1894), which affirmed *Bowlby v. Shively,* 22 Or. 410, 30 P. 154 (1892). These cases held that landowners claiming under federal patents owned seaward only to the 'high-water' line, a line that was then assumed to be the vegetation line. [3]

In 1935, the United States Supreme Court held that a federal patent conveyed title to land farther seaward, to the mean hightide line. *Borax Consolidated, Ltd. v. Los Angeles,* 296 U.S. 10, 56 S.Ct. 23, 80 L.Ed. 9 (1935). While this decision may have expanded seaward the record ownership of upland landowners, it was apparently little noticed by Oregonians. In any event, the Borax decision had no discernible effect on the actual practices of Oregon beachgoers and upland property owners.

Recently, however, the scarcity of oceanfront building sites has attracted substantial private investments in resort facilities. Resort owners like these defendants now desire to reserve for their paying guests the recreational advantages that accrue to the dry-sand portions of their deeded property. Consequently, in 1967, public debate and political activity resulted in legislative attempts to resolve conflicts between public and private interests in the dry-sand area:

ORS 390.610 '(1) The Legislative Assembly hereby declares it is the public policy of the State of Oregon to forever preserve and maintain the sovereignty of the state heretofore existing over the seashore and ocean beaches of the state from the Columbia River on the North to the Oregon-California line on the South so that the public may have the free and uninterrupted use thereof.

'(2) The Legislative Assembly recognizes that over the years the public has made frequent and uninterrupted use of lands abutting, adjacent and contiguous to the public highways and state recreation areas and recognizes, further, that where such use has been sufficient to create easements in the public through dedication, prescription, grant or otherwise, that it is in the public interest to protect and preserve such public easements as a permanent part of Oregon's recreational resources.

'(3) Accordingly, the Legislative Assembly hereby declares that all public rights and easements in those lands described in subsection (2) of this section are confirmed and declared vested exclusively in the State of Oregon and shall be held and administered in the same manner as those lands described in ORS 390.720.

The state concedes that such legislation cannot divest a person of his rights in land, *Hughes v. Washington,* 389 U.S. 290, 88 S.Ct. 438, 19 L.Ed.2d 530 (1967), and that the defendants' record title, which includes the dry-sand area, extends seaward to the ordinary or mean high-tide line. Borax Consolidated Ltd. v. Los Angeles, supra.

The landowners likewise concede that since 1899 the public's rights in the foreshore have been confirmed by law as well as by custom and usage. Oregon Laws 1899, p. 3, provided:

'That the shore of the Pacific ocean, between ordinary high and extreme low tides, and from the Columbia river on the north to the south boundary line of Clatsop county on the south, is hereby declared a public highway, and shall forever remain open as such to the public.'

The disputed area is Sui generis. While the foreshore is 'owned' by the state, and the upland is 'owned' by the patentee or record-title holder, neither can be said to 'own' the full bundle of rights normally connoted by the term 'estate in fee simple.' 1 Powell, Real Property § 163, at 661 (1949).

In addition to the Sui generis nature of the land itself, a multitude of complex and sometimes overlapping precedents in the law confronted the trial court. Several early Oregon decisions generally support the trial court's decision, i.e., that the public can acquire easements in private land by long-continued user that is inconsistent with the owner's exclusive possession and enjoyment of his land. A citation of the cases could end the discussion at this point. But because the early cases do not agree on the legal theories by which the results are reached, and because this is an important case affecting valuable rights in land, it is appropriate to review some of the law applicable to this case.

One group of precedents relied upon in part by the state and by the trial court can be called the 'implied-dedication' cases. The doctrine of implied dedication is well known to the law in this state and elsewhere. See cases collected in Parks, The Law of Dedication in Oregon, 20 Or.L.Rev. 111 (1941). Dedication, however, whether express or implied, rests upon an intent to dedicate. [4] In the case at bar, it is unlikely that the landowners thought they had anything to dedicate, until 1967, when the notoriety of legislative debates about the public's rights in the dry-sand area sent a number of ocean-front landowners to the offices of their legal advisers.

A second group of cases relied upon by the state, but rejected by the trial court, deals with the possibility of a landowner's losing the exclusive possession and enjoyment of his land through the development of prescriptive easements in the public.

In Oregon, as in most common-law jurisdictions, an easement can be created in favor of one person in the land of another by uninterrupted use and enjoyment of the land in a particular manner for the statutory period, so long as the user is open, adverse, under claim of right, but without authority of law or consent of the owner. *Feldman et ux. v. Knapp et ux.,* 196 Or. 453, 476, 250 P.2d 92 (1952); *Coventon v. Seufert,* 23 Or. 548, 550, 32 P. 508 (1893). In Oregon, the prescriptive period is ten years. ORS 12.050. The public use of the disputed land in the case at bar is admitted to be continuous for more than sixty years. There is no suggestion in the record that anyone's permission was sought or given; rather, the public used the land under a claim of right. Therefore, if the public can acquire an easement by prescription, the requirements for such an acquisition have been met in connection with the specific tract of land involved in this case.

The owners argue, however, that the general public, not being subject to actions in trespass the ejectment, cannot acquire rights by prescription, because the statute of limitations is irrelevant when an action does not lie. While it may not be feasible for a landowner to sue the general public, it is nonetheless possible by means of signs and fences to prevent or minimize public invasions of private land for recreational purposes. In Oregon, moreover, the courts and the Legislative Assembly have both recognized that the public can acquire prescriptive easements in private land, at least for roads and highways. See, e.g., *Huggett et ux. v. Moran et ux.,* 201 Or. 105, 266 P.2d 692 (1954), in which we observed that counties could acquire public roads by prescription. And see ORS 368.405, which provides for the manner in which counties may establish roads. The statute enumerates the formal governmental actions that can be employed, and then concludes: 'This section does not preclude acquiring public ways by adverse user.'

Another statute codifies a policy favoring the acquisition by prescription of public recreational easements in beach lands. See ORS 390.610. While such a statute cannot create public rights at the expense of a private landowner the statute can, and does, express legislative approval of the common-law doctrine of prescription where the facts justify its application. Consequently, we conclude that the law in Oregon, regardless of the generalizations that may apply elsewhere, [5]does not preclude the creation of prescriptive easements in beach land for public recreational use.

Because many elements of prescription are present in this case, the state has relied upon the doctrine in support of the decree below. We believe, however, that there is a better legal basis for affirming the decree. The most cogent basis for the decision in this case is the English doctrine of custom. Strictly construed, prescription applies only to the specific tract of land before the court, and doubtful prescription cases could fill the courts for years with tract-by-tract litigation. An established custom, on the other hand, can be proven with reference to a larger region. Ocean-front lands from the northern to the southern border of the state ought to be treated uniformly.

The other reason which commends the doctrine of custom over that of prescription as the principal basis for the decision in this case is the unique nature of the lands in question. This case deals solely with the dry-sand area along the Pacific shore, and this land has been used by the public as public recreational land according to an unbroken custom running back in time as long as the land has been inhabited.

A custom is defined in 1 Bouv. Law Dict., Rawle's Third Revision, p. 742 as 'such a usage as by common consent and uniform practice has become the law of the place, or of the subject matter to which it relates.'

In 1 Blackstone, Commentaries * 75 — * 78, Sir William Blackstone set out the requisites of a particular custom.

Paraphrasing Blackstone, the first requirement of a custom, to be recognized as law, is that it must be ancient. It must have been used so long 'that the memory of man runneth not to the contrary.' Professor Cooley footnotes his edition of Blackstone with the comment that 'long and general' usage is sufficient. In any event, the record in the case at bar satisfies the requirement of antiquity. So long as there has been an institutionalized system of land tenure in Oregon, the public has freely exercised the right to use the dry-sand area up and down the Oregon coast for the recreational purposes noted earlier in this opinion.

The second requirement is that the right be exercised without interruption. A customary right need not be exercised continuously, but it must be exercised without an interruption caused by anyone possessing a paramount right. In the case at bar, there was evidence that the public's use and enjoyment of the dry-sand area had never been interrupted by private landowners.

Blackstone's third requirement, that the customary use be peaceable and free from dispute, is satisfied by the evidence which related to the second requirement.

The fourth requirement, that of reasonableness, is satisfied by the evidence that the public has always made use of the land in a manner appropriate to the land and to the usages of the community. There is evidence in the record that when inappropriate uses have been detected, municipal police officers have intervened to preserve order.

The fifth requirement, certainty, is satisfied by the visible boundaries of the dry-sand area and by the character of the land, which limits the use thereof to recreational uses connected with the foreshore.

The sixth requirement is that a custom must be obligatory; that is, in the case at bar, not left to the option of each landowner whether or not he will recognize the public's right to go upon the dry-sand area for recreational purposes. The record shows that the dry-sand area in question has been used, as of right, uniformly with similarly situated lands elsewhere, and that the public's use has never been questioned by an upland owner so long as the public remained on the dry sand and refrained from trespassing upon the lands above the vegetation line.

Finally, a custom must not be repugnant, or inconsistent, with other customs or with other law. The custom under consideration violates no law, and is not repugnant.

Two arguments have been arrayed against the doctrine of custom as a basis for decision in Oregon. The first argument is that custom is unprecedented in this state, and has only scant adherence elsewhere in the United States. The second argument is that because of the relative brevity of our political history it is inappropriate to rely upon an English doctrine that requires greater antiquity than a newly-settled land can muster. Neither of these arguments is persuasive.

The custom of the people of Oregon to use the dry-sand area of the beaches for public recreational purposes meets every one of Blackstone's requisites. While it is not necessary to rely upon precedent from other states, we are not the first state to recognize custom as a source of law. See *Perley et ux'r v. Langley*, 7 N.H. 233 (1834).

On the score of the brevity of our political history, it is true that the Anglo-American legal system on this continent is relatively new. Its newness has made it possible for government to provide for many of our institutions by written law rather than by customary law.[6] This truism does not, however, militate against the validity of a custom when the custom does in fact exist. If antiquity were the sole test of validity of a custom, Oregonians could satisfy that requirement by recalling that the European settlers were not the first people to use the dry-sand area as public land.

Finally, in support of custom, the record shows that the custom of the inhabitants of Oregon and of visitors in the state to use the dry sand as a public recreation area is so notorious that notice of the custom on the part of persons buying land along the shore must be presumed. In the case at bar, the landowners conceded their actual knowledge of the public's long-standing use of the dry-sand area, and argued that the elements of consent present in the relationship between the landowners and the public precluded the application of the law of prescription. As noted, we are not resting this decision on prescription, and we leave open the effect upon prescription of the type of consent that may have been present in this case. Such elements of consent are, however, wholly consistent with the recognition of public rights derived from custom.

Because so much of our law is the product of legislation, we sometimes lose sight of the importance of custom as a source of law in our society. It seems particularly appropriate in the case at bar to look to an ancient and accepted custom in this state as the source of a rule of law. The rule in this case, based upon custom, is salutary in confirming a public right, and at the same time it takes from no man anything which he has had a legitimate reason to regard as exclusively his.

For the foregoing reasons, the decree of the trial court is affirmed.

DENECKE, Justice (specially concurring).

I agree with the decision of the majority; however, I disagree with basing the decision upon the English doctrine of 'customary rights.' In my opinion the facts in this case cannot be fitted into the outlines of that ancient doctrine. 6 Powell, Real Property 362, § 934, n. 5 (1968); 2 Thompson, Real Property, 463, § 369, n. 50 (1961); Gray, The Rule Against Perpetuities, ch. 17 (4th ed. 1942); 15 Harv.L.Rev. 329, 332 (1903).

In my opinion the doctrine of 'customary rights' is useful but only as an analogy. I am further of the opinion that 'custom,' as distinguished from 'customary rights,' is an important ingredient in establishing the rights of the public to the use of the dry sands.

I base the public's right upon the following factors: (1) long usage by the public of the dry sands area, not necessarily on all the Oregon beaches, but wherever the public uses the beach; (2) a universal and long held belief by the public in the public's right to such use; (3) long and universal acquiescence by the upland owners in such public use; and (4) the extreme desirability to the public of the right to the use [462 P.2d 679] of the dry sands. When this combination exists, as it does here, I conclude that the public has the right to use the dry sands.

Admittedly, this is a new concept as applied to use of the dry sands of a beach; however, it is not new as applied to other public usages. In *Luscher v. Reynolds,* 153 Or. 625, 56 P.2d 1158 (1963), we held that regardless of who owns the bed of a lake, if it is capable of being boated, the public has the right to boat it.

'* * * There are hundreds of similar beautiful, small inland lakes in this state well adapted for recreational purposes, but which will never be used as highways of commerce in the ordinary acceptation of such terms. As stated in *Lamprey v. State,* 52 Minn. 181, 53 N.W. 1139, 18 L.R.A. 670, 38 Am.St.Rep. 541, quoted with approval in *Guilliams v. Beaver Lake Club,* supra (90 Or. 13, 175 P. 437), 'To hand over all these lakes to private ownership, under any old or narrow test of navigability, would be a great wrong upon the public for all time, the extent of which cannot, perhaps, be now even anticipated.' Regardless of the ownership of the bed, the public has the paramount right to the use of the waters of the lake for the purpose of transportation and commerce.' 153 Or. at 635--636, 56 P.2d at 1162.

In *Collins v. Gerhardt,* 237 Mich. 38, 211 N.W. 115, 116 (1926), the defendant was wading Pine River and fishing. The plaintiff, who owned the land on both sides and the bed of Pine River, sued defendant for trespass. The court held for the defendant:

'From this it follows that the common-law doctrine, viz., that the right of fishing in navigable waters follows the ownership of the soil, does not prevail in this state. It is immaterial who owns the soil in our navigable rivers. The trust remains. From the beginning the title was impressed with this trust for the preservation of the public right of fishing and other public rights which all citizens enjoyed in tidal waters under the common law. * * *.' 237 Mich. at 48, 211 N.W. at 118.

These rights of the public in tidelands and in the beds of navigable streams have been called 'jus publicum' and we have consistently and recently reaffirmed their existence. *Corvallis Sand & Gravel Co. v. State Land Board,* 250 Or. 319, 335--337, 439 P.2d 575 (1968); *Smith Tug & Barge Co. v. Columbia-Pac. Towing,* 250 Or. 612, 638, 443 P.2d 205 (1968). The right of public use continues although title to the property passes into private ownership and nothing in the chain of title reserves or notifies anyone of this public right. *Winston Bros. Co. v. State Tax. Comm.,* 156 Or. 505, 510--511, 62 P.2d 7 (1937).

In a recent treatise on waters and water rights the authors state:

'The principle that the public has an interest in tidelands and banks of navigable waters and a right to use them for purposes for which there is a substantial public demand may be derived from the fact that the public won a right to passage over the shore for access to the sea for fishing when this was the area of substantial public demand. As time goes by, opportunities for much more extensive uses of these lands become available to the public. The assertion by the public of a right to enjoy additional uses is met by the assertion that the public right is defined and limited by precedent based upon past uses and past demand. But such a limitation confuses the application of the principle under given circumstances with the principle itself.

'The law regarding the public use of property held in part for the benefit of the public must change as the public need changes. The words of Justice Cardozo, expressed in a different context nearly a half-century ago are relevant today in our application of this law: 'We may not suffer it to petrify at the cost of its animating principle." 1 Clark (ed-in-chief), Waters and Water Rights, at 202 (1967).

NOTES:

[1] ORS 390.720 provides:

'Ownership of the shore of the Pacific Ocean between ordinary high tide and extreme low tide, and from the Oregon and Washington state line on the north to the Oregon and California state line on the south, excepting such portions as may have been disposed of by the state prior to July 5, 1947, is vested in the State of Oregon, and is declared to be a state recreation area. No portion of such ocean shore shall be alienated by any of the agencies of the state except as provided by law.'

[2] A similar, but more colorful, description of the vegetation-line, mean-high-tide-line problem on the island of Molokai may be found in Application of Ashford, Haw., 440 P.2d 76 (1968).

[3] One of the issues in the Bowlby litigation was the right to wharf out from the upland. The state had sold to various private owners certain tidelands seaward of Shively's uplands. Shively, in an early plat, had overlapped his platted lots upon state-owned tidelands. The controversy concerned the foreshore generally. The litigants were not then particularly interested in 'dry' versus 'wet' sand, because a wharf must cross both 'dry' and 'wet' sand if such topography lies between the upland and deep water. The state's title to the tidelands was confirmed to the 'high-water mark.' Most private landowners in Oregon apparently assumed after Bowlby that their patented lands run seaward only to the 'high-water mark,' which is also, for all practical purposes, the vegetation line.

[4] Because of the elements of public interest and estoppel running through the cases, intent to dedicate is sometimes 'presumed' instead of proven. But conceptually, at least, dedication is founded upon an intent to dedicate. *Security and Investment Co. of Oregon City v. Oregon City*, 161 Or. 421, 433, 90 P.2d 467 (1939); *City of Clatskanie v. McDonald*, 85 Or. 670, 674, 167 P. 560 (1917); *Portland Ry., L. & P. Co. v. Oregon City*, 85 Or. 574, 582, 166 P. 932 (1917); *Harris v. St. Helens*, 72 Or. 377, 386, 143 P. 941 (1914); *Parrott v. Stewart*, 65 Or. 254, 259, 132 P. 523 (1913); *Lownsdale v. City of Portland*, 1 Or. 397, 405 (1861). See, also, Nicholas v. Title & Trust Co., 79 Or. 226, 154 P. 391 (1916); Kuck v. Wakefield, 58 Or. 549, 555, 115 P. 428 (1911). Additional cases are collected in Parks, The Law of Dedication in Oregon, 20 Or.L.Rev. 111 (1941).

[5] See, e.g., *Sanchez v. Taylor*, 377 F.2d 733, 738 (10th Cir. 1967), holding that the general public cannot acquire grazing rights in unfenced land. Among other reasons assigned by authorities cited in *Sanchez v. Taylor* are these: prescription would violate the rule against perpetuities because no grantee could ever convey the land free of the easement; and prescription rests on the fiction of a 'lost grant,' which state of affairs cannot apply to the general public. The first argument can as well be made against the public's acquiring rights by express dedication; and the second argument applies equally to the fictional aspects of the doctrine of implied dedication. Both arguments are properly ignored in cases dealing with roads and highways, because the utility of roads and the public interest in keeping them open outweighs the policy favoring formal over informal transfers of interests in land.

[6] The English law on customary rights grew up in a small island nation at a time when most inhabitants lived and died without traveling more than a day's walk from their birthplace. Most of the customary rights recorded in English cases are local in scope. The English had many cultural and language groups which

eventually merged into a nation. After these groups developed their own unique customs, the unified nation recognized some of them as law. Some American scholars, looking at the vast geography of this continent and the freshness of its civilization, have concluded that there is no need to look to English customary rights as a source of legal rights in this country. See, e.g., 6 Powell, Real Property § 934, note 5, at 362 (1949). Some of the generalizations drawn by the text writers from English cases would tend to limit customary rights to specific usages in English towns and villages. See Gray, The Rule Against Perpetuities §§ 572--588 (1942). But it does not follow that a custom, established in fact, cannot have regional application and be enjoyed by a larger public than the inhabitants of a single village.

REFERENCES

Alfano, Paul J. *Creation and Termination of Highways in New Hampshire. New Hampshire Bar Journal.* Autumn, 2005, pages 56–67.

Beckman, James H. and David A. Thomas. *A Practical Guide to Disputes Between Adjoining Landowners – Easements.* New York: Matthew Bender & Company, Inc., 1989.

Blackstone, Sir William. *Commentaries on the Laws of England (1765–1769).* Based on the first edition printed at the Clarendon Press (Oxford, England).

Canadian Council of Land Surveyors. *Survey Law in Canada.* Toronto: The Carswell Co., Ltd., 1989.

Cartwright, John M. *Glossary of Real Estate Law.* Rochester: The Lawyers Co-operative Publishing Co. 1972.

Creteau, Paul G. *Maine Real Estate Law.* Portland: Castle Publishing Company, 1969.

Hand, Jacqueline P. and James Charles Smith. *Neighboring Property Owners.* Colorado Springs: Shepard's McGraw-Hill, Inc. 1988.

Kube, Robyn W. "Access at Last: The Use of Private Condemnation." *Colorado Lawyer,* Vol. 29, No. 2, p. 77, February 2000.

Lindstrom, C. Timothy. "Conservation Easements in Wyoming." *Wyoming Bar Journal,* vol. 28, No. 5, October, 2005.

Patton, Rufford G. and Carroll G. Patton. *Patton on Land Titles.* Volume 1. St. Paul: West Publishing Co., 1957.

Pill, Michael, Esq. *Land Title Issues including easements and derelict fees.* Seminar Notes, Massachusetts Association of Land Surveyors and Civil Engineers, Marlborough, Mass., November 17, 2006.

Robillard, Walter G., Donald A. Wilson, and Curtis M. Brown. *Brown's Boundary Control and Legal Principles.* 6th Edition. Hoboken: John Wiley & Sons, Inc., 2009.

Wait, John Cassam. *The Law of Operations Preliminary to Construction in Engineering and Architecture: Rights in Real Property, Boundaries, Easements, and Franchises for Engineers, Architects, Contractors, Builders, Public Officers, and Attorneys at Law*. New York: John Wiley & Sons., 1900.

Wattles, Gurdon H. *Writing Legal Descriptions*. Published by the author, 1976.

Wattles, William C. *Land Survey Descriptions*. Tenth Edition. Revised and Published by Gurdon H. Wattles, 1974.

Wilson, Donald A. *Easements and Reversions*. Rancho Cordova: Landmark Enterprises, 1991.

Wilson, Donald A. *Interpreting Land Records*. Hoboken: John Wiley & Sons, Inc. 2006.

Wilson, Donald A. *Locating Historical Rights of Way*. Seminar Notes, Surveyors Educational Seminars, 1992.

Wilson, Donald A. *Locating Undescribed Easements.* Seminar Notes, Surveyors Educational Seminars, 2007.

Wilson, Donald A. *Dedication of Public & Private Rights: What it means to a surveyor*. Seminar Notes, Surveyors Educational Seminars, 2001.

Wilson, Donald A. *Finding and Locating Colonial Grants.* Seminar Notes, Surveyors Educational Seminars, 2001.

FOR FURTHER REFERENCE

In American law, *American Law Reports* are a resource for finding a variety of sources concerning specific legal rules, doctrines, and principles. It is in its 6th edition, having been published since 1919, and is an important tool for legal research and useful in finding solutions to surveying-type problems.

The volumes provide an in depth coverage of legal issues in various areas of the law, and present agreements and conflicts between jurisdictions.

ALR annotations are a good place to begin when one knows little about an area of law, or a particular issue. Each annotation addresses a single issue, presenting the most important cases and state or federal statures relating to that particular issue. It also directs to other sources where the issue is discussed, such as *American Jurisprudence*, *Corpus Juris Secundum*, and numerous other reference sources. Most of them provide a list of related ALR sources.

1 ALR 884	Loss of Easement by Adverse Possession or Nonuser
1 ALR 1368	Acquisition by User or Prescription of Right of Way over Uninclosed Land
5 ALR 1325	What Will Disprove Acquiescence by Owner Essential to Easement by Prescription in Case of Known Use
30 ALR 206	Effect of Expiration of Charter of Turnpike or Toll-Road Company on Title to Road
34 ALR 233	Roadway or Pathway Used at Time of Severance of Tract as Visible Easement

38 ALR 1310	Easements: Way of Necessity Where Property Is Accessible by Navigable Water
58 ALR 824	Implied Easement in Respect of Drains, Pipes, or Sewers upon Severance of Tract
66 ALR 1099	Loss of Easement by Adverse Possession or Nonuser
68 ALR 528	Locating Easement of Way by Necessity
85 ALR 404	Boundary under Conveyance of Land Bordering on Railroad Right of Way
98 ALR 1291	Loss of Easement by Adverse Possession or Nonuser
100 ALR 1321	Roadway or Pathway Used at Time of Severance of Tract as Visible or Apparent Easement
102 ALR 472	Easement by Implication or Necessity over One Part of a Lot in Favor of Another Part in Order to Reach a Right of Way Appurtenant to Entire Lot
110 ALR 174	Locating Easement of Way Created by a Grant Which Does Not Definitely Describe Its Location
110 ALR 915	Enlargement of Easement by Use for Purpose or in a Manner Other Than That Specified in the Grant
112 ALR 1303	Right of Owner of Easement of Way to Make Improvements or Repairs Thereon
130 ALR 768	Extent and Reasonableness of Use of Private Way in Exercise of Easement Granted in General Terms
132 ALR 142	Deed to Railroad Company as Conveying Fee or Easement
136 ALR 296	Who Entitled to Land Upon Its Abandonment for Railroad Purposes, Where Railroad's Original Interest or Title Was Less Than Fee-Simple Absolute
136 ALR 379	Deed As Conveying Fee or Easement
142 ALR 467	Express Easements of Light, Air, and View
143 ALR 1402	Acquisition of Right of Way by Prescription as Affected by Change of Location or Deviation during Prescriptive Period
155 ALR 381	Title or Interest Acquired by Railroad in Exercise of Eminent Domain as Fee or Easement
156 ALR 1050	Type of Vehicle or Mode of Travel Permissible on Express Easement of Way Created in Limited Terms
164 ALR 1001	Roadway or Pathway Used at Time of Severance of Tract as Visible or Apparent Easement
165 ALR 567	Parol Evidence Rule as Applied to Question of Easement by Necessity or Visible Easement

170 ALR 776	Easement by Prescription: Presumption and Burden of Proof as to Adverse Character of Use
171 ALR 1278	Tacking As Applied to Prescriptive Easements
172 ALR 1020	Use of Streets and Highways by Co-operative Utility
6 ALR2d 205	Correlative Rights of Dominant and Servient Owners in Right of Way for Electric Line
25 ALR2d 1265	Loss of Private Easement by Nonuser or Adverse Possession
28 ALR2d 253	Width of Way Created by Express Grant, Reservation, or Exception Not Specifying Width
28 ALR2d 626	Correlative Rights of Dominant and Servient owners in Right of Way for Pipeline
46 ALR2d 1140	Acquisition by User or Prescription of Right of Way Over Uninclosed Land
46 ALR2d 1197	Foreclosure of Mortgage or Trust Deed as Affecting Easement Claimed in, over, or under Property
55 ALR2d 554	Acquisition by Adverse Possession or Use of Public Property Held by Municipal Corporation or Other Governmental Unit Otherwise Than for Streets, Alleys, Parks, or Common
55 ALR2d 1144	Easement by Prescription in Artificial Drains, Pipes, or Sewers
58 ALR2d 525	Electric Light or Power Line in Street or Highway as Additional Servitude
69 ALR2d 1236	Right of Owner of Servient Tenement Subject to Right of Way to Dedicate His Land
80 ALR2d 109	Acquisition of Right of Way by Prescription as Affected by Change of Location or Deviation during Prescriptive Period
80 ALR2d 743	Relocation of Easements (Other Than Those originally arising by necessity); Rights as between Private Parties
80 ALR2d 1095	Acquisition of Right of Way by Prescription as Affected by Change of Location or Deviation during Prescriptive Period
3 ALR3d 1256	Extent and Reasonableness of Use of Private Way in Exercise of Easement Granted in General Terms
5 ALR3d 439	Extent of, and Permissible Variations in, Use of Prescriptive Easements of Way
6 ALR3d 973	Deed to Railroad Company as Conveying Fee or Easement

9 ALR3d 600	Easements: Way of Necessity Where Property Is Accessible by Navigable Water
10 ALR3d 960	Right of Owners of Parcels into Which Dominant Tenement Is or Will Be Divided to Use Right of Way
72 ALR3d 648	Tacking as Applied to Prescriptive Easements
89 ALR3d 767	Conveyance of "Right of Way," in Connection With Conveyance of Another Tract, as Passing Fee or Easement
94 ALR3d 502	What Constitutes Unity of Title or Ownership Sufficient for Creation of an Easement by Implication or Way of Necessity
10 ALR4th 500	Way of Necessity Where Only Part of the Land Is Inaccessible
24 ALR4th 1053	Location of Easement of Way Created by Grant Which Does Not Specify Location
36 ALR4th 769	Locating Easement of Way Created By Necessity

GLOSSARY

Abutting Owner "Abutting owner," when used in relation to highways, ordinarily refers to one whose land actually adjoins the way, although it is sometimes used loosely without implying more than close proximity. *State v. Fuller*, 407 S.W.2d 215 (Texas)

Access A landowner's legal right to pass from his land to a highway and to return without being obstructed. *Webster's Third New International Dictionary* (1961), quoted in *Belluscio v. Town of Westmoreland*, 139 N.H.55 (1994). *See Easement of Access.*

Alley An alley is not necessarily a "street," and the public have not necessarily a right to its use. *Milliken v. Denny*, 47 S.E. 132, 135 N.C. 19.

The term "alley" generally relates exclusively to a way in a town or city, and an alley may be either public or private. Alleys are generally regarded as public where they have been accepted by the municipality pursuant to a valid dedication, or where authority over them has been acquired by it in some other lawful way. Private alleys are those which have not been dedicated to the public use and to which the general public is denied access, or which are set apart for some particular reason.

An "alley" is a "way of convenience," and there is no requirement that an alley have a certain type of ending or beginning. *Haab v. Moorman,* 50 N.W.2d 856, 332 Mich. 126.

An "alley" is a narrow way for the convenience of the owner of property abutting thereon and the persons dealing with owner. *Burkhard v. Bowen*, 203 P.2d 361, 365, 32 Wash.2d 613.

An "alley" is a narrow passageway, especially a walk or passage in a garden or park, bordered by rows of trees or bushes, a bordered way, a narrow passage or way in a city as distinct from a public street. *Morrow v. Whittler*, 25 Ohio N.P., N.S., 85.

An "alley" is a narrow way designed for the special accommodation of the property it reaches. *Atchison, T. & S. F. Ry. Co. v. City of Chanute,* 147 P. 836, 837, 95 Kan. 161.

An "alley" "is conventionally understood, in its relation to towns and cities, to mean a narrow street in common use." *Bailey v. Culver*, 12 Mo.App. 175, 183.

An "alley" is a narrow way, less in size than a street, and if a public one it is a highway, and in general is governed by the rules applicable to streets. *Kalteyer v. Sullivan*, 46 S.W. 288, 290, 18 Tex.Civ.App. 488.

Boulevard A boulevard is not technically a street, avenue, or highway, though a carriageway over it is a chief feature. *People v. Green*, N.Y., 52 How. Prac. 440.

"Boulevard" is a broad avenue in or around a city. *Murrell v. U.S., C.A., Fla.*, 269 F.2d 458, 461.

A "boulevard," according to the modern usage of the term, is a way of more than ordinary width and one which is construed and maintained in a more elaborate style and manner than the ordinary road or street. It is designed especially for travel for pleasure and recreation, although it is not limited to such use in the absence of express regulation to that effect. *39 Am Jur 2d, Highways, Streets and Bridges*, § 4.

In many respects, boulevards are treated in law differently from streets. *General Outdoor Adv. Co. v. Indianapolis, 202 Ind. 85*, 172 N.E. 309, 72 ALR 453.

A "boulevard" is a street constructed with park-like features, a wide street, or a street encircling a town, with sides or center for shade trees, flowers, seats, etc., and not used for heavy traffic. *State ex rel. Copland v. City of Toledo*, 62 N.E.2d 256, 258, 75 Ohio App. 378.

A "boulevard," as distinguished from ordinary street or highway, is broad avenue in or around city, especially one decoratively laid out with trees, belts of turf, etc., or improved thoroughfare for pleasure vehicles, sometimes exclusive and often a through way. *State ex rel. Copland v. City of Toledo*, 62 N.E.2d 256, 258, 75 Ohio App. 378.

Whether a certain highway is a street or "boulevard" does not depend solely upon name given to it on the plat, but is dependent upon its physical aspects, its width, its length, provision for giving it a parklike appearance by reserving spaces at the sides or center for shade trees, etc. *Campbell v. City of Detroit*, 243 N.W. 11, 12, 259 Mich. 297.

A "boulevard" is a "public way" although it is not a "state highway". *City of Medford v. Metropolitan Dist. Commission*, 22 N.E.2d 110, 111, 303 Mass. 537.

A "street" may be either a "boulevard" or a "way" and either or both may be streets according as the context requires interpretations. *City of Fargo v. Gearey*, 156 N.W. 552, 555, 33 N.D. 64.

The term "boulevard" is now sometimes extended to a street which is of especial width, is given a parklike appearance by reserving spaces at the sides or center for shade trees, flowers, seats, and the like. *Kendrick v. City of St. Paul*, 6 N.W.2d 449, 452, 213 Minn. 283.

Bridge A structure erected over a river, creek, stream, ditch, ravine, or other place to facilitate the passage thereof. *Black's Law Dictionary.*

Bridges are either public or private. Public bridges are such as form a part of the highway, common, according to their character as foot, horse, or carriage bridges, to the public generally, with or without toll. A private bridge is one which is not open to the use of the public generally, and does not form part of the highway, but is reserved for the use of those who erected it, or their successors, and their licensees. *Black's Law Dictionary.* Such a bridge will not be considered a public bridge although it may be occasionally used by the public. *Thompson v. R. Co., 3 Sandf. Ch. (N.Y.) 625*; 1 Rolle, Abr. 368, Bridges, pl. 2; 2 Inst. 701; 1 Salk. 359.

Bridle Road In the location of a private way laid out by the selectmen, and accepted by the town, a description of it as a "bridle road" does not confine the right of way to a particular class of animals or special mode of use. *Flagg v. Flagg*, 16 Gray (Mass.) 175.

Byroad The statute law of New Jersey recognizes three different kinds of roads: a public road, a private road, and a byroad. A byroad is a road used by the inhabitants, and recognized by statute, but not laid out. Such roads are often called "driftways." They are road of necessity in newly settled countries. *Van Blarcom v. Frike*, 29 N.J. Law, 516. See also, *Stevens v. Allen, 29 N.J. Law, 68.*

An obscure or neighborhood road in its earlier existence, not used to any great extent by the public, yet so far a public road that the public have right of free access to it at all times. *Wood v. Hurd*, 34 N.J. Law, 89.

Cartway Cartways are public roads in the sense that they are open to all who see fit to use them, although the principal benefit inures to the individual or individuals at whose request they were laid out. *Parsons v. Wright*, 223 N.C. 520, 27 S.E.2d 534.

Corduroy Road A road built of logs laid side by side across the roadway over sections of swampy ground to facilitate travel.

Country Road Under the Colonial Laws, a "country road" was one which belonged to the country and was under the direct charge of the country, as distinguished from the owners of the towns and manors, and it was a necessary line of communication between sparsely settled communities, and it was the better laying out, sparing, and preserving the public and general highways within the colony that legislation as to such roads was adopted. *Townsend v. Trustees of Freeholders & Commonality of Town of Brookhaven*, 89 N.Y.S. 982, 97 App. Div. 316.

County Highways Roads deeded to county before improvements were made on them were "county highways" within statutory definition. *Wine v. Boyar*, 33 Cal. Rptr. 787, 220 C.A.2d 375.

County ways County ways are roads leading from one town to another. *Inhabitants of Waterford v. Oxford County Com'rs*, 59 Me. 450.

Dedication A "dedication" is the setting apart of land for the public use. *Gore v. Blanchard*, 96 Vt. 234 (1922).

Dedication, Implied An implied dedication is one arising, by operation of law, from the acts of the owner. *Gore v. Blanchard*, 96 Vt. 234 (1922).

Driftway A road or way over which cattle are driven. *Smith v. Ladd*, 41 Me. 314.

A byroad is one which inhabitants who live some distance from a public road in a newly settled country make in going to the nearest public road—that is, it is a road which is used by the inhabitants but is not laid out—and very often it is called a "driftway," and is a road of necessity in newly settled countries. *Van Blarcom v. Rrike*, 29 N.J.L. (5 Dutch.) 516.

Driveway A path leading from a garage or house to the street, used especially by automobiles, and primarily intended for ingress and egress by vehicles. *Dolske v. Gormley*, 58 Cal 2d 513.

A "driveway" is a passage way, a travel way, a way of ingress and egress. *Frumin v. May*, 251 S.W.2d 314, 319, 36 Tenn.App. 32 (1952).

A "driveway" is a path leading from a garage or house to a street used especially by automobiles. *Dolske v. Gormley*, 25 Cal.Rptr. 270, 274, 376 P.2d 174 (1962).

Easement of Access "Easement of access" means the right of ingress and egress to and from the premises of a lot owner to a street appurtenant to the land of the lot owner. *Lang v. Smith*, 173 A. 682, 683, 113 Pa. Super. 559.

An abutting owner's "easement of access" extends to a use of public highway upon which his land fronts for purposes of ingress and egress to his property by such modes of conveyance and travel as are appropriate to the highway and in such manner as is customary or reasonable. *Rose v. State*, 123 P.2d 505, 514, 515, 19 C.2d 713.

Abutting property owners possessed as a matter of law not only right to use of streets in common with other members of the public, buy also al private right, known as "easement of access," which right may not be taken away or destroyed or substantially impaired or interfered with for public purposes without just compensation therefore. *Rose v. State*, 123 P.2d 505, 514, 515, 19 C.2d 713.

Owners of land abutting upon a highway have the right to use and enjoy the highway in common with other members of the public, and in addition they have an easement of access to their land abutting upon the highway, arising from the ownership of such land contiguous to the highway, which "easement of access" does not belong to the public generally, and which exists regardless of whether the fee of the highway is in said owners or not. This easement "includes the right of ingress, egress, and regress, a right of way from a locus a quo to the locus ad quem, and from the latter forth to any other spot to which the party may lawfully go, or back to the locus a quo." *State Highway Board v. Baxter*, 144 S.E.796, 800, 167 Ga. 124.

Farm Crossing A roadway over a railroad track at grade for the purpose of reaching tillage land cut off by the track. *True v. Maine Centr. R. Co.*, 113 Me. 375, 94 A. 183. See also, "In re Colvin Street in City of Buffalo," 155 App. Div. 808, 140 N.Y.S. 882.

Forschel A strip of land lying next to a highway. *Black's Law Dictionary*.

Freeway A "freeway" is a limited access highway where ingress and egress may be had only at certain designated points, which are determined by highway authorities. *Department of Public Works and Buildings v. Farina*, 194 N.E.2d 209, 29 Ill.2d 474.

Front Although it is true that a corner lot does front on both streets, only that portion of the lot that is opposite the rear and faces upon the street is properly designated as the "front" of such lot. *Staley v. Mears*, 142 N.E.2d 835, 13 Ill. App.2d 451.

Frontage "Frontage," within statute defining "business district" and "residence district," means space available for erection of buildings, and does not include cross streets or space occupied by sidewalk or any ornamental spaces in plat between sidewalks and curb. *Comp. Laws 1929*, § 4693 (v,w. *Wallace v. Kramer*, 296 N.W. 838, 296 Mich. 680.

Frontager Frontager is a person owning or occupying land that abuts on a highway, river, seashore, or the like. *See 27 Corpus Juris*.

Fronting Words "fronting" and "adjoining" are synonymous, and reference to lots "fronting" upon a street means that the lots touch boundary line of street. *Roach v. Soles*, D.C. Cal., 120 F.Supp. 400.

Fronting and Abutting Very often, "fronting" signifies abutting, adjoining, or bordering on, depending largely on the context. *Rombauer v. Compton Heights Christian Church*, 40 S.W.2d 545, 328 Mo. 1.

Great Roads The term "great roads" is commonly used to designate main or principal roads or highways. *Ex parte Withers*, S.C., 3 Brev. 83; State v. Mobley, S.., 1 McMul. 28.

Ground of Railroad The phrase "the ground of the railroad," used in describing a parcel of land in a mortgage as being "north of the ground of the railroad," evidently means its right of way. *Pence v. Armstrong*, 95 Ind. 191.

Highway "Highway" is defined as a main road or thoroughfare; hence a road or way open to use of public, including in broadest sense the term ways upon water as well as upon land; "highway" is generic term for all kinds of public ways, whether it be carriageways, bridleways, footways, bridges, turnpike roads, railroads, canals, ferries, or navigable rivers, and includes county and township roads, railroads and tramways, bridges and ferries, canals, and navigable rivers. In short, every thoroughfare is a "highway." *City of Long Beach v. Payne*, 44 P.2d 305 (1935).

A highway is a way open to the public at large, for travel or transportation, without distinction, discrimination, or restriction, except such as is incident to regulations calculated to secure to the general public the largest practical benefit therefrom and enjoyment thereof. *39 Am Jur 2d, Highways, Streets, and Bridges*, § 1.

It is a generic term and includes all public roads and ways. In its broad or general sense, it covers every common way for travel in any ordinary mode or by any ordinary means, which the public has the right to use either conditionally or unconditionally, and thus may include turnpikes and toll roads, bridges, canals, ferries, navigable waters, lanes, pent roads, and crossroads. In a limited sense, however, the term means a way for general travel which is wholly public. *Ibid.*

Intersection The word "intersection" as applied to a highway, means the space common to two highways. *City of Birmingham v. Young*, 22 So.2d 169, 246 Ala. 650; *Rodgers v. Commercial Casualty Ins. Co.*, 186 So. 684, 237 Ala. 301.

Laid Out "Laid out," as used in reference to a highway, has a well-known meaning, under our statutes, and plainly includes the doing of those things by the proper local officers which are essential in creating a public highway, to authorize it to be worked and traveled, and especially the surveying, marking the course or boundaries, and ordering it to be established as a highway. The affirmative action of the public authorities is indispensable in such case. *Chicago Anderson Pressed Brick Co. v. City of Chicago, 28 N.E. 756, 137 Ill. 628.*

The phrase "laid out," as used in the statute defining highways, does not imply the building or construction of a structure, and hence a way need not have a road on it in order to be a county highway. *Watson v. Greely*, 232 P. 475, 69 Cal.App. 643.

Land Service Road "Land-service road" is term used to describe ordinary road or highway which is intended primarily to enable abutting landowners to have access to outside world as distinguished from limited-access road, which is a "traffic-service road" designed primarily to move through traffic. *State ex rel. State Highway Commission v. Vorhof-Duenke Co.*, Mo. 366 S.W.2d 329.

Lay Out Term "lay out" of a highway embraces all measures for creation or establishment of the way, and from earliest colonial time, has had reference to new highways within a township and between towns. *Town of Woodbridge v. Merwin*, 244 A.2d 57, 27 Conn.Sup. 469.

The term "lay out" well expresses the work to be done by the viewers in establishing upon the ground the lines and angles of the road. *Feagins v. Wallowa County*, 123 P. 902, 62 Or. 186.

"To lay out a highway" means to locate the highway and to define its limits, which is accomplished by a survey and particular description of the highway. *Hough v. City of Bridgeport*, 18 A. 102, 57 Conn. 290.

Term "to lay out," when used with reference to highways, means taking necessary legal steps for establishment of and looking toward construction of highway, but does not include actual construction. *French v. Lewis and Clark County*, 288 P. 455, 87 Mont. 448.

Layout a Highway To "lay out a highway" is to locate it and define its limits. *U.S. v. Certain Parcels of Land in Riverside County*, Cal., 67 F.Supp. 780.

Lay out Roads "Laying out roads" across improved property is the designation of boundary lines of such roads which, in such natural state, are nothing more than dirt covered by natural growth, including grass, brush, trees, and holes, within such boundary lines. *Pasco County v. Johnson, Fla.*, 67 So.2d 639.

Line of Railroad "On the line of a railroad," as used in a deed describing the land as lying "on the line of a certain railroad," means next to or bounded by the railroad. The line of a road must mean, as used in the contract, the boundary of the land appropriated for its use. *Burnam v. Banks*, 45 Mo. 349.

Locate To ascertain or fix the position of something, the place of which was before uncertain or not manifest; as to locate the calls in a deed.

To decide upon the place or direction to be occupied by something not yet in being; as to locate a road. *Black's Law Dictionary.*

The words "locate" and "establish" are not synonymous. *Givens v. Woodward, Tex. Civ. App.*, 207 S.W.2d 234.

"Locate" means to designate the site or place of; to define the location or limits of, as by a survey; as, to locate a public building, a mining claim; to locate the land granted by a land warrant. *Delaware, L. & W. R. Co. v. Chiara*, C.C.A.N.J., 95 F.2d 663.

Locate a Road Strictly speaking, to "locate a road" is to build it, because until it is completed it is not a road. When a subscription book speaks of locating a road, everybody understands it to man less than a finished road—a mere line surveyed and established as the locus where the road is to be built. It must be supposed and held that it was the location of the road, and not its structure, that was intended to be specified. *Warner v. Callender*, 20 Ohio St. 190.

Paper Street A street appearing on the recorded plat, but which has never been opened, nor prepared for use, nor used as a street, is known as a "paper street." *Riaolo v. Northern Pac. R. Co.*, 122 N.W. 489, 108 Minn. 431.

Parkway A parkway is essentially a "boulevard," giving to the term its modern meaning. *Kleopfert v. City of Minneapolis*, 95 N.W. 908, 909, 90 Minn. 158, citing Cent. Dict. *See Boulevard.*

Passage The word "passage" means a way, road, path, route, channel, an entrance or exit, or means of passing, and a "passageway" affords passage. *Wright v. Cherry, Tex. Civ. App.*, 284 S.W.2d 273.

Passageway The term "passageway" is descriptive of a way used for public passage. *People ex rel. Mather v. Field*, 266 Ill. 609, 107 N.E. 864.

Path "'Path' is constantly used in our old acts as synonymous with 'road,' and until within a few years it was the custom for large portions of the population of this state to use the term 'path' to the exclusion of 'road.'" *Singleton v. Commissioner of Roads*, S.C., 2 Nott. & McC. 526.

Plank Road Plank roads, which are open to use by the whole public, with their own teams and vehicles, are highways in the strictest sense. *Flint v. P.M. Ry. Co. v. Gordon*, 2 N.W. 648, 41 Mich. 420.

A plank road is unquestionably a public highway. By the act of constructing it and opening it for use, and being used on payment of tolls, it was as emphatically dedicated to the public as it could have been by a deed or dedication acknowledged and recorded. The elements of a complete dedication are found in the permitted use of the road by the public, and building it for such purposes, and the subsequent use of it by the public without denial or interruption. The very purpose of constructing the road carries with it, when constructed, a dedication. This being so, it would be unjust, in view of the privileges granted, that the parties finding it not remunerative should, on the strength of the claim that the road is private property, suddenly close it up and deprive the public of its use. *Craig v. People,* 47 Ill. 487.

Post Roads The roads or highways, by land or sea, designated by law as the avenues over which the mails shall be transported. *Railway Mail Service Cases,* 13 Ct. Cl. 204. A "post route," on the other hand, is the appointed course or prescribed line of transportation of the mail. *U.S. v. Kochersperger,* 26 Fed. Cas. 803; *Blackham v. Gresham (C.C.),* 16 Fed. 611.

Private Path A "private way" is an individual right, while a "private path" is a neighborhood road. *Earle v. Poat,* 41 S.E. 525, 63 S.C. 439.

A "private path" is a neighborhood road running from one public road to another; from a public place to another public place. *Kirby v. Southern Ry.,* 41 S.E. 765, 63 S.C. 494.

Private Road A "private road" has been defined as a road established by public authority chiefly for the accommodation of an individual or individuals, and at his or their instance and expense. 72 *C.J.S. Private Roads,* § 2.

Private Street Literally speaking, this is an impossibility, for no way can be both private and a street. It may be one or the other, but not both. *Greil v. Stollenwerck, 201 Ala. 303, 78 So. 79.*

Private Way A "private way" is a way over which only a limited number of persons have the right to pass. *Stabola v. Palmer,* 73 A.2d 831, 136 Conn. 670.

Road dedicated to, but never accepted by public remains "private," as distinguished from "public" way. *Wyatt v. Chesapeake & Potomoc Tel. Co. of Virginia, Va.,* 163 S.E. 370.

Projected Street "Projected street," as used in a deed of land by the owner while making a street, bounding the land conveyed upon the projected street, does not mean a designated, intended, or contemplated street merely, but a street already projected and then in process of construction. *Greenhood v. Carroll,* 114 Mass. 588.

Public Highway Roadway open at one end only may be a "public highway." *Foster v. Bullock,* 50 P.2d 892, 184 Wash. 254.

The distinction between "public highway" and "private road or driveway" is not based on ownership of fee or nature of controlling authority, but upon whether road or highway or driveway in question is open to use of public for purposes of vehicular traffic. *State v. O'Connor,* 184 A.2d 83, 76 N.J. Super. 246.

Public Road A "public road" is one that the public generally, and not merely a portion thereof, is privileged to use. *In re Penn Ave.,* 126 A.2d 715, 386 Pa. 403.

Public Way "Streets" and "alleys" are "public ways." *Western Union Tel. Co. v. Dickson,* 173 S.W.2d 714, 27 Tenn App. 752.

"Public ways," as applied to ways by land, are usually termed "highways" or "public roads" and are such ways as every citizen has a right to use. *Kripp v. Curtis, 11 P. 879, 71 C. 62.*

Railroad "Railroad" may mean a corporation rather than a track. *Nicolson v. Brown*, 135 F.2d 245, 77 U.S. App. D.C. 314.

The word "railroad" when used to describe a boundary of land, means the railroad right of way, and not the steel rails on the right of way, a all property essential to the operation of trains constitutes a railroad and not any individual part of railroad property. *Cross v. Bernstein*, 8 La. App. 380.

Railroad Track "Railroad track," as used in a deed bounding the land by the line of a railroad track, should be construed to mean the line of the rails, if the grantor could convey so far, and not simply to the right of way. *Reid v. Klein*, 37 N.E. 967, 138 Ind. 484.

Release of a right of way The phrase "release of a right of way," in an instrument in writing by which the absolute owner of a tract of land released to a steam railroad company a right of way 100 feet in width and passing through the land, leaving a portion thereof on each side of the right of way, conveys an easement in the land for railroad purposes, leaving the fee subject to such servitude in the general owner. *Cincinnati, H. & D. Ry. Co. v. Wachter*, 70 N.E. 974, 70 Ohio St. 113.

Relocate To "relocate" a highway means to locate again and contemplates a change of course. *City of Lakewood v. Thormyer, Ohio Com. Pl.*, 154 N.E.2d 777.

Reserving the Highway "Reserving the highway," as used in a conveyance of land containing the usual covenants of warranty, but reserving the highway, means the reservation of the highway for its use and purpose as a highway, and does not prevent the fee of the lands included in the highway from passing to the purchaser, subject only to the easement of the public. *Peck v. Smith, 1 Conn.*, 103, 6 Am. Dec. 216.

Reversion The term "reversion" has two meanings, first, as designating the estate left in the grantor during the continuance of a particular estate and also the returning of the land to the grantor or his heirs after the grant is over. *Davidson v. Davidson*, 167 S.W.2d 641, 350 Mo. 639.

Revert The ordinary meaning of the word "revert" is to return or to go back. *Reaney v. Wall*, 60 A.2d 505, 134 Conn. 663.

Right of Way "Right of way" is a right of passage over another person's land. *Almada v. Superior Court in and for Napa County,* Cal. App., 149 P.2d 61.

The term "right of way" has a two fold signification. It is sometimes used to describe a right belonging to a party—a right of passage over any tract, and it is also used to describe that strip of land which railroad companies have taken upon which to construct their roadbed. *Maysville & B.S.R. Co. v. Ball*, 56 S.W. 188, 108 Ky. 241.

A "right of way" may be defined generally to be a right to pass over the land of another. It may be a private way or a public way, and it may belong to one or several persons or to the entire community. *Poole v. Greer, Del.*, 65 A. 767, 6 Penneuill, 220.

Term "right of way" in deed is frequently used to describe not only easement, but the strip of land occupied by such use. *Moakley v. Los Angeles Pac. Ry. Co.*, 34 P.2d 218, 139 Cal. App. 421.

A "right of way" may consist of either of the fee or merely of a right of passage and use or servitude. *Brightwell v. International-Great Northern R. Co.* (Tex.) 41 S.W.2d 319.

A right of way is not a road in the legal sense of the term, nor can any one have a right of way over his own land. A way, in law, is the right of going over another man's ground. *Green v. Morris & E.R. Co.*, 24 N.J.L. (4 Zab.) 486.

A conveyance of a "right of way" is an easement only and title does not pass, and grantee acquiring only the right to a reasonable and usual enjoyment thereof with owner of the soil retaining all rights and benefits of ownership consistent with the easement, including the right to make any incidental changes consistent with the easement. *Kleih v. Van Schoyck*, 27 N.W.2d 490, 250 Wis. 413.

"Right of way," whether public or private, is mere right of limited or special use, and differs from ownership in fee of land over which way passes. *West v. Maryland Gas Transmission Corp.*, 159 A. 758, 162 Md. 298.

Right of Way by Custom "A "right of way by custom" in favor of the inhabitants of a particular locality might be set up on the common law of England. It could be proved by immemorial usage. From such proof a presumption was deemed to arise that the usage was founded on a legal right. This right was not assumed to arise from a grant by an owner of land of an easement in it. * * * A 'right of way by custom' appertains to a certain district of territory, but not to any particular tenement forming part of that territory. It belongs to the inhabitants of that territory, whether landowners or not. To a fluctuating body of that kind, no estate in lands can be granted. If, therefore, an easement be claimed to exist in their favor, a title cannot be made out by prescription, on the theory of a lost grant. It must have come, if at all, from some public act of a governmental nature. Theory of English law was that, if there had been a usage from time immemorial (that is, so far as could be ascertained, from the coronation of Richard I), affecting the use of real estate by those not able to show any paper title to warrant it, it might be fairly presumed that it arose under an act of Parliament or other public act of governing power, the best evidence of which had perished. A charter from some feudal lord or ecclesiastical corporation might be such an act. Of such charters there were no public records. That the accidental destruction of the parchment on which one was written should annul the privileges which it gave would be plainly unjust." The laws of Connecticut do not recognize personal rights of way or other easements resting on local custom. *Graham v. Walker, 61 A. 98*, 78 Conn. 130, 2 L.R.A., N.S., 983, 112 Am. St. Rep. 93, 3 Ann. Cas. 641.

Road A highway; an open way or public passage; a line of travel or communication extending from one town or place to another; a strip of land appropriated and use for purposes of travel and communication between different places.

Private road. This term has various meanings: (1) A road, the soil of which belongs to the owner of the land which it traverses, but which is burdened with a right of way, (2) A neighborhood way, not commonly used by others than the people of the neighborhood, though it may be used by any one having occasion, (3) A road intended for the use of one or more private individuals, and not wanted nor intended for general public use, which may be opened across the lands of other persons by statutory authority in some states, (4) A road which is only open for the benefit of certain individuals to go from and to their homes for the service of their lands and for the use of some estates exclusively.

Public road. A highway; a road or way established and adopted (or accepted as a dedication) by the proper authorities for the use of the general public, and over which every person has a right to pass and to use it for all purposes of travel or transportation to which it is adapted and devoted.

The word "road" is a term of general use to designate a passageway in the country outside of municipality, and word "street" is term of general use designating passageway in municipality. *Howard v. State*, 49 S.E.2d 684, 77 Ga.App. 712.

"Road," in its popular sense, is a generic term, including overland ways of every character. *Johnston v. Wortham Machinery Co., 151 P.2d 89, 60 Wyo. 301.*

A "road" or "highway" is nothing more than a strip of ground set aside, improved, and dedicated to the public for use as a passageway. *State ex rel. Wabash Ry. Co. v. Public Service Commission of Missouri, 100 S.W.2d 522, 340 Mo. 225, 109 ALR 754.*

"Road" means any piece of land used or appropriated for travel, whether by an individual, a corporation, or the public. *39 Am Jur 2d, Highways, Streets, and Bridges*, § 2.

"Roads" and highways are generic terms, embracing all kinds of public ways, such as county and township roads, streets, alleys, township and plank roads, turnpike or gravel roads, tramways, ferries, canals, navigable rivers including also railroads, a street is a highway but a highway is not necessarily a street. *Shannon v. Martin*, 139 S.E. 671, 164 Ga. 872.

The word "road" is uniformly taken to mean a public highway in both legal and common acceptation. It is synonymous with "highway." *Heiple v. City of East Portland, 8 P. 907, 13 Or. 97.*

A "road" is a way actually used in passing from one place to another. A mere survey or location of a route for a road is not a road. A mere location for a road falls short of a road as much as a house lot falls short of a house. *Brooks v. Morrill, 42 A. 357, 92 Me. 172.*

"Road," as used in a reservation in a grant of the range way if ever wanted for a road, means a public highway, and not a private way. *Morgan v. Palmer, 48 N.H. 336.*

Settlement Road The designation of a road as a "settlement road" does not necessarily mean it is not also a "public road" as distinguished from a "private road," it being the character rather than the quantum of use that controls, and although the chief users be a few families having a special need therefore, this does not necessarily stamp it as a "private way." *Moore v. Cruit*, 191 So. 252, 238 Ala. 414.

Shell Road A road constructed of crushed sea shells, such as oyster shells.

Shunpike A shunpike is a road intended to furnish a way of evading a tollgate, and constructed for that special purpose. *Elliott, Road and S. p. 74.* The county road which would be of great convenience to people living between two existing roads, one of which is a turnpike, by shortening their line of travel to the town in which the roads terminated, and which would in time be of great utility, is not a shunpike, though the effect would be to decrease the business of the turnpike so as to entitle such company to enjoin its location. *Clarkesville & R. Turnpike Co. v. City of Clarksville, Tenn.*, 36 S.W. 979.

Side The side lines of a road or railroad are the lines which include the territory covered by the road. Public roads and highways and railroads are regarded as having three lines, the side lines and the center line equidistant between the side lines. *Maynard v. Weeks*, 41 Vt. 617.

Sidewalk The "sidewalk" is usually recognized as that portion of the street on each side thereof which has been arranged for pedestrians and is not intended for use by vehicles, and though devoted to use of pedestrians it is nevertheless a portion of the public highway. *Nikiel v. City of Buffalo*, 165 N.Y.S.2d 592 (1957).

Street The term "street" is ordinarily applied to a public way in a city, town or village, and the word "road" to a free public way in the country, while the two terms include all public highways by land, whether designated as highway, road, street, alley, lane, place, or boulevard. *Parsons v. Wright*, 27 S.E.2d 534, 223 N.C. 520.

While the term "street" or "avenue" commonly applies to a public highway in a village, town, or city, and the term "road" to suburban highways, there may be roads in a city or town and streets and avenues in the country. *City of Spokane v. Spokane County*, 36 P.2d 311, 179 Wash. 130; Murphy v. King County, 88 P. 1115, 45 Wash. 587.

A "street" is a way upon land, more properly a paved way, lined, or proposed to be lined, by houses on each side. *United States v. Bain*, 24 Fed. Cas. 940.

A "street" is a public thoroughfare and includes sidewalks. *Nikiel v. City of Buffalo*, 165 N.Y.S.2d 592 (1957).

A "street" is a highway in an urban district. *Fischer v. Shasta County*, 299 P.2d 222, 46 C.2d 771.

Strictly speaking, a "street" is a public thoroughfare in an urban community such as a city, town or village, and the term is not ordinarily applicable to roads and highways outside of municipalities. *39 Am Jur 2d, Highways, Streets, and Bridges*, § 8.

A "street" is a city road. *Muesig v. Harz*, 283 Ill. App. 115.

Land, though owned by city in fee-simple, does not acquire status of street until it is not only defined and laid out by survey, but also actually opened by work. *City of Jacksonville v. Padgett, 108 N.E.2d 460*, 413 Ill. 189.

The term "street," in ordinary legal signification, includes all parts of the way, the roadway, the gutters, and sidewalks. *Thomas v. City of Fremont*, 12 Ohio Dec. 605.

Street Railroad A "street railroad" is a road constructed on a street or highway for purposes of conveying passengers living upon or having business on such street or highway, its main object being to accommodate street travel, and for this purpose the cars make frequent stops to take on and discharge passengers along street or highway. *Goddard v. Chicago, M. & St. P. Ry. Co.*, 104 Ill. App. 533.

Taken Private property is "taken for public use" when it is appropriated to the common use of the public at large. *Craighall v. Lambert*, 18 S. Ct. 217, 168 U.S. 611, 42 L. Ed. 599.

Taking The "taking" of land under the right of eminent domain occurs when the land within the highway is condemned for that public use. *Gilpin v. City of Ansonia*, 35 A. 777, 68 Conn. 72.

Thoroughfare The term means, according to its derivation, a street or passage *through* which one can *fare*, (travel); that is, a street or highway affording an unobstructed exit at each end into another street or public passage. If the passage is closed at one end, admitting no exit there, it is called a "*cul de sac*." *Black's Law Dictionary*.

A "thoroughfare" is a passage through, an unobstructed way open to the public, a public road, a frequented street. *Jewett v. State, Ohio*, 22 O.L.A. 37.

Toll Road A "toll road" is a public highway, differing from ordinary public highways chiefly in the fact that the cost of its construction in the first instance is borne by individuals or by a corporation having authority from the state to build it, and, further, in the right o the public to use the road after its completion, subject only to the payment of toll. *Virginia Cañon Toll Road Co. v. People, 45 P. 398*, 22 Colo. 429, 37 L.R.A. 711.

Town Way "Town ways" are roads within the territorial limits of a particular town. *Inhabitants of Waterford v. Oxford County*, 59 Me. 450.

"Town way," as used in the statutes relating to such ways, means a way which was originally laid out by the selectmen of the town. *Inhabitants of Blackstone v. Worcester County*, 108 Mass. 68.

A "town way" is a term used in some statutes signifying a highway laid out in proceedings in which a town or city has original jurisdiction, and which is within the territorial limits of a particular town. *39A C.J.S. Highways*, § 1(3).

Trace Course or path followed. A mark or line left by anything that has passed; footprint; tract; also, a beaten path; trail. *Webster's Dictionary*.

Traveled Part The "traveled part of the highway" means the part traveled by vehicles. *Westlund v. Iverson*, 191 N.W. 253, 154 Minn. 52.

Turnpike A turnpike road is a road having tollgates or bars on it, and on which tolls are collected. *Northam Bridge Co. v. London Ry.*, 6 M. & W. 428.

The words "turnpikes," "tollgates," and "tollbars," in the act relating to the taking of tolls, are used synonymously. *Company of Proprietors of Northam Bridge & Roads v. London & S. Ry. Co.*, 6 Mees. & W. 428.

Whether a road is a turnpike road is a question depending on whether it is a road maintained by tolls payable by passengers, and not whether it is an important road or not. No amount of traffic can make a road a turnpike road. *Reg. v. East & West I.D. Ry. Co.*, 2 Ellis & Bl. 464.

Turnpikes and Toll Roads Originally the term had a different connotation than today, as a road built by public authority but at private expense, usually by a turnpike company, which thereafter was entitled to exact a charge for the use of the road. *39 Am Jur 2d, Highways, Streets, and Bridges*, § 7.

Turnpike Road These are roads on which parties have by law a right to erect gates and bars, for the purpose of taking toll, and of refusing the permission to pass along them to all persons who refuse to pay.

A turnpike road is a public highway, established by public authority for public use, and is to be regarded as a public easement, and not as private property. The only difference between this and a common highway is that, instead of being made at the public expense in the first instance, it is authorized and laid out by public authority and made at the expense of individuals in the first instance, and the cost of construction and maintenance is reimbursed by a toll, levied by public authority for the purpose. *Com. v. Wilkinson*, 16 Pick. (Mass.) 175, 26 Am. Dec. 654. *Black's Law Dictionary*.

Way A passage, path, road, or street. In a technical sense, a *right* of passage over land.

A right of way is the privilege which an individual, or a particular description of persons, as the inhabitants of a village or the owners or occupiers of certain farms, have of going over another's ground. It is an incorporeal hereditament of a real nature, entirely different from a public highway.

The term "way" is derived from the Saxon, and means a right of use for passengers. It may be private or public. By the term "right of way" is generally meant a private way, which is an incorporeal hereditament of that class of easements in which a particular person, or particular description of persons, have an interest and a right, though another person is the owner of the fee of the land in which it is claimed. *Wild v. Deig*, 43 Ind. 455, 13 Am. Rep. 399. *Black's Law Dictionary*.

A way is the right of going over another man's ground. *Boyd v. Hand*, 65 Ga. 468.

Word "way" means that along which one passes to reach some place, a road, street, track, path, or that along which one passes to reach some place. *Roach v. Soles, D.C. Cal.*, 120 F.Supp. 400.

The word "way" is more generic than the word "road," referring to many things besides roads. *Kister v. Reeser*, 98 Pa. 1, 42 Am. Rep. 608.

Strictly speaking, the words "road" and "way" are not synonymous, although in one case they have been so held. As a matter of fact, however, the words are often used interchangeably in general conversation. "A way" is an incorporeal hereditament, and imports ex vi termini the right of passing over the land of another along a particular line or route, and it may arise either from grant, necessity, or prescription. *Washb. Easem.* 256. *Webster*, as one of the definitions of a "road," says: "It is the ground appropriated for travel, forming a communication." Bouvier defines a "road" as "a passage through the country for the use of the people." Anderson, in his dictionary, says: "A road is an open way or public passage ground appropriated for travel, and generally it includes highways, streets and lanes." As a general proposition, the word "road" is applied to a highway, street, or land, but the term is often applied to a private way or a pathway. The word "road" includes in some sort the sense of "way," although the latter word is more generic, and refers to many things besides roads. The grant of a way over one's land, nothing appearing in the grant to restrict it, will be understood to be a general way for all purposes. *Newsom v. Newsom, Tenn.*, 56 S.W. 29.

Although the term "way" is sometimes used in the same sense as "road," the two are not synonymous. "Way" in its legal, technical sense means nearly the same thing as "right of way," or in other words, the right of one person, of several persons, or of the community at large, to pass over the land of another. *39 Am Jur 2d, Highways, Streets, and Bridges*, § 2.

Way Appurtenant A "way appurtenant" is an easement running with a dominant tenement, and is necessarily comprised of two distinct tenements, the dominant, to which the right of possession belongs, and the servient, on which the burden rests, and one terminus of the way must lie on the land to which it is claimed to be appurtenant. *Downey v. Sklebar*, Mo., 261 S.W. 697.

Way by Dedication A "way by dedication" is created by the permission or gift of the owner, and upon the acceptance of such gift by the public authorities, it becomes a way, and the owner cannot withdraw his dedication. *Commonwealth v. Coupe*, 128 Mass. 63.

Way by Prescription A "way by prescription" is established on evidence of user by the public, adverse and continuous, for a period of 20 years or more, *Schwerdtle v. Placer County*, 41 P. 448, 108 Cal. 589; from which use arises a presumption of a reservation or grant and the acceptance thereof, or that it has been laid out by the proper authorities, of which no record exists. *Commonwealth v. Coupe*, 128 Mass. 63.

Way of Necessity A purchaser of land, excluded from the public highway except by passing over the vendor's land, takes a way over the vendor's land as a "way of necessity." *Bentley v. Hampton, Ky.*, 91 S.W. 266.

"Way of necessity" is a right of way; the privilege of going over another's land; a legal way, to use which one has a legal right, which may be enforced, and which may not be rightfully interfered with. *Cox v. Tipton*, 18 Mo. App. 450.

Way Opened and Dedicated to Public Use A way by dedication is a way over land which owner thereof has dedicated to use of public for a way. *Dakin v. City of Somerville*, 160 N.E. 260, 262 Mass. 514.

Way Reserved A "way reserved," as used in a popular sense, is strictly an easement newly created, by way of a grant from the grantee in the deed, of the estate to the grantor. *Winston v. Johnson*, 45 N.W. 958, 42 Minn. 398; *2 Washb. Real Prop.* (5th Ed.) 465; *Hagerty v. Lee*, 25 A. 319, 54 N.J.L. (25 Vroom) 580, 20 L.R.A. 631.

INDEX

Printed and bound by CPI Group (UK) Ltd, Croydon, CR0 4YY

27/10/2024

14580319-0001